Nanoscale Materials in Chemistry: Environmental Applications

ACS SYMPOSIUM SERIES **1045**

Nanoscale Materials in Chemistry: Environmental Applications

Larry E. Erickson, Editor
Kansas State University

Ranjit T. Koodali, Editor
University of South Dakota

Ryan M. Richards, Editor
Colorado School of Mines

Sponsored by the
ACS Division of Industrial & Engineering Chemistry

American Chemical Society, Washington, DC

Library of Congress Cataloging-in-Publication Data

Nanoscale materials in chemistry : environmental applications / Larry E. Erickson, Ranjit T. Koodali, Ryan M. Richards, editors ; sponsored by the ACS Division of Industrial & Engineering Chemistry.
p. cm. -- (ACS symposium series ; 1045)
Includes bibliographical references and index.
ISBN 978-0-8412-2555-8 (alk. paper)
1. Metallic oxides--Environmental aspects. 2. Nanocrystals--Industrial applications. I. Erickson, L. E. (Larry Eugene), 1938- II. Koodali, Ranjit T. III. Richards, Ryan. IV. American Chemical Society. Division of Industrial and Engineering Chemistry.
QD181.O1N36 2010
628--dc22

2010029297

The paper used in this publication meets the minimum requirements of American National Standard for Information Sciences—Permanence of Paper for Printed Library Materials, ANSI Z39.48n1984.

Distributed by Oxford University Press

PRINTED IN THE UNITED STATES OF AMERICA

Foreword

The ACS Symposium Series was first published in 1974 to provide a mechanism for publishing symposia quickly in book form. The purpose of the series is to publish timely, comprehensive books developed from the ACS sponsored symposia based on current scientific research. Occasionally, books are developed from symposia sponsored by other organizations when the topic is of keen interest to the chemistry audience.

Before agreeing to publish a book, the proposed table of contents is reviewed for appropriate and comprehensive coverage and for interest to the audience. Some papers may be excluded to better focus the book; others may be added to provide comprehensiveness. When appropriate, overview or introductory chapters are added. Drafts of chapters are peer-reviewed prior to final acceptance or rejection, and manuscripts are prepared in camera-ready format.

As a rule, only original research papers and original review papers are included in the volumes. Verbatim reproductions of previous published papers are not accepted.

ACS Books Department

Contents

Indexes

Preface

At the 237th National American Chemical Society meeting in Salt Lake City, many of the contributors to this book presented their work in the symposium "Nanoscale Materials in Chemistry: Environmental Applications: In Honor of Professor Klabunde." The symposium honored 30 years of research by Professor Kenneth Klabunde and his coworkers. Dr. Klabunde has authored two books and edited two other books on the topic of nanoscale materials in chemistry, and he has started a company, NanoScale Corporation, Inc. that produces and markets products for environmental applications. This book describes research on the development of catalysts and adsorbents based on nanoscale materials.

The book includes new fundamental research and applications. It starts with a review of research on the development of nanoscale metal oxides that have environmental applications. Information on product development is described for selected products that have been developed and commercialized.

This book is for scientists and engineers who are engaged in research, development, and commercialization of nanoscale materials for environmental applications. Those interested in the pathway from idea to product will find this book valuable to them. Those interested in sustainable indoor environments will find new information on in room devices that may be able to reduce energy use in buildings. Toxicology and product safety are included as well.

The editors wish to thank all of the reviewers that assisted with the peer review effort that has improved the quality of the manuscripts. We also wish to thank those at ACS who have helped manage the peer review process and production of this book.

Larry E. Erickson
Department of Chemical Engineering, Kansas State University
Manhattan, KS 66506
lerick@ksu.edu (e-mail)

Ranjit T. Koodali
Department of Chemistry, University of South Dakota
Vermillion, SD 57069
Ranjit.Koodali@usd.edu (e-mail)

Ryan M. Richards
Department of Chemistry, Colorado School of Mines
1500 Illinois Street
Golden, CO 80401
rrichard@mines.edu (e-mail)

Chapter 1

Review of Nanoscale Materials in Chemistry: Environmental Applications

Kenneth J. Klabunde,[*,1] Larry Erickson,[2] Olga Koper,[3] and Ryan Richards[4]

[1]Department of Chemistry, Kansas State University, Manhattan, KS 66506
[2]Department of Chemical Engineering, Kansas State University, Manhattan, KS 66506
[3]NanoScale Corporation, Manhattan, KS 66502
[4]Department of Chemistry and Geochemistry, Colorado School of Mines, Golden, Colorado 80401
[*]kenjk@ksu.edu

This chapter provides historical background and a foundation for the other chapters in this book by reviewing the most closely related research and developments in nanoscale materials in chemistry and their environmental applications. The review includes nanoscale sorbents, destructive sorbents, nano-catalysts, and photocatalytic nanomaterials.

Environmental safety of nanoscale materials is considered and the focus of the review is on substances that have potential commercial applications in environments where health and safety considerations are evaluated.

Introduction

The discussion of engineered nanomaterials for environmental applications could include many very old and historic aspects of chemistry, including heterogeneous catalysis, carbon sorbents, and air purification (*1*). Thus, nanomaterials include nanostructured porous solids as powders, pellets, or even stand-alone monoliths (*2*).

For the purposes herein, we will deal only with recent discoveries that complement these older, important fields, and considerations of the environmental safety of deploying nanoscale materials in remediation technologies (Chapter 13 herein).

Clearly, there are safety issues whenever new technologies are deployed, and nanomaterials are no exception (*3*). In this regard, nanocrystalline solids should be considered new chemicals, and development, manufacture, and use should follow appropriate protocols.

Sorbents for Environmental Remediation

First, we will deal with new discoveries in the field of sorbents. As mentioned above, activated high surface area carbon is the "gold standard" in sorbent technology. A wide variety of carbon sorbents made from coal, wood, coconut shells, fruit seeds, polymers, etc. have been used for centuries for purification of chemicals, water, and air. Chemical additives to the carbon sorbents often enhance their abilities to be more selective for certain sorbates. And, recent developments in the field of fullerenes, carbon nanotubes, graphene, and carbon fibers have added greatly to the usefulness of carbon as a whole. Carbon is certainly an intriguing and amazing material especially in nanostructured solid form.

Nonetheless, there are some drawbacks to carbon sorbents. Since it mainly operates by physisorption (rather than strong chemisorptions), many volatile sorbates are not trapped very well by carbon. Also, sorbates are usually not destroyed or detoxified by carbon, and eventual leaching or bleeding-off of the sorbates is a common problem. And, one other disadvantage is that carbon is black, and so it does not lend itself well to behaving as a colorimetric sensor.

There are, of course, many other solids that serve as high surface area (high capacity) sorbents. This has become especially the case in recent years when new synthetic methods have allowed nanoscale metal oxides to be prepared, such as sol-gel (*2–5*), aerogel, and aerosol methods.

The periodic table of the elements presents us with at least 60 metallic elements that can be used to obtain stable metal oxides. On the other hand, when considering environmental safety issues, the list diminishes, and from the beginning of our work in the 1980s, we have centered our interest in only a few, including magnesium oxide (MgO), calcium oxide (CaO), titanium oxide (TiO_2), aluminum oxide (Al_2O_3), iron oxide (Fe_2O_3), and zinc oxide (ZnO). These oxides in high surface area form, and their physical mixtures and intimate (molecular or nanoscale) mixtures have proven to be excellent sorbents for many applications.

Destructive Adsorbents

Elevated Temperature Chemistry

Our first foray into this field was the use of metal oxides as "destructive adsorbents" of organophosphorus compounds (chemical warfare agent mimics and pesticides). The metal oxide (MgO) was contained in a fixed bed reactor tube, and heated to temperatures hot enough to very rapidly destroy a series of organophosphorus chemicals. As an example, triethylphosphate [$(CH_3CH_2O)_3P{=}O$] adsorbed strongly on a MgO surface, and about four surface

MgO moieties adsorbed one phosphate molecule (*6*, *7*). This is close to a full monolayer, and slight bandshifts in the IR to higher energy suggest a net electron loss. Upon heating ,the adsorbed species evolved ethene and diethylether, leaving a $[PO_4H]_{ads}$ fragment strongly bound.

Numerous organophosphorus compounds behaved similarly; strong adsorption and destructive adsorption at slightly elevated temperatures (100-200°C).

Following this success, we began to investigate synthetic methods for preparing much higher surface area oxides, and this resulted in the discovery of the "nano-effect" on destructive adsorption. Again, looking at organophosphorus reagents, it was found that much higher reactivities and capacities for destructive adsorption were realized, even after correcting for surface area. For example, 0.48 mole of dimethylmethyl phosphate $CH_3PO(OCH_3)_2$, DMMP could be destroyed (essentially mineralized) for one mole of nano-MgO (which we dubbed AP-MgO for "aerogel prepared") (*8*). This finding indicated that the reaction of solid MgO with gaseous DMMP was almost stoichiometric, which meant that the 4nm MgO crystallites were providing even the inner MgO moieties for reaction at 500°C.

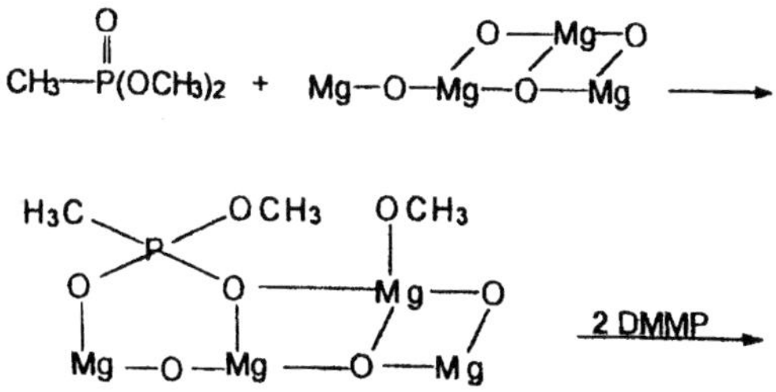

$$[CH_3(CH_3O)PO]_{(a)} + 2\,[CH_3(CH_3O)P]_{(a)} + 2\,HCOOH_{(g)} + CH_3O_{(a)} + 2\,H_{(a)}$$

The very high capacity was attributed to a combination of initial high surface area plus an enhanced proportion of edge/corner and defect sites on AP-MgO (*8*).

Chlorocarbon destructive adsorption was also found to be very efficient when hot MgO, CaO, or Fe_2O_3 nanoscale materials were employed. These reactions, such as with carbon tetrachloride, are exothermic, as the $\Delta H°_{rxn}$ based on $\Delta H°_f$ values show:

$$2\,CaO_{(s)} + CCl_{4(g)} \longrightarrow 2CaCl_{2(s)} + CO_{2(g)} \quad \Delta H^{o}_{rxn} = -573kJ$$

$$2MgO_{(s)} + CCl_{4(g)} \longrightarrow 2MgCl_{(s)} + CO_{2(g)} \quad \Delta H^{o}_{rxn} = -334kJ$$

In these reactions, high surface areas and reactivities for our nanoscale oxides were again found to be very beneficial.

However, the inner core of the nanocrystals did not react. Therefore, efforts were made to find a "second-generation" of destructive adsorbents, and core-shell mixed oxides were found to be surprisingly effective (*8*).

The high exothermicities of the MO/CCl_4 reactions were noted above. This is not generally the case for transition metal oxides. Perhaps there could be synergism between transition metal oxides and alkaline earth oxides, where transition metal ions react fast, but due to thermodynamics, transfer of chloride to Mg or Ca ions would occur. In fact, Berty and coworkers had reported $CaCO_3$ or Na_2CO_3 impregnated with Mn, Ni, or Cu ions did enhance capacities for oxidative destruction of chlorocarbons (*9*, *10*).

We explored core/shell structures of $[M_xO_y]MgO$ and CaO in the direct conversion of CCl_4 to CO_2 in the absence of oxygen where M = Mn, Fe, Co, and Ni. To our pleasant surprise, these core/shell nanoscale powders performed very well, so that a stoichiometric reaction was achieved at a reaction temperature of 425°C (*11*, *12*):

$$[Fe_2O_3]CaO + CCl_4 \longrightarrow [FeCl_3]CaCl_2 + CO_2$$

Performance Efficiency = 0.5
(Theoretical Efficiency = 0.5)

For comparison, normal CaO (microscale particles) gave 0.01 and AP-CaO gave 0.31 efficiencies. These results could not be attributed solely to higher surface areas, but the nanocrystalline materials of MgO and CaO exhibited inherently high surface reactivities. Then, by adding small amounts of transition metals, a catalytic ion-mixing took place. It was rationalized that the lower melting $FeCl_3$ was mobile in the CaO or MgO matrix, and, due to a thermodynamic driving force, exchanged the Cl^-, forming an iron oxide species capable of reaction again. In this way, complete reaction of the bimetallic oxide took place.

These favorable findings proved to hold for other high temperature destructive adsorption reactions for a series of chlorocarbons and for acid gases such as SO_2 (*14*).

These results suggest that nanoscale crystalline solids can be viewed as a new class of near stoichiometric chemical reagents. Large surface areas, coupled with enhanced surface reactivities, coupled with strategic addition of small amounts of catalytic transition metal ions make this a reality for a large number of environmental remediation chemistries.

Lower Temperature Chemistry

It soon became apparent that nanoscale oxides were capable of destructive adsorption of organohalides and organophosphorus compounds at much lower temperatures, in many cases, at room temperature. This observation opened up a large number of environmental remediation possibilities.

One interesting observation, made by an undergraduate researcher (*15*) using FT-IR and TG-MS techniques, was that methyl iodide (CH_3I and CD_3I) reacted spontaneously at 100°C to form monodentate methoxide (due to surface –OH reaction), and methyl cation on oxide surface moieties:

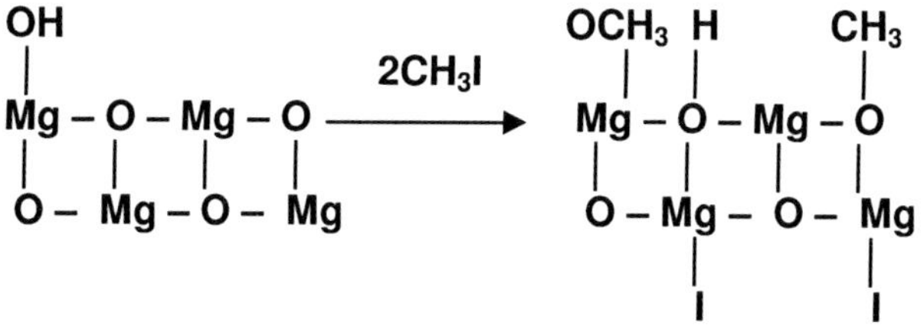

Furthermore, a large increase in hydrogen bonded hydroxyl groups and a disappearance of isolated hydroxyl groups occurred at 3750 cm^{-1}, which suggested that the CH_3 group was being partially stripped of protons. Experiments with CD_3I supported that the C – H (or C – D) bonds were also being broken, and substantial H – D exchange on the methyl group took place.

Clearly, extensive bond breaking occurred upon exposure of AP-MgO to methyl iodide at 100°C.

When even more reactive organohalides were studied, such as 2-Chloroethylethyl sulfide (2-CEES, a mimic of the chemical warfare agent mustard gas $ClCH_2CH_2SCH_2CH_2Cl$), AP-MgO caused destructive adsorption at room temperature.

Extensive studies by FT-IR of the volatile products evolved and extracted products from the 2-CEES exposed AP-MgO showed that the reaction took place as shown below (*16–18*):

$$CH_3CH_2SCH_2CH_2Cl + \underset{\text{(surface moieties)}}{2MgO} \xrightarrow{25^{\circ}C} CH_3CH_2SCH = CH_2 + MgCl_2 + H_2O$$

Comparisons of AP-MgO, CP-MgO, CM-MgO (high surface, moderate surface and low surface areas, respectively) and high surface $Mg(OH)_2$ clearly showed the superiority of the AP-MgO samples, based on the same mass of each sample (see Figure 1). Of even more interest were studies of differing masses that equaled 2:1 moles <u>surface</u> oxide to one mole 2-CEES (in order to correct for surface area effects). It was shown that AP-MgO was still clearly the best destructive adsorbent, once again showing the inherent high surface reactivity/capacity of the smallest nanocrystals (porous aggregates of 4nm crystallites) (*19*).

Further reaction with a second 2-CEES molecule, followed by proton migration on the MgO surface, would yield H_2O and $MgCl_2$.

These results and others (*20*) led to in-depth studies of nano-MgO, CaO, Al_2O_3 with real chemical warfare agents (*21–23*), and these are covered in Chapter 7 by Dr. George Wagner. Also, commercial products have been developed by NanoScale Corporation, based on the accumulation of data on the efficacy and

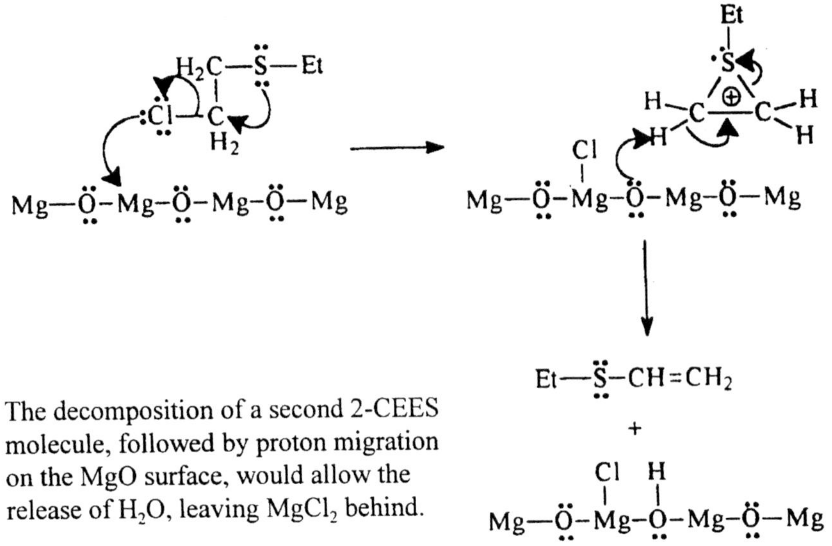
The decomposition of a second 2-CEES molecule, followed by proton migration on the MgO surface, would allow the release of H_2O, leaving $MgCl_2$ behind.

Figure 1. Proposed mechanism for destructive adsorption of 2-CEES on AP-MgO (19).

the safety of these nano-metal oxides in a variety of environmental remediation technologies. These products are discussed in chapters 5 and 14, including FASTACT® (First Applied Sorbent Treatment Against Chemical Threats).

A wide variety of toxic industrial chemicals (TICs) have been found to be susceptible to treatment by these destructive adsorbents. Remediation of chemical spills, contaminants in commercial gases, and odors in air are now commonplace with this class of safe-to-use nano-metal oxides (*3*).

Biocidal Nanoscale Materials

Biological toxins, including bacteria, bacterial spores, such as anthrax, and viruses are probably the most serious threats. The need is growing for broad-spectrum biocides that can decontaminate tough-to-kill spores, as well as other biological threats, such as prions.

There are numerous effective biocides available, such as bleach solution, chlorine dioxide gas, volatile halogens (Cl_2, Br_2), methyl bromide gas, and high energy radiation. However, there are very few effective, solid biocides. Therefore, we considered the high sorption capacities of these nanocrystalline metal oxides, and did some experiments where gaseous biocides were adsorbed on their surfaces, with the idea that thereby biocidal solids would be produced. This led to the formation of metal oxide-halogen and metal oxide-mixed halogen adducts, where the halogens were chemisorbed rather strongly, but in a zero-valent state, thus allowing the solid oxides to provide active halogen toward bacteria, spores, and viruses (*24–28*). Indeed, these high surface area solids were very effective for destroying *E-Coli, Bacillus Cerus, and Bacillus Subtilis* spores, Sterne strain anthrax spores, and several viruses. The most stable halogen adducts were formed with MgO, Al_2O_3, TiO_2, and CeO_2, and the most effective biocidal action was noted for adducts with Br_2, ICl, and ICl_3, although all adducts showed good activity.

There were four modes of action discerned: (1) electrostatic attraction of the spores to the nanoparticles, (2) high local pH weakening the spore protective outer layer, (3) abrasive action, and (4) oxidative reactions with the attached zero-valent halogens.

Figure 2 demonstrates how Al_2O_3·halogen dry contact with *Bacillus Anthracis* (Sterne strain) disrupts the spore cell completely (*29*).

These nano-oxide-halogen adducts are effective in dry contact with the biothreat, or as water slurries.

One set of experiments was very interesting. In a level 3 biocidal chamber, *Bacillus Subtillus* spore-water solution was sprayed as a mist. Five minutes later, MgO-Cl_2 yellow-green adduct was sprayed as a dry, fine powder. Air samples pulled air down and around agar plates. In this way the sticky agar trapped airborne spores and nanopowders aggregates. These plates were incubated, and showed, with the help of control experiments, that the trapped spores were completely decontaminated (see Figure 3) (*30*).

All of these results were quite positive, and the halogen adducts are strongly biocidal as powders or in pellet form. Indeed, they are very active zero-valent halogen sources, even capable to selective halogenation of organic alkenes and alkanes (*31*).

On the negative side, these adducts are too reactive to allow exposure to animals or humans. On the positive side, they are very effective, and in the environment rather rapidly degrade to mineral and salt substances, and so are not long-term pollutants.

Nonetheless, the search for easier to handle and safer biocides must go on, and work with nano-oxides containing silver as a biocide is discussed in Chapter 11.

Nano-Catalysts for Environmental Applications

Nanoscale metal particles dispersed onto high surface area supports (Al_2O_3, SiO_2, C, etc.) actually are the first "nanoscale materials in chemistry" and have made a huge impact on society. Fuels, textiles, chemicals, and reduced automobile emissions have all been profoundly affected by such heterogeneous catalysts (*2*). As discussed in Chapter 15, indoor air quality is another area of application.

In more recent years, three related areas of interest in catalysis have been explored, as discussed below.

Catalytic Transition Metal Ions in Destructive Adsorption

Solid-gas chemical reactions at elevated temperatures can often be enhanced to stoichiometric reactions by the use of transition metal ions in metal oxides. Five toxic chemicals (CCl_4, $CHC{=}CCl_2$, ortho-$C_6H_4Cl_2$, $CH_3P(O)(OCH_3)_2$, and SO_2)

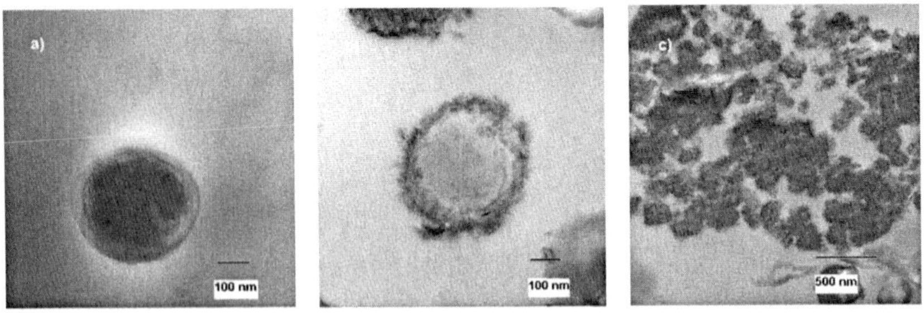

Figure 2. TEM micrograph of untreated (a) and treated (b and c) B. anthracis spores.

Level 3 Chamber

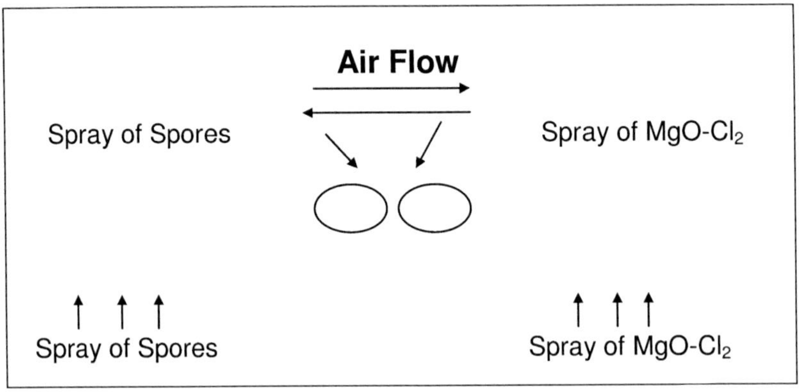

Figure 3. Room size chamber for mixing and sampling air-borne spores and air-borne nanoparticles.

(*12–14*) were exposed to MgO, CaO, and core/shell M_xO_y/MgO and M_xO_y/CaO solids of varying surface areas.

The transition metal oxides incorporated onto MgO or CaO were Se_2O_3, TiO_2, V_2O_5, Cr_2O_3, and ZnO. The loading ranged from 1 to 43 weight %. Only very small amounts were needed (1 to 5%) to yield dramatic results, and order of effectiveness was V > Mn > Co > Fe > Zn > Ti > Cu > Ni > Cr > none > Se.

For example, DMMP [$CH_3P(O)(OCH_3)_2$) destructive adsorption was enhanced such that breakthrough was 30 μl to 60 μl comparing AP-MgO with [Fe_2O_3]AP-MgO (see Figure 4) (*12*).

$$CH_3O\overset{\overset{\displaystyle O}{\|}}{P}\!-\!(OCH_3)_2 \xrightarrow{MgO_{(s)}} [CH_3O-P-CH_3]_{ads} + HCOOH + CH_3OH + H_2O$$

Mole/mole ratios of DMMP/MgO ranged from 0.1 to 0.5 for [Fe_2O_3]MgO. Since earlier work indicated that two surface MgO were required to destructively

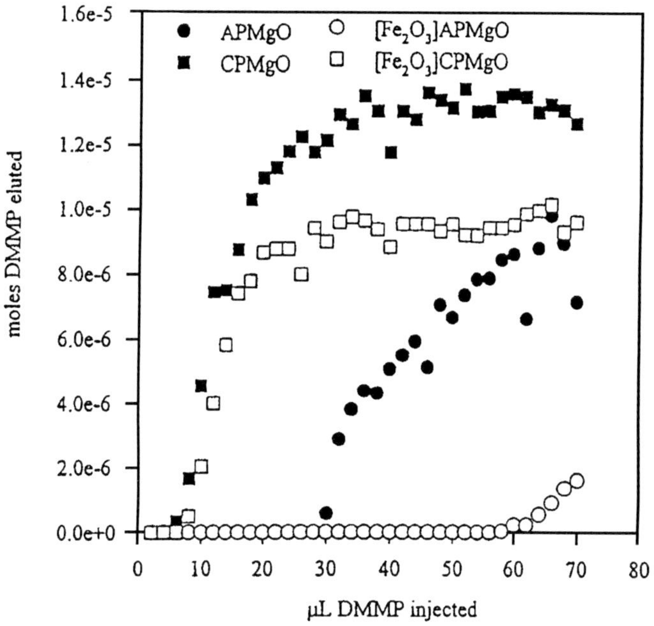

Figure 4. Destructive adsorption of DMMP with AP-MgO, [Fe_2O_3]AP-MgO, and [Fe_2O_3]CP-MgO. (Reproduced with permission from reference (12).)

adsorb one DMMP molecule, a mole/mole ratio of 0.50 is stoichiometric chemistry.

Similar enhancements were observed for CCl_4 (0.50 mole/mole) and dichlorobenzene 0.80 mole/mole MgO when Fe_2O_3 was present.

Interestingly, this beneficial effect of V_2O_5 or Fe_2O_3 was not observed with $CHCl{=}CCl_2$, and carbon deposition appeared to be a problem.

The catalytic effect on enhancing solid-gas reactions appears to depend on the generation of mobil transition metal compounds (chlorides or hydroxides/alkoxides) that tend to exchange with anions with MgO, until the MgO is consumed. Fairly clear evidence of this was found for the [V_2O_5]MgO system with CCl_4. As shown in Figure 5, upon reaction VCl_3 exchanged chloride to eventually yield V_2O_3, which could migrate back to the surface for further reaction with CCl_4 (*12*).

Additional studies showed that quite a wide range of environmentally problematic compounds can be destroyed stoichiometrically by utilizing small amounts of Fe_2O_3 as a catalyst coupled with MgO, CaO, ZnO, or Al_2O_3 nanocrystalline materials (*13*). A particularly interesting result was that hydrogen sulfide (H_2S) could be converted to metal sulfides under very mild conditions, and the best nanoscale materials for this were [Fe_2O_3]CaO and ZnO (*32*, *33*). Once again, the presence of Fe_2O_3 as a catalyst was found to be very effective, even at room temperature. Nano-ZnO, however, was effective even without the Fe_2O_3.

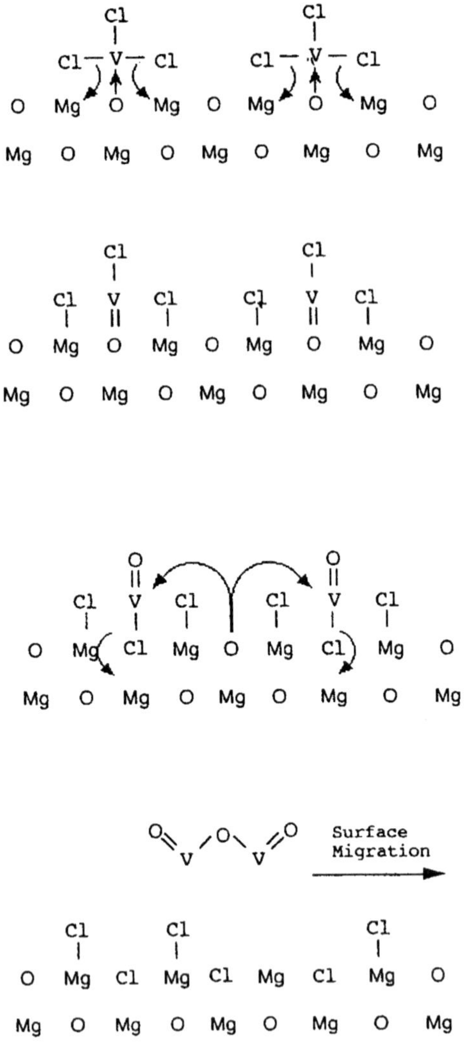

Figure 5. Proposed process by which VCl_x exchanges Cl^-/O^{2-}. (Reproduced with permission from reference (12).)

Super-Base Catalysts

Although not strictly related to environmental remediation, improved catalytic methods for alkene conversions to higher alkenes is beneficial in clean chemical/fuel production (*34*). For example, nanocrystalline MgO doped with small amounts of potassium metal possess highly basic reaction sites (super-base sites) in large amounts. The total number of basic sites closely correlates with the estimated number of ions that exist on corners/edges on the polyhedral MgO crystallites. These results represent another example of nanocrystalline materials possessing unique surface chemistry.

Isomerization of alkenes, and alkylation of toluene and propene by ethane were quite facile on these K/MgO nanocrystalline catalysts:

$$C_6H_5CH_3 + H_2C{=}CH_2 \rightarrow C_6H_5CH_2CH_2CH_3$$

$$C_6H_5CH_3 + CH_3CH{=}CH_2 \rightarrow C_6H_5CH_2\text{-}CH(CH_3)_2$$

$$CH_3CH{=}CH_2 + H_2C{=}CH_2 \rightarrow \text{pentenes and heptanes}$$

Turnover numbers (TON) were estimated to be $0.59s^{-1}$ for isomerization of 2,3-dimethyl-1-butene to tetramethylethylene for a K/AP-MgO catalyst with a 10% loading of elemental potassium. The pK_{BH} of these samples were as high as 0.95 compared with 0.40 for AP-MgO without potassium, according to an indicator-color change procedure in hexane solvent (*35*).

Hydrocarbon Processing

Another example where nano-MgO proved beneficial in hydrocarbon processing (*36*, *37*) was with vanadium oxide/MgO used for dehydrogenation of butane to form butadiene. With vanadium ion as a catalyst promoter, and iodine as a catalyst mediator, the following reactions take place, aided by the high surface area and reactivity of MgO:

$$C_4H_{10} + I_2 + MgO \rightarrow C_4H_8 + MgI_2 + H_2O$$

$$C_4H_8 + I_2 + MgO \rightarrow C_4H_6 + MgI_2 + H_2O$$

$$2MgI_2 + O_2 \rightarrow 2\,MgO + 2I_2$$

This sequence shows that there are really three "catalysts" operating in this system: vanadium ions, iodine, and magnesium oxide nanocrystals. Only oxygen and butane are consumed.

Photocatalytic Nanomaterials

Regarding the historical buildup of technology for the use of nanomaterials for environmental application, photocatalysis has evolved as an important area. Indeed, the combination of destructive adsorption properties with photo-activity allows a synergistic "photochemical boosting" whereby increased capacities are possible for destruction of toxins.

In order to bring photo-activity into the picture, semiconductor metal oxides need to be employed. Therefore, titanium oxide (TiO_2) has emerged as a favored material, since it is ubiquitous in our environment and is used in many products, such as paints and even processed food. It can be prepared in high surface area form, and is efficient in photon absorption in the ultra-violet (UV) range.

A drawback is that TiO_2 does not normally absorb light in the visible range, although TiO_2 prepared with numerous defect sites (especially Ti^{3+} ions) has been shown to be active under visible light (*38*).

As an introduction to this area, which is covered in more detail in Chapter 10, the use of TiO_2 as a photocatalyst for 2-CEES destruction bears mention. In this study, at 25°C and 100°C, 2-CEES suffered extensive bond cleavages, and eventually, in the presence of air, was mineralized to SO_2, HCl, H_2O, CO_2, and sulfate ions immobilized on the surface of the TiO_2 (*39*).

Numerous alkenes, chloroalkenes, aldehydes, disulfides, and alcohols were observed as intermediates. Apparent quantum yields at the beginning of the degradation process ranged up to 0.15.

Conclusions

This brief historical review of the discoveries regarding Nanoscale Materials in Chemistry: Environmental Applications leads to several conclusions:

A. Nanocrystalline oxides provide advantages due to a combination of higher surface areas and intrinsically higher reactivities.
B. Addition of small amounts of transition metal ions can increase destructive adsorption capacities, in many cases such that stoichiometric chemistries are observed.
C. Photocatalysis can complement and further enhance technologies for remediation of a wide variety of toxic chemicals.
D. Nanoscale metal oxides that have been identified as environmentally safe to use are MgO, CaO, TiO_2, ZnO, Al_2O_3, Fe_2O_3.

References

1. Interrante, L. V., Hampden-Smith, M. J., Eds. *Chemistry of Materials*; Wiley-VCH: New York, NY, 1998; pp 1–18.
2. Klabunde, K. J., Ed. *Nanoscale Materials in Chemistry*; Wiley Interscience: New York, NY, 2001; pp 1–14, 85–120, 223–62.
3. Klabunde, K. J., Richards R., Ed. *Nanoscale Materials in Chemistry*, 2nd ed.; Wiley: New York, NY, 2009; pp 629–768.
4. Brinker, C. J.; Scherer, G. W. *Sol-Gel Science*; Academic Press: San Diego, CA, 1990; pp 1–403.
5. Grassian, V., Ed. *Environmental Catalysis*; CRC Press: Boca Raton, FL, 2005; pp 1–3, 391–420.
6. Lin, S. T.; Klabunde, K. J. *Langmuir* **1985**, *1*, 600–5.
7. Ekerdt, J. G.; Klabunde, K. J.; Shapley, J. R.; White, J. M.; Yates, J. T. *J. Phys. Chem.* **1988**, *92*, 6182–8.
8. Klabunde, K. J.; Stark, J. V.; Koper, O.; Mohs, C.; Park, D. G.; Decker, S.; Jiang, Y.; Lagadic, I.; Zhang, D. *J. Phys. Chem.* **1996**, *100*, 12142–53.

9. Berty, J. U.S. Patent 5,021,383, June 1991.
10. Stenger, H. G., Jr.; Buzan, G. E.; Berty, J. M. *Appl. Catal. B* **1993**, *2*, 117–30.
11. Klabunde, K. J.; Khaleel, A.; Park, D. *High Temp. Mater. Sci.* **1995**, *33*, 99–106.
12. Jiang, Y.; Decker, S.; Mohs, C.; Klabunde, K. J. *J.Catal.* **1998**, *180*, 24–35.
13. Decker, S.; Klabunde, J. S.; Khaleel, A.; Klabunde, K. J. *Environ. Sci. Technol.* **2002**, *36*, 762–68.
14. Decker, S.; Klabunde, K. J. *J. Am. Chem. Soc.* **1996**, *118*, 12465–6.
15. Pauzauski, P.; Richards, R. M. Unpublished work.
16. Lucas, E.; Klabunde, K. *Nanostruct. Mater.* **1999**, *12*, 179–82.
17. Klabunde, K. J.; Koper, O.; Khaleel, A. U.S. Patent 6,093,236, July 25, 2000.
18. Klabunde, K. J. U.S. Patent 5,990,373, November 23, 1999.
19. Lucas, E. Ph.D. Thesis, Kansas State University, 2000; p 104.
20. Martin, M. E.; Narske, R. M.; Klabunde, K. J. *Microporous Mesoporous Mater.* **2005**, *83*, 47–50.
21. Wagner, G. W.; Bartram, P. W.; Koper, O.; Klabunde, K. J. *J. Phys. Chem. B* **1999**, *103*, 3225–8.
22. Wagner, G. W.; Koper, O. B.; Lucas, E.; Decker, S.; Klabunde, K. J. *J. Phys. Chem. B* **2000**, *104*, 5118–23.
23. Wagner, G. W.; Procell, L. R.; O'Connor, R. J.; Munavalli, S.; Carnes, C. L.; Kapoor, P. N.; Klabunde, K. J. *J. Am Chem. Soc.* **2001**, *123*, 1636–44.
24. Stoimenov, P. K.; Klinger, R. L.; Marchin, G. L.; Klabunde, K. J. *Langmuir* **2002**, *18*, 6679–86.
25. Koper, O.; Klabunde, J. S.; Marchin, G.; Klabunde, K. J.; Stoimenov, P.; Bohra, L. *Curr. Microbiol.* **2002**, *44*, 49–55.
26. Häggström, J. A. Ph.D. Thesis, Kansas State University, 2008.
27. Stoimenov, P. K.; Zaikovski, V.; Klabunde, K. J. *J. Am. Chem. Soc.* **2003**, *125*, 12907–13.
28. Smetana, A.; Klabunde, K. J.; Marchin, G. R.; Sorensen, C. M. *Langmuir* **2008**, *24*, 7457–64.
29. Häggström, J. A., Klabunde, K. J.; Marchin, G. In press.
30. Private communications with Midwest Research Institute.
31. Sun, N.; Klabunde, K. J. *J. Am. Chem. Soc.* **1999**, *121*, 5587–8.
32. Carnes, C. L.; Klabunde, K. J. *Langmuir* **2000**, *16*, 3764–72.
33. Carnes, C.; Klabunde, K. J. *Chem. Mater.* **2002**, *14*, 1806–11.
34. Sun, N.; Klabunde, K. J. *J. Catal.* **1999**, *185*, 506–12.
35. Take, J.; Kikuchi, N.; Yoneda, Y. *J. Catal.* **1971**, *21*, 164–70.
36. Chesnokov, V. V.; Bedilo, A. F.; Heroux, D. S.; Mishakov, I. V.; Klabunde, K. J. *J. Catal.* **2003**, *218*, 438–46.
37. Pak, C.; Bell, A. T.; Tilley, T. D. *J. Catal.* **2002**, *206*, 49–59.
38. Martyanov, I. N.; Uma, S.; Rodrigues, S.; Klabunde, K. J. *Chem. Commun.* **2004**, 2476–7.
39. Martyanov, I. N.; Klabunde, K. J. *Environ. Sci. Technol.* **2003**, *37*, 3448–53.

Chapter 2

Size-Dependent Properties and Surface Chemistry of Oxide-Based Nanomaterials in Environmental Processes

Vicki H. Grassian*

Department of Chemistry, University of Iowa, Iowa City, IA 52246
***vicki-grassian@uiowa.edu**

Both natural and engineered oxide-based nanomaterials play important roles in environmental processes. In the case of engineered nanomaterials, e.g. nanocyrstalline zeolites with high external and internal surface areas, the properties can be tailored for a number of different environmental applications including carbon dioxide removal and conversion. On the nanoscale, titanium dioxide particles show size-dependent adsorption behavior that suggests nanoscale titanium dioxide may also exhibit size dependent behavior in their toxicity and/or environmental impacts. For naturally occurring oxide-based nanomaterials, e.g. iron oxyhydroxides, size-dependent properties and surface chemistry have the potential to impact the cycling of iron in the global environment. In this chapter, some specific examples of size-dependent properties and surface chemistry of both natural and engineered oxide nanomaterials in environmental processes are presented.

1. Introduction

There is a great deal of interest in engineered nanomaterials because these materials exhibit unique properties, especially surface properties, that can be exploited in heterogeneous catalysis, environmental remediation, biomedical diagnostics and drug delivery. The surface chemistry of nanoparticles for various environmental applications has been pioneered by Kenneth Klabunde, Distinguished Professor at Kansas State University. As noted by Klabunde and co-workers (*1–3*), since most inorganic nanoparticles are not spherical in shape but in fact more cubic or octahedral in nature and they will have edge and corner sites. Klabunde and co-workers suggested that nanoparticles can exhibit enhanced reactivities because of these corner and edge sites that may render these particles more reactive (*1–3*). In particular, nanoparticles less than 10 nm may show unique surface chemistry due to a high surface density of corner and edge sites.

Figure 1 shows conceptually what occurs when a large cube is cut into smaller-sized cubes. From geometric arguments alone, it can be seen that corner and edge sites increase as the cube is made into smaller cubes. This pictorial representation is in some ways the conceptual framework put forth by Klabunde and co-workers. Additionally, in the case of inorganic materials, nanoparticle phases and shapes will be dictated by thermodynamics and differences in surface free energies become an increasingly important component of the total free energy as the particle size decreases. For example, metal oxide-based nanomaterials not only have an abundance of corner and edge sites present but distinct surface planes become predominant as well (see Section 2.3).

Besides the applications of nanomaterials because of their unique properties, there has been an increasing interest recently in understanding the implications of engineered nanomaterials in the natural and human-impacted environment (*4*). This includes the potential of engineered nanomaterials to get into water systems and increase the concentration of heavy metals or the potential for airborne nanoparticles to be toxic and an occupational hazard for those working in the area. These implications are briefly touched upon here.

2. Specific Examples of Size-Dependent Properties and Surface Chemistry of Both Natural and Engineered Oxide Nanomaterials in Environmental Processes

In this section, specific examples are shown to demonstrate the unique and interesting behavior found for nanoscale oxide-based materials. Figure 2 shows the transmission electron micrograph of the three oxide-based nanomaterials discussed in this chapter. These materials include nanocrystalline zeolites, titanium dioxide and an iron hydroxide, FeOOH. Nanocrystalline zeolites are aluminosilicates with internal pore sizes on the order of 1 nm and particle sizes on the order of 20 nm. The ability to synthesize zeolites with particle sizes below 100 nm gives these crystalline materials both high internal and external surface areas and adsorption studies for carbon dioxide show both of these sites are active toward carbon dioxide uptake. Nanoscale titanium dioxide can be made much smaller in size than nanocrystalline zeolites. Figure 3 shows ca. 5 nm TiO_2 nanoparticles. These particles are pure anatase and at this size nearly 40% of the atoms in the particle are at the surface. Unique adsorption sites are observed in FTIR experiments of these small TiO_2 nanoparticles. The third nanomaterial to be discussed here is that of the mineral goethite, α-FeOOH, which is a rod-shaped mineral. As shown in Section 2.3, nanorods exhibit enhanced dissolution behavior compared to larger microrods.

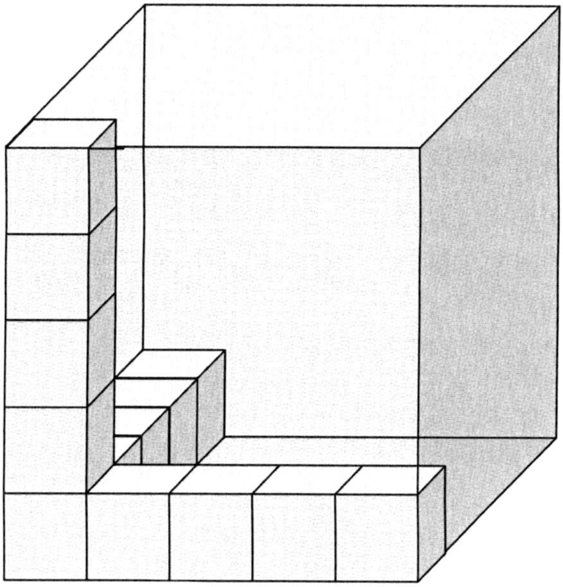

Figure 1. The number of edge and corner sites increases with decreasing size as shown geometrically for cube being divided into smaller cubes. Take for example, a one 5 cm x 5 cm cube has length of edges = 60 cm and one hundred twenty five-1cm x 1cm cubes have length of edges = 1500 cm (25 times greater edge length). In general, edge length increases by 1/(factor of size change) (2) and number of corners increases by 1/(factor of size change) (3).

2.1. Nanocrystalline Zeolites

Zeolites are well-known crystalline, aluminosilicate molecular sieves with pores of molecular dimensions (*5*). While zeolites have been studied extensively for many years due to their adsorptive and catalytic properties, the ability to control the crystal size has only recently been realized. As such, there is a great deal of interest in the unique properties of nanocrystalline zeolites (zeolites with crystal sizes of 100 nm or less) relative to conventional micron-sized zeolite crystals. A key feature of nanocrystalline zeolites is that the external surface can be utilized as a reactive or adsorptive surface resulting in materials with a wide variety of physical and chemical characteristics (*5*). The surface properties of the nanocrystalline materials can be easily functionalized (*6*). Other advantages of nanocrystalline zeolites include improved mass-transfer properties and optical transparency due to the small crystal size (*7*).

Recent studies indicate there is a potential to use nanocrystalline zeolites for a number of environmental applications (*8*). Faujasite zeolites (e.g. NaY) are considered promising environmental catalysts because of their cation exchange capacity and acid–base properties and nanocrystalline NaY zeolites have shown enhanced performance for the selective catalytic reduction of NO_x (NO_x = NO + NO_2) with urea (*9*). Recent studies indicate that zeolites have the potential to adsorb, store and convert CO_2 into more useful products (*10–14*). As CO_2 plays

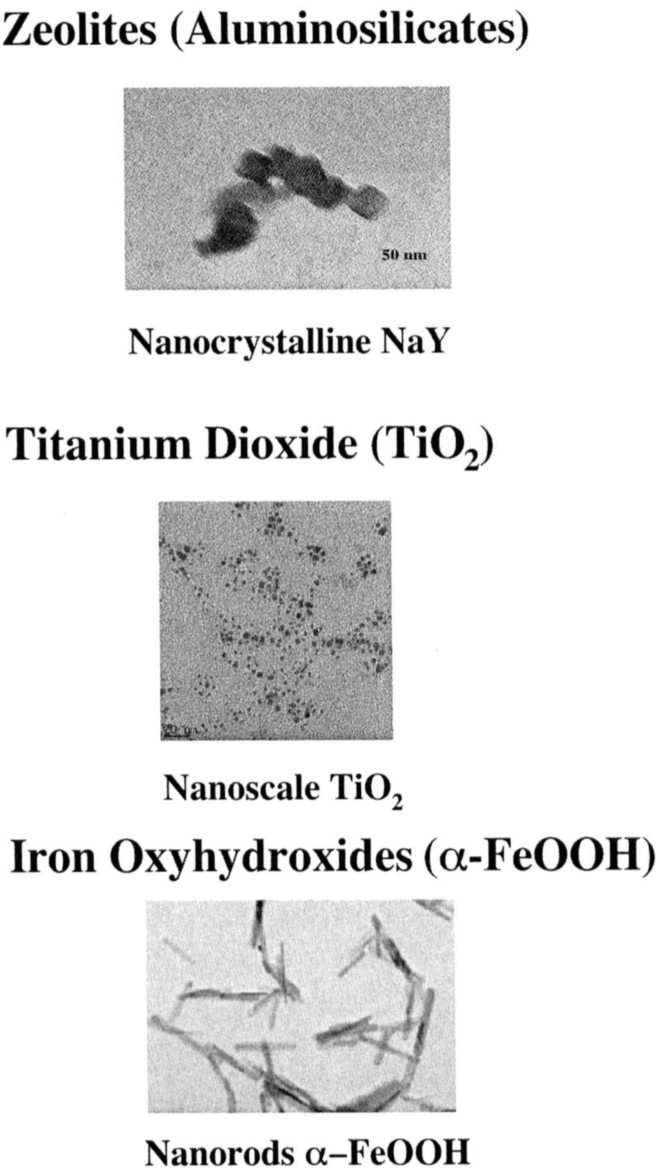

Figure 2. Transmission electron micrograph of three oxide-based nanomaterials discussed in this chapter: NaY nanocrystalline zeolites, nanoscale TiO_2 and α-FeOOH nanorods.

an important role in global warming, it has become necessary to think about new approaches and novel ideas for CO_2 management.

In this chapter, FTIR studies of CO_2 adsorption in commercial NaY and nanocrystalline NaY; herein referred to as NaY, and nano-NaY, are compared. FTIR spectroscopy is a useful probe of CO_2 and can be used to determine if CO_2 adsorption in nanocrystalline zeolites with particle size below 100 nm, show

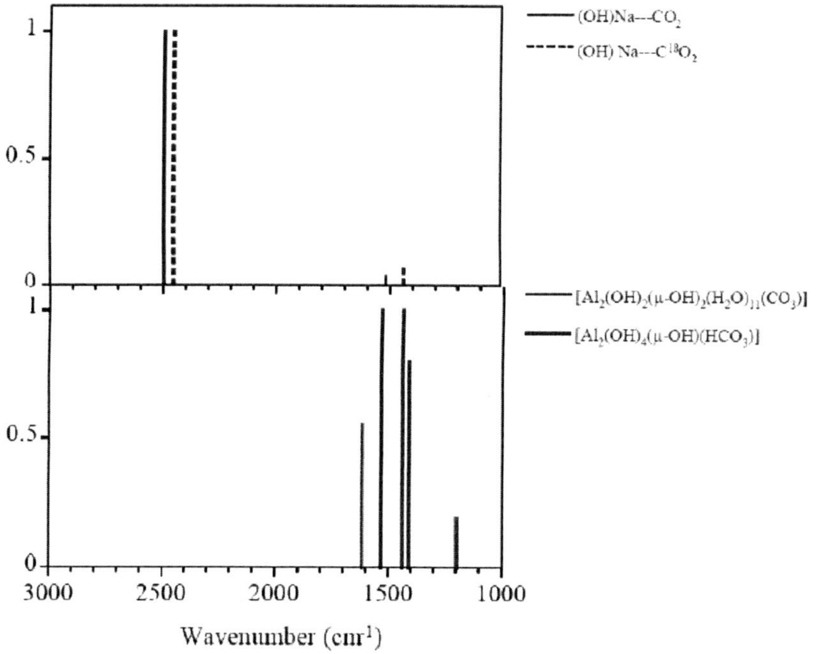

Figure 3. Top panel represents the calculated spectra of (OH)Na clustered with $C^{16}O_2$ unlabeled (solid blue line) and $C^{18}O_2$.Bottom panel represents the vibrational modes for aluminum oxide carbonate complexes (red line) and aluminum oxide bicarbonate complexes (green line). Energy Environ. Sci., ***2009****, 2, 401 – adapted and reproduced by permission of the PCCP Owner Societies.*

unique carbon dioxide adsorption sites due to the high external surface area compared with commercial NaY, with larger particle size.

2.1.1. Zeolite Characterization

Commercial zeolite, NaY (Aldrich) and synthesized nano-NaY were used in these studies. The crystal sizes were ~1 mm and 38 nm for NaY (Aldrich) and nano-NaY, respectively. The most important difference between the two samples is particle size and the difference in external surface area. The external surface area for the two sizes is 1 m^2/g and 100 m^2/g respectively (*14*). Thus, the external surface area for the 38 nm nano-NaY is 100 times greater than the commercial sample. The external surface provides for additional reaction sites and functionality in these nanocrystalline zeolite materials.

2.1.2. Calculated Vibrational Frequencies for Adsorbed CO_2

Figure 3 summarizes and provides in stick format the vibrational frequencies expected for different adsorption modes of carbon dioxide – linear bonded CO_2

and carbonate and bicarbonate, possible adsorption products. These data are used as a guide for the interpretation of the experimental data for CO_2 adsorbed in NaY and nano-NaY (*15*).

2.1.3. Transmission FTIR Data

Figure 4 shows the spectral region, from 1200 to 4000 cm^{-1} for $C^{16}O_2$ adsorbed in NaY at a pressure of 20 Torr. As seen in Figure 4, there is an intense absorption band at 2351 cm^{-1}. This band is assigned to the asymmetric stretching mode (ν_3) of CO_2. The ν_3 peak is shifted from the gas-phase value of 2347 cm^{-1} due to the interaction with the exchangeable cation. This band increases in intensity as a function of $C^{16}O_2$ pressure (*data not shown*). A sharp peak is observed near 1382 cm^{-1} due to the symmetric stretch of CO_2 as is observed in the calculated spectra shown in Figure 3. In the gas-phase this peak is IR inactive. However, this peak becomes IR active when CO_2 is adsorbed in the zeolite due to interaction with the exchangeable cation and the symmetry is lowered on upon interaction, because of two unequal bond lengths of C—O bond as indicated by theoretical calculations (*15*), giving rise to an IR active mode. Additional bands are observed in the spectrum and are assigned to other vibrational modes of adsorbed carbon dioxide, some of which are overtones and some are combination bands. One of them is a small peak observed at around 1276 cm^{-1} for these zeolites, which is a combination band arising from Fermi resonance between the Raman-active ν_1 symmetric vibration and the overtone of the bending mode ν_2 ($2\nu_2$) . As already discussed, due to the reduction in symmetry of carbon dioxide on interaction with the zeolite surface, this mode becomes IR active as well. The combination bands of $\upsilon_3+\upsilon_1$ and $\upsilon_3+2\upsilon_2$ at 3713 and 3605 cm^{-1} are also present.

In addition to $C^{16}O_2$ adsorption in NaY, Figure 5 shows $C^{16}O_2$ adsorption in nano-NaY zeolites. Interestingly, the ν_3 vibrational mode for the commercially available NaY as well as nanocrystalline NaY zeolite is observed at exactly the same wavenumber. Besides the intense ν_3 vibrational band, a number of other vibrational bands are observed including the nano-NaY spectra that are associated with $C^{16}O_2$ adsorbed inside the zeolite cage. Of particular interest are the differences observed in these spectra for the different zeolite materials. In comparing the 1200 to 1800 cm^{-1} spectral region for the NaY and nano-NaY zeolites, there is evidence for additional bands observed in nano-NaY spectra.

Figure 6 shows the IR spectra collected upon adsorption of $C^{16}O_2$ in nano-NaY zeolite as a function of pressure in the spectral region between 1200 and 4000 cm^{-1}. In particular, for nano-NaY there are several broad features near 1640, 1461 and 1381 cm^{-1}. As the carbon dioxide equilibrium pressure increases, one can see these broad features grow in intensity. The FWHM (full width half maximum) for these peaks are greater than the other absorption bands in the spectrum by more than 20 cm^{-1}. Although the peak at 1381 cm^{-1} has some contribution to the symmetric stretching mode as already discussed, there is a broad band underlying the sharper feature.

Based on earlier studies, these new absorption bands are proposed to be due to the formation of bicarbonate and carbonate species on extra framework

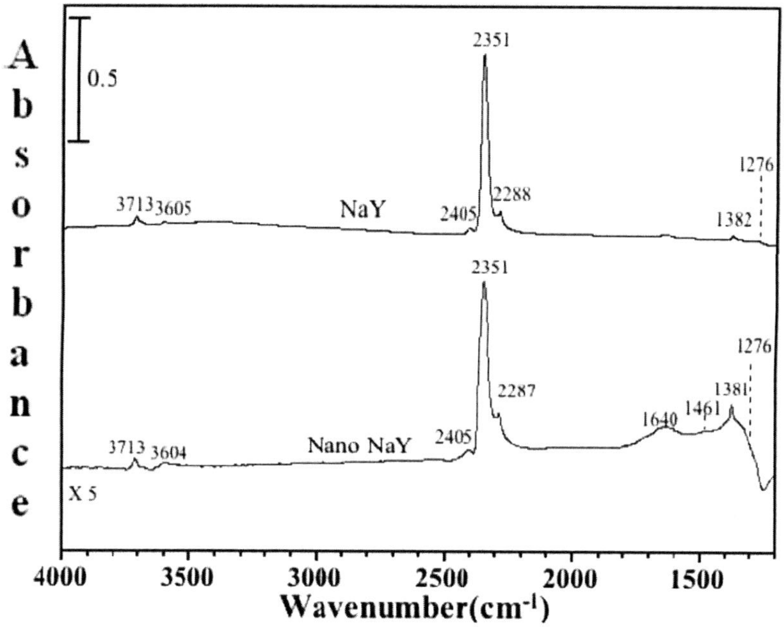

*Figure 4. FTIR spectra of adsorbed $C^{16}O_2$ on dehydrated NaY and nano-NaY zeolite at a CO_2 pressure of 20 Torr and temperature of 296 K. Energy Environ. Sci., **2009**, 2, 401 –adapted and reproduced by permission of the PCCP Owner Societies.*

aluminum (EFAL) sites (*5, 16*) known to be present on the external surface of nanocrystalline zeolites. As has been shown previously, nanocrystalline zeolites not only possess high surface area but also have a high concentration of reactive surface sites because of the increased external surface area due to the smaller particle size. EFAL sites are most notably active catalytic sites in nanocrystalline zeolites that cause these zeolites to demonstrate higher activity as compared to commercial zeolites with large particle size. For example, it has been previously shown that molecules, e.g. NO_2, can adsorb to these sites yielding surface complexes unique to nanocrystalline NaY zeolite (*16*). We propose here that these EFAL sites play a similar role in the adsorption of CO_2 in the nanocrystalline NaY zeolite and result in the formation of carbonate and bicarbonate species.

2.1.4. Summary of CO_2 Adsorption in Nano-NaY.

These studies show that nano-NaY zeolites have two modes of adsorption for carbon dioxide. These zeolites can store carbon dioxide in the internal pores similar to zeolites with larger particle size and carbonate/bicarbonate species formation on the external surface sites. Thus nanocrystalline zeolites possess adsorption sites that are efficient in conversion of carbon dioxide. Further studies should investigate how modified and/or functionalized nanocrystalline zeolites can be used to convert carbon dioxide to more useful products such as methanol.

Because of the high external surface area and active sites present on the external surface, nanocrystalline zeolites in particular may be useful zeolite materials in CO_2 recycling and conversion processes.

2.2. Nanoscale TiO_2

Titanium dioxide manufactured nanomaterials are useful materials in a number of applications including photocatalysts, solar cells, biomaterials, memory devices and as environmental catalysts. Although some studies have investigated size effects in the interactions of molecules and ions adsorbed on TiO_2 particle surfaces, there is no clear consensus on the impact of particle size on surface adsorption and surface chemistry.

From thermodynamic considerations, smaller nanoparticles are predicted to show enhanced adsorption due to an increase in interfacial tension and surface free energy with decreasing particle size (*17*, *18*). Solution phase adsorption studies with a series of organic acids onto TiO_2 nanoparticles seemed to confirm this prediction (*17*). However in another study, the surface adsorption of Cd^{2+} on TiO_2 was found to decrease with decreasing particle size (*19*). Using sum frequency generation (*20*), Shultz and co-workers showed that for substrate-deposited nanoparticles, smaller TiO_2 nanoparticles were more reactive than larger ones as determined by the relative amounts of dissociative versus molecular adsorption of methanol from the gas phase. Smaller nanoparticles were shown to enhance dissociative adsorption (*20*). From the above examples, it remains unclear as to the expected size-dependent trends in the adsorption and surface chemistry of TiO_2 nanoparticles.

2.2.1. TiO_2 Nanoparticle Surface Adsorption of Oxalic Acid

To better understand the adsorption of oxalic acid on 5 and 32 nm TiO_2 particles, both macroscopic solution phase adsorption isotherms and spectroscopic measurements were done to quantify coverages and to probe the molecular structure of adsorbed oxalic acid on these two different-sized nanoparticles (*21*). Using a Langmuir adsorption model fit to the data,

$$\theta = \frac{N}{N_s} = \frac{CK_{ads}}{1 + CK_{ads}} \qquad (1)$$

where θ is the fractional surface coverage, N is the number of adsorbed molecules or ions, N_s is the maximum number of adsorbed molecules or ions, C is the solution phase concentration, and K_{ads} is the Langmuir adsorption equilibrium constant, the following adsorption parameters are determined: K_{ads} values of 4900 ± 400 and 2900 ± 400 M^{-1} for 5 and 32 nm particles, respectively and saturation surface coverages, N_s, of 7.2 ± 0.2 x 10 (*13*) molecules cm^{-2} compared to 6.6 ± 0.6 x 10 (*13*) molecules cm^{-2} for 5 and 32 nm particles, respectively, where N_s has been normalized to BET surface area. Within error, the maximum surface coverage is the same for the different-sized nanoparticles, however, K_{ads} differs by nearly a factor of two indicating some differences in the adsorption of oxalic acid on the smaller nanoparticles.

Spectroscopic studies can probe any molecular-level differences found in the adsorption of oxalic acid on TiO_2 nanparticles. ATR-FTIR spectra for oxalic acid adsorbed on 5 and 32 nm TiO_2 particles, as well as solution phase oxalic acid, at pH 6.5 are shown Figure 6 (*21*). These spectra show that speciation in solution phase differs from that on the surface. In particular, at pH 6.5, the oxalic acid is completely deprotonated in solution to yield oxalate, $C_2O_4^{2-}$, as seen by the characteristic absorption bands at 1577 and 1309 cm^{-1}, consistent with previous results. However, the spectrum for adsorbed oxalate clearly shows that these absorption bands are not present. Instead there are characteristic oxalic acid peaks around 1287 cm^{-1}, 1431 cm^{-1}, seen in the 32 nm particle spectrum which have been previously assigned in the literature to the symmetric stretch, $\nu_s(CO_2)$ and a combination band, respectively, whereas the bands at 1695 cm^{-1} and 1719 cm^{-1} are assigned to an asymmetric stretch, $\nu_{as}(CO_2)$ (*22*, *23*). These same absorption bands are present for the 5 nm particle spectrum, however, some are shifted in frequency (for example the band at 1703 cm^{-1} is now centered at 1695 and the band at 1287 is now at 1300). A comparison of the spectra for oxalic acid adsorption on 5 versus 32 nm particles shows that besides slight differences in the frequencies of some of the absorption bands, there is an additional absorption band present in the spectrum for oxalic acid adsorbed on 5 nm TiO_2 particles. In particular, a unique band near 1630 cm^{-1}, which is either completely absent in the spectrum for oxalic acid adsorption on 32 nm particles or relatively weak compared to the other absorption bands in the spectrum, is observed. This peak is assigned to a red-shifted asymmetric stretching carboxyl region, $\nu_{as}(CO_2)$ for oxalic acid adsorption and it is proposed to be associated with a unique species adsorbed on 5 nm particle surfaces, potentially adsorption on to low coordination edge and/or corner sites which will be in greater abundance for 5 versus 32 nm particles. Although edge and corner sites have been proposed to increase reactivity in nanoparticles, it has been difficult to attribute unique adsorption complexes to these sites as seen here for the adsorption of oxalic acid on the smaller 5 nm TiO_2 particles.

2.2.2. TiO_2 Nanoparticle Toxicity

Several earlier studies have suggested that for a given mass of particles, nanoparticle toxicity increases with a decrease in size of the primary particle size due to the greater surface area of these smaller particles. It was shown that for titanium dioxide, the increase in the number of neutrophils, a measure of inflammation and toxicity, was directly correlated with particle surface area for two different sized particles, 20 and 200 nm (*24*, *25*). Specifically it was shown that the data collected for these two different sized particles fell on the same dose-response curve when plotted as a function of surface area. Thus, these earlier studies suggested that the most relevant and important dose metric in nanoparticle toxicity is nanoparticle surface area determined from either geometric consideration or BET measurements.

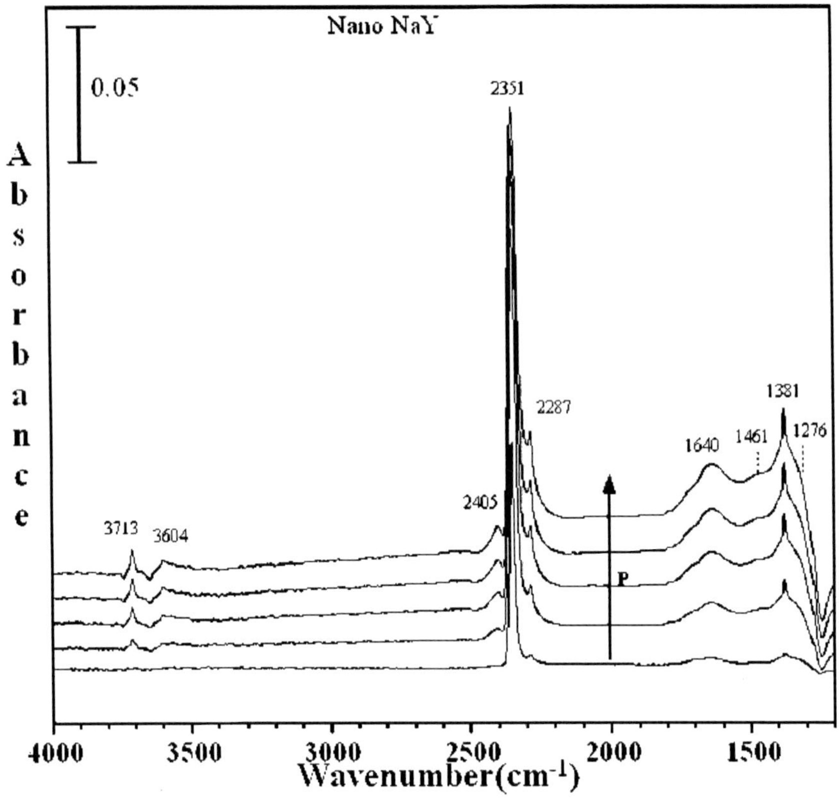

Figure 5. FTIR spectra for of adsorbed $C^{16}O_2$ as a f(P) on nano-NaY. (P_{eq} = 1.0, 5.2, 10.6, 14.4 and 19.8 Torr). Energy Environ. Sci., ***2009****, 2, 401 –adapted and reproduced by permission of the PCCP Owner Societies.*

More recently there have been additional studies that suggest that TiO_2 nanoparticle toxicity in particular, and potentially nanoparticle toxicity in general, may in fact be more complex. For example, Sayes et al. have suggested the phase of the nanoparticle plays an important role in nanoparticle toxicity (*26*). Grassian et al. have shown that for smaller TiO_2 nanoparticles below 10 nm, the expected increase in toxicity relative to 20 nm particles did not occur (*27*). Some of these data from Grassian et al. are shown in Figure 7. In particular, Figure 7 shows the analysis of the broncheoalvelolar lavage fluid for macrophages, neutrophils and lymphocytes in mice following exposure to 5 and 20 nm particles at high (H) and low (L) concentrations of particles, 7.7 mg/m^3 and 0.7 mg/m^3, respectively, compared to sentinels and mice exposed to aerosolized water. These are acute exposures (4 hrs) shown in Figure 7 and mice are analyzed immediately after 4 hours or 24 hours after exposure began. The main point here is that the largest response, as shown by the increase in macrophages, is seen for the larger particles and not the smaller ones at the same mass even though the 5 nm particles have a specific surface area between 5-6 times greater than the larger 20 nm particles. Furthermore, it was shown in Grassian et al. that when the number of

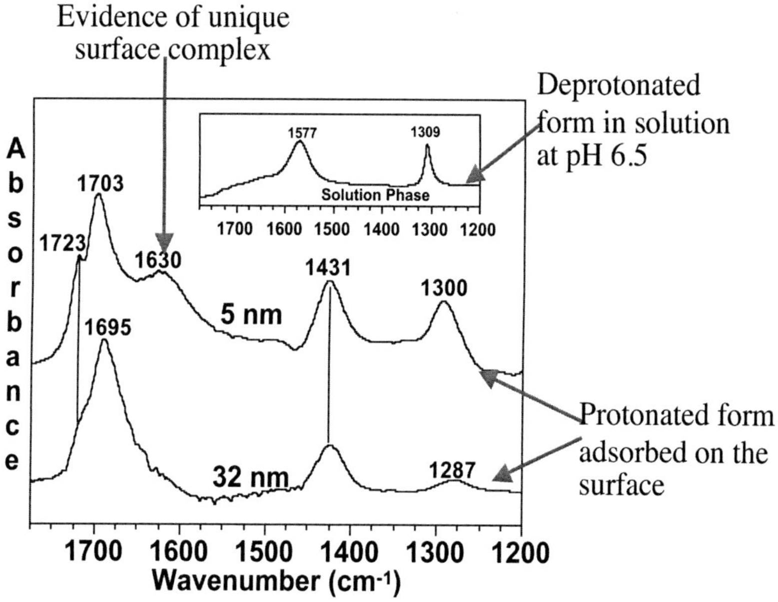

Figure 6. ATR-FTIR spectra for oxalic acid adsorption onto 5 and 32 nm TiO_2 particles at pH 6.5 compared to the solution phase spectrum shown in the inset. Adapted and reproduced with permission from Langmuir ***2008****, 24, 6659. Copyright 2008 American Chemical Society.*

neutrophils are plotted as a function of total surface area for the two different sized nanoparticles, 5 nm and 20 nm, the data did not fall on the same dose-response curve as expected from the earlier instillation studies for larger particles (20 and 200 nm). A possible cause for this unexpected result comes from a very recent study that investigated the formation of reactive oxygen species (ROS), often the cause of nanoparticle toxicity toward cells, as a function of TiO_2 particle size (*28*). It was determined that ROS formation is a maximum at a particle size near 20 nm and that for smaller nanoparticles ROS formation actually decreases. Although for the most part TiO_2 nanoparticles show moderate if very limited toxicity, these more recent studies indicate that nanoparticle toxicity is complex and will not be just a function of surface area but will depend more specifically on nanoparticle physicochemical properties which are size-dependent.

2.2.3. Summary of Surface Adsorption and Toxicity of Nanoscale TiO_2

It is clearly shown in this section, through very different types of studies, that there is size-dependent adsorption and toxicity for titanium dioxide on the nanoscale that cannot be interpreted based on surface area effects alone, thus making generalization and predictive conclusions more difficult. From a chemist's perspective, this makes the surface chemistry of nanoscale TiO_2 and nanomaterials in general, very interesting and worthy of additional attention.

From an environmental, health and safety perspective, this makes challenges associated with these issues more difficult.

2.3. α-FeOOH Nanorods

Iron oxides and iron hydroxides are a natural and reactive component of air, water and soils and play an important role in biogeochemical cycles and the global cycling of iron. Fe-containing mineral dust aerosol provides iron to regions of the ocean where it is a limiting nutrient. There has been much discussion on the implications of size-dependent reactivity of iron oxides and hydroxides on global iron cycling and although recent evidence suggests that particle size may be an important and perhaps a controlling factor in the dissolution of iron from mineral dust aerosol, a fundamental understanding of the influence of particle size on iron oxide dissolution remains unclear. More specifically, what, if any, is the role of particle size in Fe dissolution and bioavailability? Size-dependent behavior for iron dissolution can manifest itself in several ways – through adsorption, proton and ligand-promoted dissolution and photochemical reductive dissolution. In this chapter, the size-dependent surface reactivity of α-FeOOH nanorods and microrods with oxalic acid, an abundant water soluble organic found to be associated with Fe-containing mineral dust aerosol, and the ligand-promoted dissolution of nano versus micro rods is discussed.

2.3.1. Characterization of α-FeOOH Nanorods and Microrods

Nanorods and microrods were characterized using powder X-ray diffraction (XRD) performed on a Bruker D-5000 diffractometer with a Cu Kα source. Powder X-ray diffraction patterns were consistent with that expected for goethite, with patterns for nanorods exhibiting line broadening as is typically observed with decreasing particle size. Nanorod and microrod dimensions were obtained from single particle analysis with TEM. Nanorods were 81 (± 27) nm by 7 (± 2) nm, and microrods were 670 (± 370) nm by 25 (± 9) nm (uncertainties represent one standard deviation) (*29, 30*). The specific surface area of nanorods and microrods were 110 (± 7) and 40 (± 3) m^2/g, respectively.

2.3.2. Oxalate-Promoted Dissolution of α-FeOOH Nanorods and Microrods

At pH 3, proton-promoted goethite dissolution was not observed over 30 h (data not shown). In the presence of oxalate, however, the rate of ligand-promoted dissolution was appreciable over this time scale and was considerably greater for nanorods compared to microrods (Figure 8). From linear regression analysis, the initial rate of dissolved Fe(III) production in nanorod suspensions was 52 (± 10) $\mu mole \cdot g^{-1}h^{-1}$. For microrods, this rate was 4.6 (± 1.4) $\mu mole \cdot g^{-1}h^{-1}$, indicating roughly an 11 (± 4)-fold difference in the initial rate of oxalate-promoted dissolution on the basis of goethite mass. Surface-area-normalized rates of goethite dissolution were 0.47 (± 0.09) and 0.12 (± 0.04) $\mu mole \cdot m^{-2}h^{-1}$ for

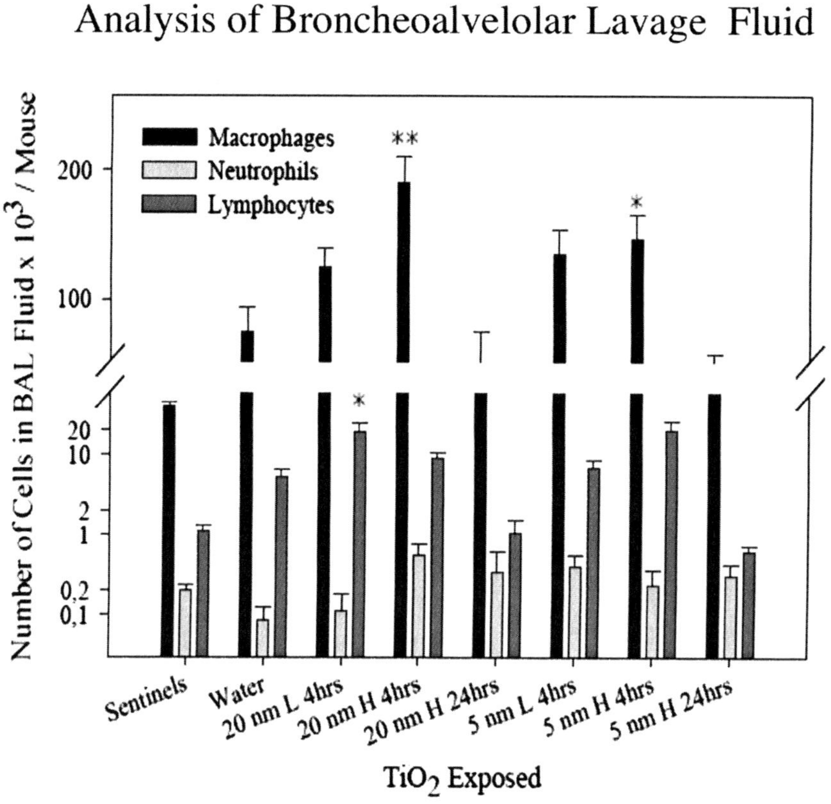

Figure 7. Number of cells in BAL fluid in animals exposed to low (L) and high (H) concentration of TiO_2 particles by inhalation (animals were necropsied 4 hours or 24 hours from the beginning of exposure). Asterisks represent significant increase ($p<0.05$, ** $p<0.01$, and *** $p<0.001$) in parameter measured, compared to controls. Reprinted by permission from Reference (27) by the publisher Taylor & Francis Ltd., http://www.tandf.co.uk/journals, "Inflammatory response of mice to manufactured titanium dioxide nanoparticles: Comparison of size effects through different exposure routes", 2009, Informa Healthcare.*

nanorods and microrods, respectively. On the basis of particle surface area, therefore, nanorods exhibit a 4.1 (± 1.3)-fold increase in their rate of dissolution relative to microrods. To the best of our knowledge this is one of the first demonstrating such size-dependent behavior for ligand-promoted dissolution of iron oxides.

The difference in the inherent surface reactivity of nanorods and microrods is likely greater than that estimated simply from consideration of specific surface area. Specifically, the production rate of dissolved Fe(III), which is equal to the rate of goethite dissolution, is described by a second-order rate law, where the rate is proportional to the surface concentration of adsorbed oxalate ($[Ox]_{ads}$) in the form of a bidentate mononuclear complex and the concentration of Fe(III) surface sites ($[Fe(III)]_{surf}$) (eq 1) (*31*).

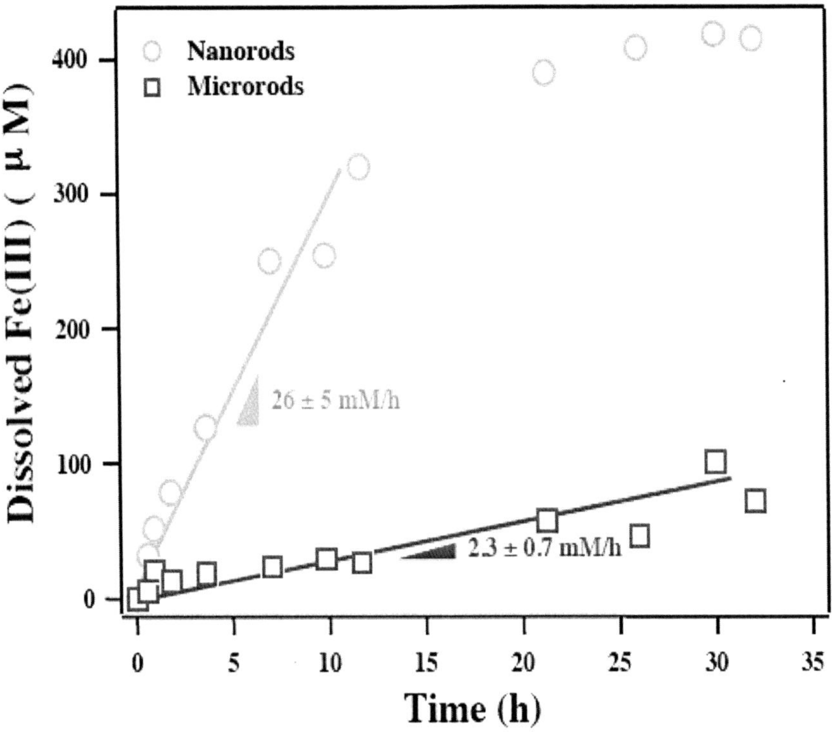

Figure 8. Dissolved Fe(III) production via ligand-promoted dissolution in suspensions containing 0.5 g/L goethite and 1 mM oxalate. Experiments were conducted in the absence of light in pH 3 solutions containing 5 mM $NaClO_4$. Adapted and reproduced with permission from Journal of Physical Chemistry C ***2009****, 113, 2175. Copyright 2009 American Chemical Society.*

$$\textit{Dissolution Rate} = d[Fe(III)]/dt = k[Ox]_{ads}[Fe(III)]_{surf} \quad (2)$$

Note that that adsorption isotherms show that that nanorods adsorbed ca. 30% less oxalate per unit BET surface area than microrods at pH 3. Thus, the rate of goethite dissolution per unit of adsorbed oxalate is roughly 6-fold greater for nanorods relative to microrods. It appears, therefore, that the oxalate complexes responsible for goethite dissolution are more reactive and promote dissolution to a greater extent when formed on nanorod surfaces.

2.3.3. TEM Analysis of α-FeOOH Nanorods Following Oxalate-Promoted Dissolution

Changes in nanorod morphology after exposure to oxalate-containing solutions at pH 3 (Figure 9) were also observed here. TEM images of nanorods prior to oxalate addition (Figure 9a) show sharp, well-defined ends characteristic of their (021) faces. In contrast, images collected 24 h (not shown) and 200 h (Figure 9b) after addition of 1 mM oxalate revealed changes in particle morphology over time; nanorods became cigar-shaped, developing narrow, rounded end. Over the course of 200 h, the average length of nanorods also decreased from an initial value of 81 (± 27; n = 530) nm to 66 (± 33; n = 100) nm. Although the standard deviations associated with these averages are relatively large, particle size analysis clearly illustrates that dissolution yields a greater percentage of nanorods with smaller lengths relative to unreacted particles (Figure 9c). In contrast, little change was observed in the width of nanorods over long-term dissolution experiments (Figure 9d). Although similar morphological changes were expected for microrods, they were far more difficult to discern, presumably due to the lower rate of oxalate-promoted dissolution observed in these systems.

Size-dependent trend for oxalate-promoted dissolution also reflects differences in the relative amount and reactivity of crystal faces present on nanorod and microrod surfaces. Specifically, changes in nanorod morphology are consistent with oxalate-promoted dissolution being most pronounced on the (021) faces and along their intersections with (110) faces. Thus, anisotropic dissolution is not limited to the proton-promoted processes explored by Cornell *et al.* (*32*), but also applies to ligand-mediated dissolution as well.

Cornell *et al.* (*32*) proposed that the unique reactivity of the (021) faces may result from differences in the coordination and concentration of the surface hydroxyl groups present on each face. Consistent with this hypothesis, Barón and Torrent (*33*) calculated that the highest density of singly and doubly coordinated surface hydroxyl groups occurs on the (021) faces of goethite (8.2 groups/nm^2, respectively). Lower densities (~ 3 groups/nm^2) were determined for the (110) faces. At first, this scenario would seem consistent with our results, particularly because ATR-FTIR characterization suggests a greater density of singly coordinated surface hydroxyl groups on nanorods relative to microrods. However, a greater concentration of surface sites would also be anticipated to yield higher concentrations of adsorbed oxalate per unit surface area, which was not observed in nanorod suspensions. This apparent contradiction may reflect the role that particle aggregation plays in our systems. It is possible that more extensive aggregation in nanorod systems limits the number of surface sites available for oxalate uptake, but that the surface hydroxyl groups on the (021) faces of nanorods are considerably more reactive than their counterparts on microrod surfaces, thereby yielding higher rates of goethite dissolution. Indeed, molecular simulations suggest that the charging behavior of (021) faces is influenced by goethite particle size (*34*), behavior that would affect the interfacial reactivity of nanorods and microrods.

2.3.4. Summary of Nanorod versus Microrod Dissolution

Results from the study discussed here indicate that molecular-level differences exist in the surface chemistries of goethite nanorods and microrods, and that these differences have implications for their interactions with oxalate. Most notably, ATR-FTIR analysis suggests that nanorods exhibit a greater density of surface hydroxyl groups relative to microrods and that differences

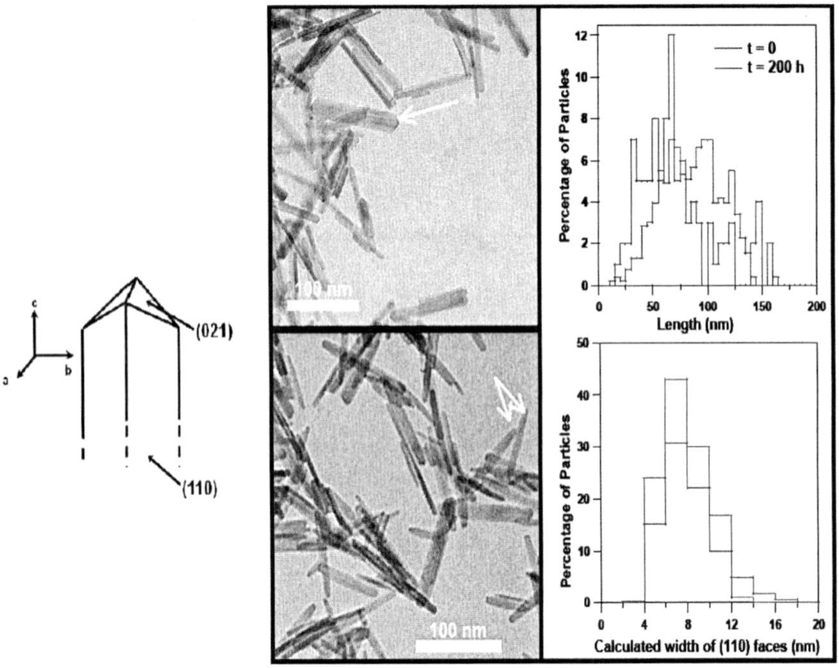

Figure 9. (a) TEM images of acicular goethite rods prior to reaction with oxalate. Inset illustrates the dominant crystal faces of goethite nanorods, whereas the arrow highlights their well-defined ends terminated by (021) faces. (b) TEM images of goethite nanorods collected 200 h after the addition of 1 mM oxalate. The arrows highlight the narrow, rounded ends resulting from dissolution. Similar images were collected for nanorods after photochemical reductive dissolution mediated by oxalate. TEM analysis was used to construct distributions for the (c) length and (d) width of nanorods before and after reaction with oxalate for 200 h. For unreacted nanorods, distributions were constructed from the results of sizing 530 nanorods. Distributions for reacted nanorods were constructed after analysis of 100 particles. Note that the widths in (d) represent the width of the 110-type faces, as explained earlier in the text. Adapted and reproduced with permission from Journal of Physical Chemistry C ***2009****, 113, 2175. Copyright 2009 American Chemical Society.*

likely exist in the oxalate complexes formed on the surfaces of different sized particles. It is challenging to quantitatively assess the impact of these molecular-level differences on macroscopic reactivity because oxalate induces extensive aggregation, introducing much uncertainty as to the available surface area in each suspension. However, results of ligand-promoted and photochemical reductive dissolution reactions suggest that nanorods tend to be more labile than microrods, a result with fairly important implications for the role that nanoscale iron oxides may play in iron solubilization processes and the cycling of iron in the global environment. In particular, ligand-promoted, non-reductive dissolution reactions appear to occur at a much greater rate on nanorod surfaces, and the difference in reactivity far exceeds that expected from simple considerations of nanorod specific surface area.

3. Oxide-Based Nanomaterials in Environmental Processes

The data shown in Section 2 provide several examples of unique surface chemistry of nanoscale materials compared to larger particles. These examples are for nanocrystalline zeolites, nanoscale titanium dioxide and nanorods of iron hydroxyoxides. Specifically we have shown that:

A. Nanocrystalline NaY zeolites unlike zeolites with larger particle size shows two modes of CO_2 adsorption – a linear complex in the pores of the zeolite with carbon dioxide interacting with the sodium cation and the formation of carbonate and bicarbonate on the external surface.
B. Titanium dioxide shows size-dependent adsorption properties and unique adsorption sites following oxalic acid adsorption on 5 nm particles and nanoparticle toxicity that is less for smaller nanoparticles compared to larger ones.
C. Iron oxyhydroxide nanorods (α-FeOOH) show size-dependent dissolution due to increased density of surface planes in nanorods that are more reactive with respect to dissolution.

Acknowledgments

I would like to thank my collaborators and students at the University of Iowa for their contributions to this work. In particular, I would like to acknowledge Professors Sarah Larsen, Dr. Juan Navea and Pragati Galhotra for contributing to the nanorystalline zeolite studies, John Pettibone for contributing to the nanoscale TiO_2 studies and Professor Michelle Scherer and David Cwiertny for contributing to the α-FeOOH dissolution studies. This work was supported by grants from the National Science Foundation and Environmental Protection Agency.

References

1. Utamapanya, S.; Klabunde, K. J.; Schlup, J. R. *Chem. Mater.* **1991**, *3*, 175–181.
2. Stark, J. V.; Park, D. G.; Lagadic, I.; Klabunde, K. J. *Chem. Mater.* **1996**, *8*, 1904–1912.
3. Klabunde, K. J.; Stark, J.; Koper, O.; Mohs, C.; Park, D. G.; Decker, S.; Jiang, Y.; Lagadic, I.; Zhang, D. J. *J. Phys. Chem.* **1996**, *100*, 12142–12153.
4. Pettibone J. B.; Elzey, S.; Grassian, V. H. An Integrated Approach Toward Understanding the Environmental Fate, Transport, Toxicity and Health Hazards of Nanomaterials. In *Nanoscience and Nanotechnology: Environmental and Health Impacts*; Grassian, V. H., Ed.; 2008; pp 43–68.
5. Larsen, S. C. *J. Phys. Chem. C* **2007**, *111*, 18464–18474.

6. Song, W.; Woodward, J. F.; Grassian, V. H.; Larsen, S. C. *Langmuir* **2005**, *21*, 7009–7014.
7. Alwy, H.; Li, G.; Grassian, V. H.; Larsen, S. C. Development of Nanocrystalline Zeolites as Environmental Catalysts. In *Nanotechnology and the Environment*; ACS Symposium Series 890; Karn, B., Ed.; American Chemical Society: Washington, DC, 2005; pp 277−283.
8. Song, W.; Li, G. H.; Grassian, V. H.; Larsen, S. C. *Environ. Sci. Technol.* **2005**, *39*, 1214–1220.
9. Li, G.; Jones, C. A.; Grassian, V. H.; Larsen, S. C. *J. Catal.* **2005**, *234*, 401–413.
10. Angell, C. L.; Howell, M. V. *Canadian J. Chem.* **1969**, *47* (20), 3831.
11. Chan, B.; Radom, L. *J. Am. Chem. Soc.* **2008**, *130* (30), 9790–9799.
12. Bonenfant, D.; Kharoune, M.; Niquette, P.; Mimeault, M.; Hausler, R. *Sci. Technol. Adv. Mater* **2008**, *9* (1), 013007.
13. Khelifa, A.; Derriche, Z.; Bengueddach, A. *Microporous Mesoporous Mater.* **1999**, *32*, 199–209.
14. Goj, A.; Sholl, D. S.; Akten, E. D.; Kohen, D. *J. Phys. Chem. B* **2002**, *106* (33), 8367–8375.
15. Galhotra, P.; Navea, J.; Larsen, S. C.; Grassian, V. H. *Energy Environ. Sci.* **2009**, *2*, 401–409.
16. Li, G.; Larsen, S. C.; Grassian, V. H. *Catal. Lett.* **2005**, *103*, 23–32.
17. Zhang, H. Z.; Penn, R. L.; Hamers, R. J.; Banfield, J. F. *J. Phys. Chem. B* **1999**, *103*, 4656.
18. Lu, H. M.; Wen, Z.; Jiang, Q. *Chem. Phys.* **2005**, *309*, 303.
19. Gao, Y.; Wahi, R.; Kan, A. T.; Falkner, J. C.; Colvin, V. L.; Tomson, A. B. *Langmuir* **2004**, *20*, 9585.
20. Wang, C. Y.; Groenzin, H.; Shultz, M. J. *J. Am. Chem. Soc.* **2005**, *127*, 973620.
21. Pettibone, J. B.; Cwiertny, D. M.; Scherer, M.; Grassian, V. H. *Langmuir* **2008**, *24*, 6659–6667.
22. Hug, S. J.; Sulzberger, B. *Langmuir* **1994**, *10*, 3587.
23. Duckworth, O. W.; Martin, S. T. *Geochim. Cosmochim. Acta* **2001**, *65*, 4289.
24. Oberdörster, G; Oberdörster, E; Oberdörster, J. *Environ Health Perspect* **2005**, *113*, 823.
25. Oberdörster, G. *Int. Arch. Occup. Environ. Health* **2000**, *74*, 1.
26. Sayes, C. M.; Wahi, R.; Kurian, P. A.; Liu, Y.; West, J. L.; Ausman, K. D.; Warheit, D. B.; Colvin, V. L. *Toxicol. Sci.* **2006**, *92*, 174.
27. Grassian, V. H.; Adamcakova-Dodd, A.; Pettibone, J. M.; O'Shaughnessy, P. T.; Thorne, P. S. *Nanotoxicology* **2007**, *1*, 211.
28. Jiang, J.; Oberdorester, G.; Alder, A.; Gelein, R.; Mercer, P.; Biswas, P. *Nanotoxicology* **2008**, *2*, 33.
29. Cwiertny, D. M.; Handler, R. M.; Schaefer, M. V.; Grassian, V. H.; Scherer, M. M. *Geochim. Cosmochim. Acta* **2008**, *72*, 1365–1380.
30. Cwiertny, D. M.; Hunter, G. J.; Pettibone, J. M.; Scherer, M. M.; Grassian, V. H. *J. Phys. Chem. C* **2009**, *113*, 2175–2186.
31. Zinder, B.; Furrer, G.; Stumm, W. *Geochim. Cosmochim. Acta* **1986**, *50*, 1861–1869.

32. Cornell, R. M.; Posner, A. M.; Quirk, J. P. *J. Inorg. Nucl. Chem.* **1974**, *36*, 1937–1946.
33. Barron, V.; Torrent, J. *J. Colloid Interface Sci.* **1996**, *177*, 407–410.
34. Rustad, J. R.; Felmy, A. R. *Geochim. Cosmochim. Acta* **2005**, *69*, 1405–1411.

Chapter 3

Nanoparticle Solutions

C. M. Sorensen*

Departments of Physics and Chemistry, Kansas State University, Manhattan, KS 66506-2601
***sor@phys.ksu.edu**

This chapter describes novel synthetic methods for preparation of macroscopic quantities of a large variety of nearly monodispersed nanoparticles. Suspensions of these nanoparticles act as thermally reversible solutions. Experiments and theory to understand the solution properties, nucleation and self assembly via precipitation of 2d and 3d superlattices are also described.

Introduction

We have been fortunate to have been funded by an NSF/NIRT grant "Nanometer Stoichiometric Particle Compound Solutions and Control of their Self-Assembly into the Condensed Phase" for the past 2.5 years, beginning in October, 2006. The work we are doing is at the boundary between synthetic and supramolecular chemistry and chemical physics. It involves essentially an equal number of physics and chemistry faculty and graduate students. There are two major goals of this project:

1. Develop novel synthetic methods for preparation of macroscopic quantities of a large variety of nearly monodispersed nanoparticles. Such narrowly dispersed systems can be considered very large molecules or "**stoichiometric particle compounds**."
2. Understand their solution properties to allow for control of self assembly of 2d and 3d superlattices of the nanoparticles. Control will be gained by nanoparticle material and surface ligand type and functionality. In some cases molecular engineering techniques will be employed.

Below I will describe some of the techniques we have developed and some of the findings we have obtained.

Synthetic Methods

The Solvated Metal Atom Dispersion Method

The Solvated Metal Atom Dispersion method, or SMAD method, was developed by Klabunde and his students (*1–4*). It involves evaporation of a solid precursor, such as a metal or compound, in a closed reactor while simultaneously spraying a volatile liquid, typically an organic solvent, from a "shower head" within the reactor. See Figure 1. The reactor is immersed in liquid nitrogen at 77K. The evaporated metal and the sprayed liquid co-deposit on the cold inner surface of the reactor to form a frozen matrix of solvent molecules and metal atoms. When the reactor is removed from the liquid nitrogen, the matrix melts and the metal atoms find each other and form particles which are limited in their growth in some manner by the solvent. A picture of an as-prepared particle system is shown in Figure 2.

The SMAD method has important favorable attributes. It can be used for a great variety of materials, and it can be scaled up to large quantities.

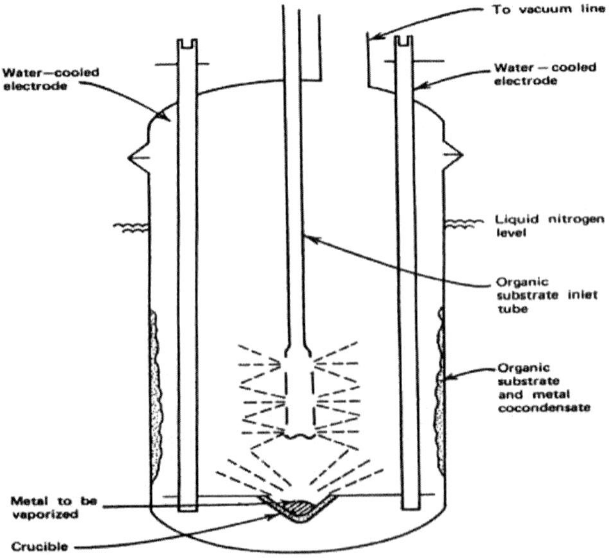

Figure 1. Schematic diagram of the SMAD reactor.

The Inverse Micelle Method

The inverse micelle method works on the concept that amphiphilic molecules will form inverse micelles in an organic solvent (*5*). An inverse micelle is a closed bundle of amphiphilic molecules that have all their hydrophilic groups clustered

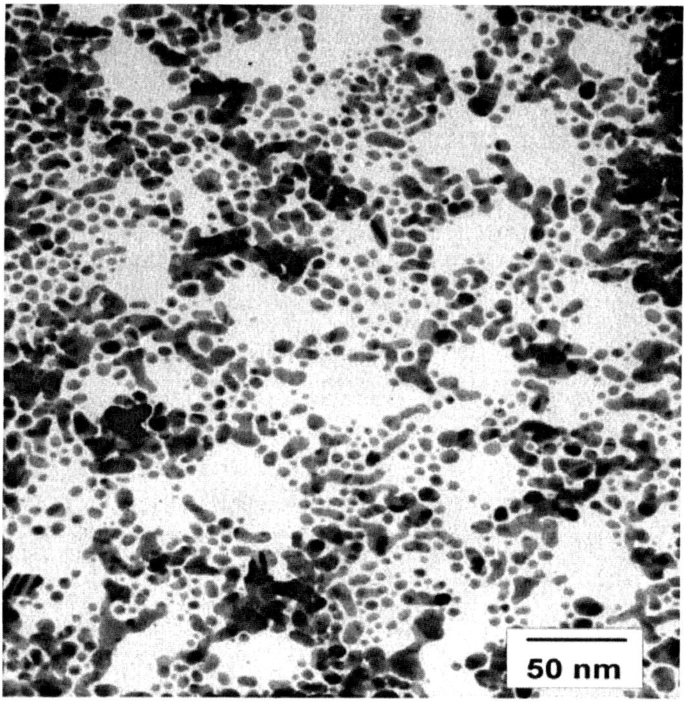

Figure 2. Gold SMAD as-prepared in acetone.

on the inside of the bundle while their hydrophobic, hence alkylphillic, tails reside on the outer circumference of the bundle sticking out into the "friendly" organic phase as drawn in Figure 3 (opposite to a regular micelle formed in water in which the hydrophilic groups stick out). As such, one can dissolve water and other hydrophilic entities into the small inner regions of the inverse micelles. In this way the inverse micelles can act as nano-reactors for inorganic reactions. This is a pretty picture, but this researcher has developed the opinion that it is rather idealistic. A more realistic picture views the organic solvent plus amphiphillic molecule solutions as able to solubilize the inorganic compounds, such as $AuCl_3$; a perfect micelle may or may not form. Once solubilized, these compounds can be reduced by reducing agents that have also been solubilized. The growing metal particle is limited in its growth by the amphiphillic surfactant, probably by surface ligation to the metal particle and probably not via constraint in the inverse micelle. Regardless, a polydisperse system of nanoparticles can be formed as illustrated in Figure 4.

Digestive Ripening

About 10 years ago our group discovered a very important process whereby polydisperse nanoparticle colloids can be made nearly monodisperse (*6*). Working with colloidal gold, we discovered that by boiling the colloid under reflux in the presence of excess dodecane thiol for about one hour, a nearly monodisperse

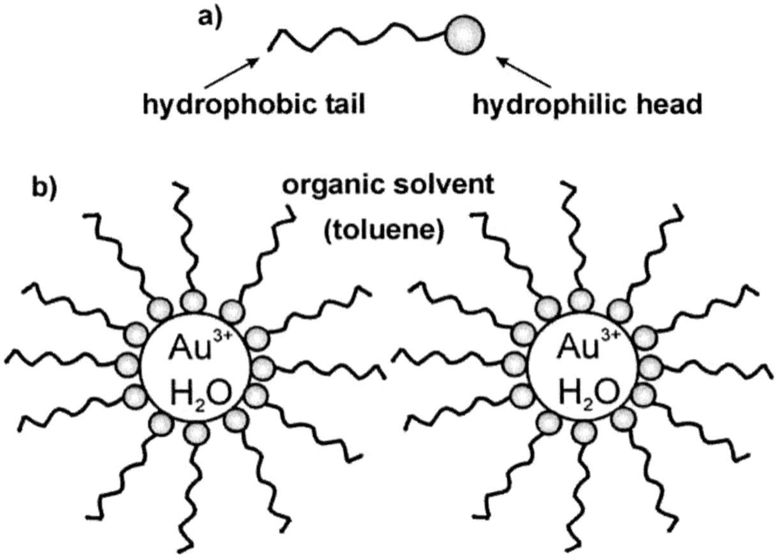

Figure 3. a) An amphiphilic molecule. b) Inverse micelles filled with an aqueous solution of gold ions.

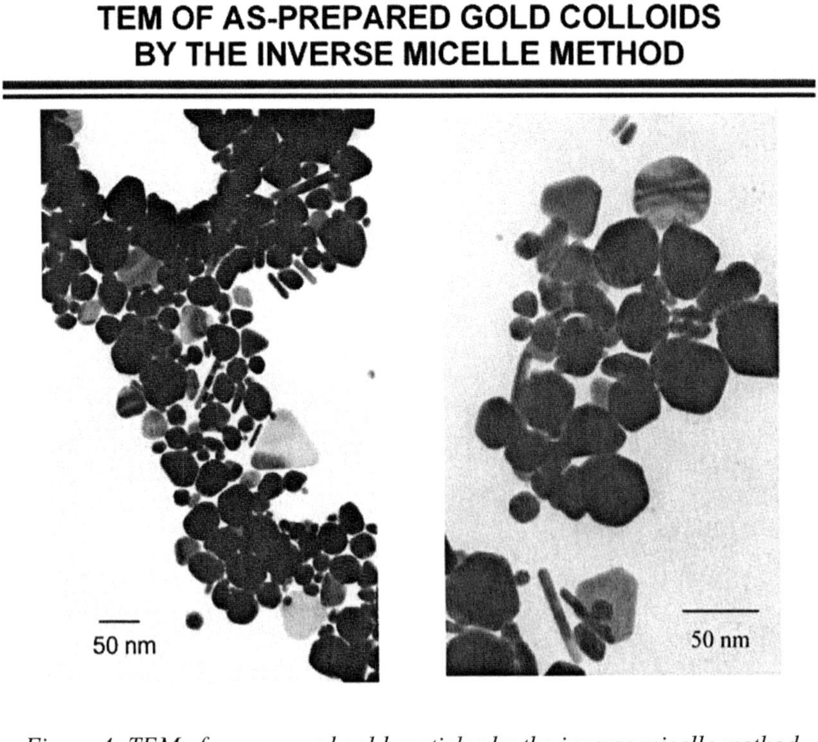

Figure 4. TEM of as-prepared gold particles by the inverse micelle method.

colloid was obtained. The transformation is illustrated in Figure 5. We named this process **Digestive Ripening**. Subsequently, we have applied this method to a variety of materials using a variety of ligands as described in Table 1 (*7–15*).

We can define digestive ripening as the heating of a nano-colloid in the presence of an excess of surface active ligand. In strong contrast to Ostwald ripening where the particles continue to grow, the particles in digestive ripening evolve to an equilibrium size and then stop. Indeed, in many cases the particles get smaller as seen in Figure 5. As a further example, we found that the quasi-monodisperse system can be digestively ripened to polyhedra as shown in Figure 6 by first removing the excess alkylthiol ligand and then adding excess DDAB (didodecyl ammonium bromide) ligand. Although the mechanism of digestive ripening is unknown, it likely involves ligand solvation of the material of the particles. This has been demonstrated by combining separately prepared gold and silver nanoparticle colloids and digestively ripening them together. The yield was a quasi-monodisperse system of gold/silver alloy particles.

Nanoparticles as Molecules

Stoichiometry

The great size uniformity afforded by the digestive ripening process (and other modern synthetic procedures) means that the well known size dependent properties of nanoparticles will be rather homogeneous across a quasi-monodisperse colloid. Colloids of the same material but of different size will have uniformly different properties as sure as they would if the were made of different chemical compounds. Moreover, consider that a ligated 5.0 nm Au nanoparticle can be represented by

$$Au_{3800}\,(C_{12}SH)_{365}$$

with ~10% variation on the numbers 3800 and 365. This is almost stoichiometry! These considerations lead us to recognize that with quasi-monodisperse nanoparticle systems, we have a new class of macromolecules. We have chosen to call these **Stoichiometric Particle Compounds**. A picture of single molecule of a stoichiometric particle compound (SPC), also known as a ligated nanoparticle, is given in Figure 7.

Superlattices

Because of their size uniformity, SPC molecules can form molecular crystals that are typically called superlattices, i.e. a lattice of nanoparticles. Note the two dimensional superlattice in Figure 6; a three dimensional superlattice is shown in Figure 8.

Solutions of Nanoparticles. Equilibrium Properties

We all know that when mixed with a liquid, particles form colloidal suspensions and molecules form solutions. But how is a suspension different

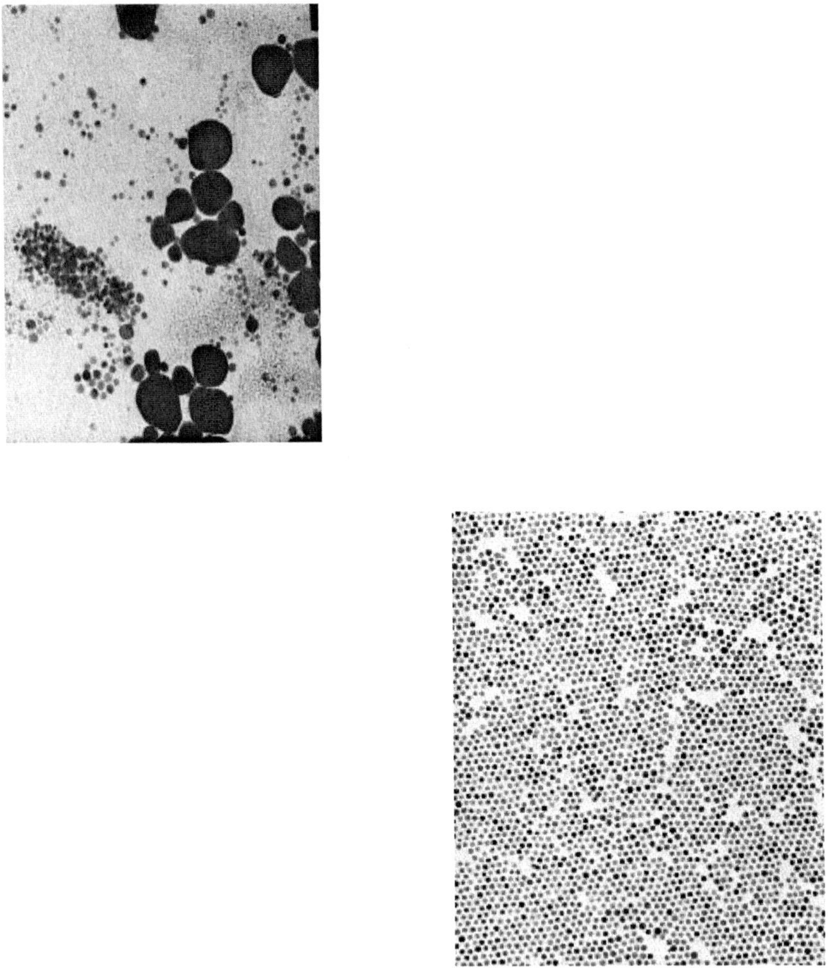

Figure 5. Illustration of the transformation of a polydisperse colloid into a nearly monodisperse colloid by cooking under reflux with excess ligand, a process we call ***Digestive Ripening****. (Reproduced with permission from reference (6). Copyright 2000 Springer Press).*

Table 1. Materials and Ligands for Digestive Ripening

Materials:	Gold, silver, copper, CdS, CdSe, CdTe, Pd, In, Fe_2O_3.
Ligands:	Alkane thiols, e.g. $(C_{12}H_{25})SH$ or C_8 through C_{16}
	amines RNH_2
	phosphines e.g. TOP
	acids RCOOH

than a solution? If suspensions eventually settle out, then the magnitude of gravity is involved in whether a system is a suspension or a solution, and that can't be correct because gravity is not intrinsic to either. We propose that a suspension is a solution if the particles are all the same size and hence interact both with each other and the solvent in the same, homogeneous manner. The laws of statistical mechanics insist that if these interactions exist, the system will exhibit temperature dependent bulk phenomena. Then given the same size with concomitant homogeneous interactions stronger than the gravitational interaction, it follows that there will exist definite, temperature dependent bulk properties. Our nanometer size particles are in colloidal suspension. But their same size leads to the same interactions throughout the colloid and hence the colloid acts like a solution with definite, temperature dependent phase boundaries. Said in another way, our nanoparticles are molecules of stoichiometric particle compounds. Suspensions of these particles are solutions of these molecules.

We have preliminary results for solubility phase diagrams for nanoparticle solutions. Figure 9 shows an example for 5 nm gold ligated with dodecane thiol. The solvent is a 4/96 vol/vol mixture of a poor solvent, 2-butanone, and a good solvent, t-butyl toluene. We see classic exothermic dissolution behavior with solubility increasing with temperature. In another preliminary example, Figure 10 shows solubility at room temperature for 5 nm gold nanoparticles ligated with dodecane thiol in a series of n-alkanes from hexane to decahexane. The non-monotonic "hump" behavior suggests a like-dissolves-like interpretation.

Solution theory gives the enthalpy of dissolution ΔH in terms of the phase boundary as

$$\Delta H = nRT^2 \, (d \ln x / dT) \qquad (1)$$

In eq 1, n is the number of moles of particles, R is the ideal gas constant, T the temperature, and x is the mole fraction concentration. The derivative term implies we need merely to use the slope of x vs. T phase boundary to determine the enthalpy. We have done this for the data in Figure 9 to find ΔH = 3.3kJ/mole of nanoparticles = 0.43 eV/nanoparticle.

To make sense of this result, we need a theory for interparticle interactions. This has not yet been accomplished due, we propose, to both its novelty and its difficulty. The difficulty lies in the many interactions that sum to make the complete nanoparticle-nanoparticle interaction. There is the van der Waals interaction between the core metal particles. There are also ligand-ligand interactions that that can be both attractive and, at closer distances, repulsive. And there are ligand-solvent interactions.

We have made some preliminary progress in this area. Figure 11 shows our current best calculation for the nanoparticle interaction. The depth of the potential, which is $\varepsilon = 3.2$ kT, where k is Boltzmann's constant, can be used for a simple comparison to the enthalpy data above.

The spherical nanoparticles interact isotropically and form close packed superlattices. Thus we make the simplifying assumption that the physics of inert gas solids can be applied to the nanoparticle superlattices. For such solids the lattice cohesive energy per particle is (*16*)

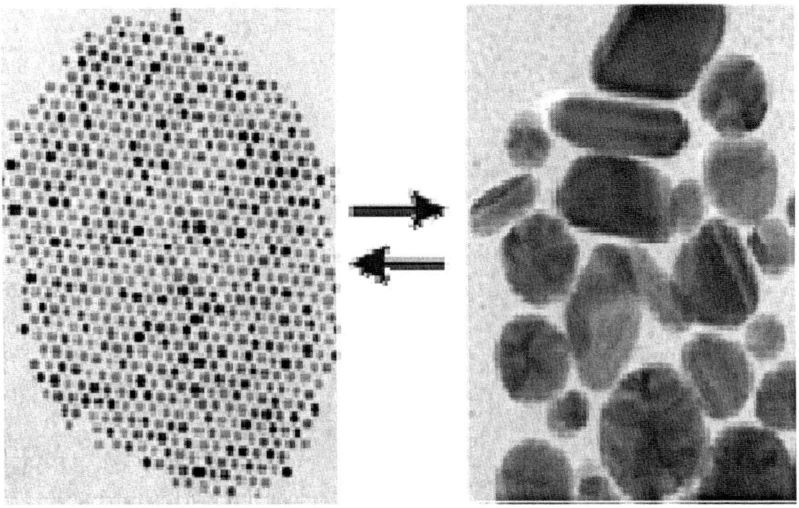

Figure 6. Reversing the morphology of a gold colloid with digestive ripening. The colloid on the left (5nm diameter particles) was made by digestively ripening with excess dodecane thiol, the one on the right had excess DDAB. This reversal can be performed any number of times. (Reproduced with permission from reference (13). Copyright 2005 American Chemical Society).

$$U_{coh} = 8.6\varepsilon \quad (2)$$

From the interparticle potential in Figure 10 we have

$$U_{coh} = 27.5\ kT \quad (3)$$

At room temperature kT = 0.25 eV. Thus the estimated superlattice cohesive energy per particle is

$$U_{coh} = 0.69\ eV \quad (4)$$

This value is close to the experimental enthalpy of dissolution. This result gives promise that our method, with refinement, will eventually lead to quantitative description of nanoparticle solutions.

Solutions of Nanoparticles. Non-Equilibrium Properties

Given the possibility of two phases in equilibrium, such as the dissolved state in equilibrium with a solid precipitate, there must be a mechanism for material to transform between pure states. For the dissolved state the mechanisms are nucleation and growth.

We have temperature quenched equilibrium solutions of 5 nm gold nanoparticles with dodecanethiol ligands from the high temperature, single phase into the low temperature, two-phase regime (*17*). The one-phase regime is the

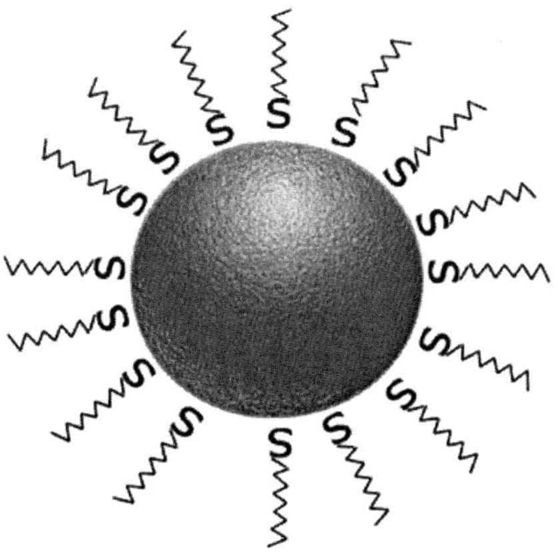

Figure 7. A ligates nanoparticle or a single molecule of a stoichiometric particle compound.

dissolved state; the two-phase regime is the supernatant in equilibrium with the precipitate phase. A schematic of this quenching is given in Figure 12.

We found that after a short growth period of about 5 minutes the precipitated phase was a systems of clusters of nanoparticles. We used dynamic light scattering to determine the diameter of these superclusters. We found that the deeper the quench, the smaller the diameter. This is a classic result from nucleation; deep quenches make fine precipitates, and now we see it for the first time in nanoparticle solutions. In a supersaturated metastable system homogeneous nucleation theory describes embryonic nuclei of the precipitate which form by monomer-monomer aggregation. Only the nuclei larger than the critical nucleation size will continue to grow. The theory gives the number density n^* of growable nuclei as (*18*, *19*)

$$n^* = n_0 \exp\left\{ \frac{-16\pi v^2 \gamma^3}{3k_B^3 T^3 \left[\ln\left(C/C^*\right)\right]^2} \right\}, \qquad (5)$$

where n_0 is the initial number density of monomers, v is the volume occupied by one monomer in the supercluster, not in solution, γ is the surface tension of the supercluster, k_B is Boltzmann's constant, T is the temperature, C is the solution concentration, and C^* is the equilibrium solubility of monomers at the given temperature.

The initial number density of monomers can be calculated from the solution mass concentration

$$n_0 = \frac{C}{\rho v} \qquad (6)$$

where ρ is the mass density of the nanoparticle supercluster.

The growth model of LaMer and Dinegar assumes a nucleation burst followed by diffusional growth (*19*). Each nucleus is influenced only by the monomers in a

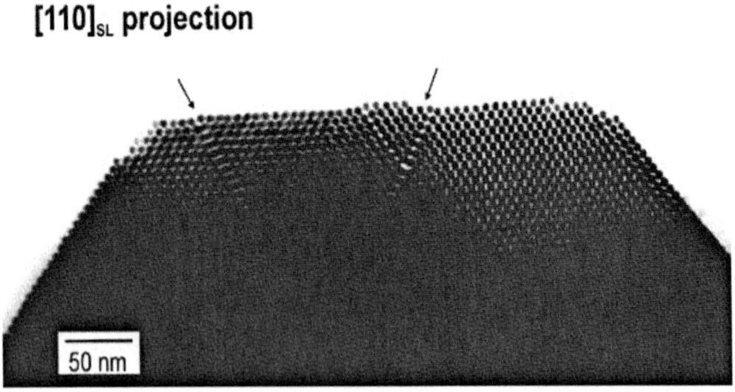

Figure 8. Three dimensional superlattice composed of 5.0 nm Au/C12SH nanoparticles. (Reproduced with permission from reference (11). Copyright 2003 American Chemical Society).

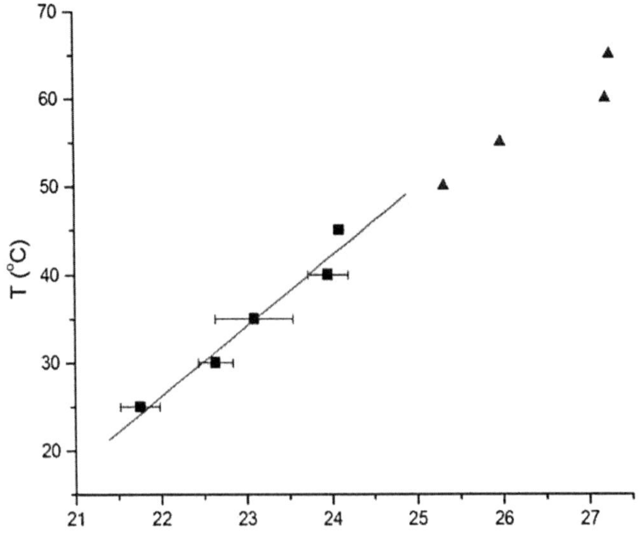

Figure 9. Solubility phase diagram for 5.0 nm gold nanopartilces ligated with dodecane thiol dissolved in a 4/96 mixture of t-butyl toluene/2-butanone.

spherical shell, centered on the nucleus, with the volume $1/n^*$. Once all the excess monomers, equal to $(C - C^*)$, are exhausted, the superclusters with number density n^* have grown to their maximum size V

$$V = \frac{4\pi}{3}R^3 = \frac{C - C^*}{n^* \rho}. \tag{7}$$

Combining these equations, we have

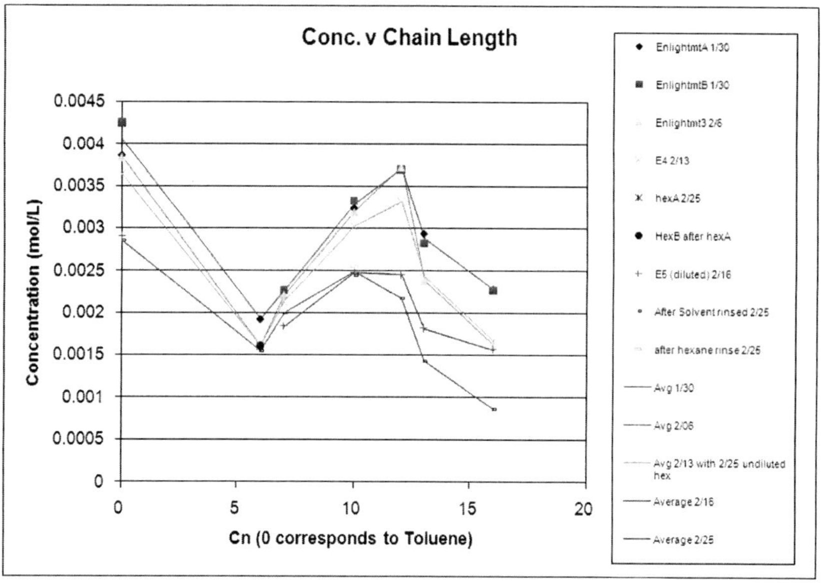

Figure 10. 5nm Au/C12SH nanoparticle supernatant concentration at room temperature versus solvent n-alkane chain length. Cn = 0 is solvent toluene.

$$\frac{4\pi}{3}R^3 = v\left(1-\frac{C^*}{C}\right)\exp\left\{\frac{16\pi v^2\gamma^3}{3k_B^3T^3\left[\ln\left(C/C^*\right)\right]^2}\right\}. \tag{8}$$

The solubility curve C^* was measured as above. All the other quantities in eq 8 are experimentally measurable except the superlattice surface tension γ. The measured radii of Figure 13 were fitted with eq 8 with the surface tension as the single fit parameter. Figure 13 shows an excellent fit can be obtained with a surface tension of 0.042 *erg/cm*^{2}.

This fit is successful in two aspects. First, the functionality of decreasing supercluster size with increasing quench depth is obtained. Second, the fit parameter gamma represents a new quantity: the surface tension of a solid phase of aggregated nanoparticles, most likely a superlattice. From atomic and molecular perspectives, the value we obtain from our fit is quite small, nearly three orders of magnitude smaller than those found for atomic and molecular liquids and solids which are typically in the range 10 to 30 *erg/cm*^{2}. However, nanoparticle solids are new materials composed of large stoichiometric particle compound "molecules". The surface tension of a hard sphere condensed phase is entirely entropic, and one would expect entropy to be the dominant contribution to the surface tension for the weakly interacting nanoparticle molecules where the interactions are on the order of the thermal energy, kT. The entropic surface tension for a close packed fcc lattice is given by *(20)*

$$\gamma = 0.61k_BT/\sigma^2 \tag{9}$$

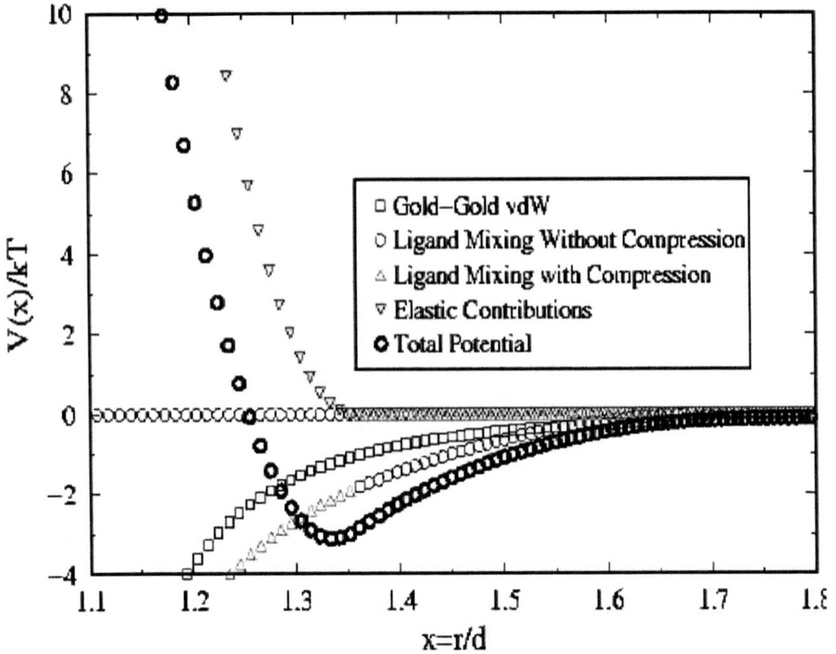

Figure 11. Various Components of the interaction potential between two d = 5.0 nm gold nanoparticles with dodecanethiol ligands.

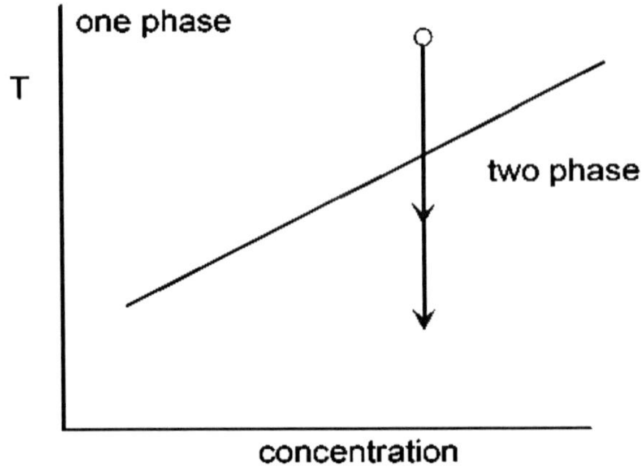

Figure 12. Schematic diagram of the temperature quench experiments.

where σ is the diameter of the "molecule". The nanoparticle diameter with ligand shell was measured with dynamic light scattering to be 8.4+/-1.0 *nm*. Then we find, at room T, $\gamma = 0.033 +/- 0.08$ erg/cm^2, which has surprising consistency with the fit value.

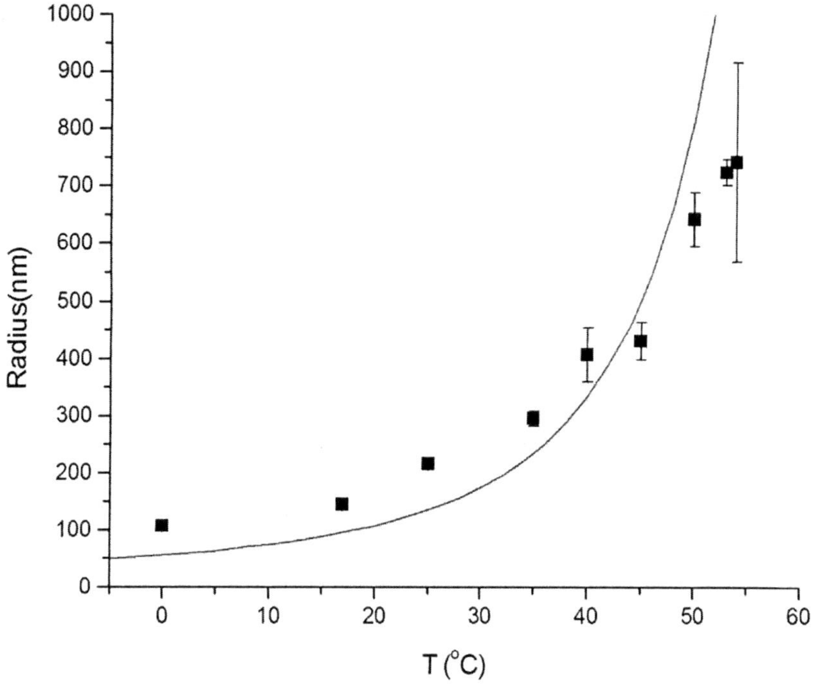

Figure 13. The sizes of superclusters formed by quenching the nanoparticle solution from 65C to the temperature indicated. The line is the best fit to the nucleation theory described in the text with the surface tension of the solid phase as the fit variable to yield a value of 0.042erg / cm. (Reproduced with permission from reference (17). Copyright 2009 American Physical Society).

Conclusions

The theme developed in this work is that current synthetic methods have led to nanoparticles with great size uniformity. With this uniformity, they represent a new class of macromolecules which we have named stoichiometric particle compounds. Thus nanoparticle colloidal suspensions act like solutions with thermally reversible, solvent dependent solubility and nucleation phenomena.

Acknowledgments

I have enjoyed my collaboration with my good friend, colleague and extraordinary chemist Ken Klabunde for some 20 years. I also thank the other members of the NIRT team including Professors Christer Aakeroy, Amit Chakrabarti, and Bruce Law and our students Hao Yan, Brandon Lohman, Jeffery Powell, Sreeram Cingarapu, Siddique Khan, Haeng Sub Wi, Evan Pugh and Ben Scott. This work was supported by NSF/NIRT grant 0609318.

References

1. Klabunde, K. J. *Nanoscale Materials in Chemistry*; Wiley Interscience: New York, NY, 2001.
2. Klabunde, K. J.; Timms, P. L.; Skell, P. S.; Ittel, S. Introduction to Metal Vapor Synthesis. *Inorg. Synth.* **1979**, *19*, 59.
3. Stoeva, S.; Klabunde, K. J.; Sorensen, C. M.; Dragieva, I. Gram-Scale Synthesis of Monodisperse Gold Colloids by the Solvated Metal Atom Dispersion Method and Digestive Ripening and their Organization into Two- and Three-Dimensional Structures. *J. Am. Chem. Soc.* **2002**, *124*, 2305–2311.
4. Prasad, B. L. V.; Sorensen, C. M.; Klabunde, K. J. Gold Nanoparticle Superlattices. *Chem. Soc. Rev.* **2008**, *37*, 1871–1883.
5. Pelini, M. P. Semiconductor Nanocrystals. In *Nanoscale Materials in Chemistry*; Klabunde, K. J., Ed.; Wiley Interscience: New York, NY, 2001; Vol. 3, pp 61–84.
6. Lin, X. M.; Sorensen, C. M.; Klabunde, K. J. Digestive Ripening, Nanophase Segregation and Superlattice Formation in Gold Nanocrystal Colloids. *J. Nanopart. Res.* **2000**, *2*, 157.
7. Lin, X. M.; Jaeger, H. M.; Sorensen, C. M.; Klabunde, K. J. Formation of Long-Range-Ordered Nanocrystal Superlattices on Silicon Nitride Substrates. *J. Phys. Chem. B* **2001**, *105*, 3353–3357.
8. Stoeva, S.; Sorensen, C. M.; Klabunde, K. J.; Dragieva, I. Gram Scale Synthesis of Monodisperse Gold Colloids by the Solvated Metal Atom Dispersion (SMAD) Method and their Organization into 2D- 3D-Structures. *J. Am. Chem. Soc.* **2002**, *124*, 2305–2311.
9. Prasad, B. L. V.; Stoeva, S. I.; Sorensen, C. M.; Klabunde, K. J. Digestive Ripening of Thiolated Gold Nanoparticles: The Effect of Alkyl Chain Length. *Langmuir* **2002**, *18*, 7515.
10. Prasad, B. L. V.; Stoeva, S. I.; Sorensen, C. M.; Klabunde, K. J. Digestive Ripening Agents for Gold Nanoparticles: Alternatives to Thiols. *Chem. Mater.* **2003**, *15*, 935.
11. Stoeva, S.; Prasad, B.; Uma, S.; Stoimenov, P.; Zaikovski, V.; Sorensen, C. M.; Klabunde, K. J. Face-Centered Cubic and Hexagonal Close-Packed Nanocrystal Superlattices of Gold Prepared by Different Methods. *J. Phys. Chem. B* **2003**, *107*, 7441–7448.
12. Smetana, A. B.; Klabunde, K. J.; Sorensen, C. M. Synthesis of Spherical Silver Nanoparticles Stabilized with Various Agents and their 3D and 2D Superlattice Formation. *J. Colloid Interface. Sci.* **2005**, *284*, 521.
13. Stoeva, S.; Zaikovski, V.; Prasad, B.; Stoimenenov, P. S.; Sorensen, C. M.; Klabunde, K. J. Reversible Transformations of Gold Nanoparticle Morphology. *Langmuir* **2005**, *21*, 10280–10283.
14. Smetana, A. B.; Klabunde, K. J.; Sorensen, C. M.; Ponce, A. A.; Mwale, B. Low-Temperature Metallic Alloying of Copper and Silver Nanoparticles with Gold Nanoparticles through Digestive Ripening. *J. Phys. Chem. B* **2006**, *110*, 2155–2158.

15. Stoeva, S. I.; Smetana, A. B.; Sorensen, C. M.; Klabunde, K. J. Gram Scale Synthesis of Aqueous Gold Colloids Stabilized by Various Ligands. *J. Colloid Interface Sci.* **2007**, *309*, 94–98 (invited).
16. Kittel, C. *Introduction to Solid State Physics*; Wiley: New York, 2005.
17. Yan, H.; Cingarapu, S.; Klabunde, K. J.; Chakrabarti, A.; Sorensen, C. M. Nucleation of Gold Nanoparticle Superclusters from Solution. *Phys. Rev. Lett.* **2009**, *102*, 095501.
18. Abraham, F. F. *Homogeneous Nucleation Theory; the Pretransition Theory of Vapor Condensation*; Academic Press: New York, 1974. Mullin, J. W. *Crystallization*; Butterworth-Heinemann: Oxford: 2001.
19. LaMer, V. K.; Dinegar, R. H. *J. Am. Chem. Soc.* **1950**, *72*, 4847.
20. Laird, B. B. *J. Chem. Phys.* **2001**, *115*, 2887.

Chapter 4

Chemistry of Rocksalt-Structured (111) Metal Oxides

April Corpuz and Ryan Richards*

Colorado School of Mines, 1500 Illinois St., Golden, CO 80401
***rrichard@mines.edu**

Rocksalt-type metal oxides with the (111) surface are considered novel catalysts because their surface is comprised of a layer of entirely oxygen anions or entirely metal cations. (111) surfaces terminated by oxygen anions have theoretically been surmised to promote catalytic reactions which need strong base sites. Since the (111) surface for rocksalt metal oxides is less thermodynamically stable than the (100) surface, most of the catalysis with rocksalt metal oxides has been studied with the (100) surface. Also for this reason, most work carried out with the (111) facet has been conducted under high vacuum or ultra high vacuum conditions. A wet chemical synthesis of two rocksalt metal oxides, MgO(111) and NiO(111), has been discovered, paving the way for new catalysis.

Setting the Scene

The focus of this chapter is the development of the wet chemical synthesis of (111) rocksalt metal oxides, namely MgO(111) and NiO(111), pioneered by the Richards group. It is not meant to be a comprehensive text on all (111) metal oxides. There are many extensive texts and review articles on metal oxides (*1–3*), including a book edited by Kenneth Klabunde (*4*).

This chapter will begin with a short introduction to (111) rocksalt metal oxides. It will continue with some of the combined experimental and theoretical studies on (111) rocksalt metal oxides, as well as some completely theoretical studies. After a brief mention of work on (111) structures grown under ultra high vacuum, an in-depth discussion of the wet chemical synthesis, characterization, and chemistry of the MgO(111) and NiO(111), inspired by theoretical studies and the work done by Kenneth Klabunde's group, will ensue. Lastly, an example of chemistry of a non-rocksalt (111) metal oxide, ceria, will round out the chapter.

Introduction

Metal oxides, as the second most abundant class of material on Earth, are widely applied in many different areas. From insulators in cables to semiconductors in diodes, from glass to gel, as a doping agent or catalyst, the use of metal oxides spans a wide range of different electronic and physical properties. And, as science branches into the nano realm, even more properties of metal oxides are emerging, and many studies have been performed on nanoparticle metal oxides. Since nanoparticles generally possess high surface areas, some studies focus on the potential applications of metal oxides as novel catalysts. Some examples are the catalytic effect of MgO(100) on the combustion of methane (*5*), the catalytic abilities of MgO(111) nanosheets and NiO(111) nanosheets (*6*, *7*), and the catalytic conversion of NO to NO_2 on nickel oxide (*8*, *9*). With these studies comes a need to better understand the processes that occur on the surface of these metal oxide nanoparticles.

Complete understanding of the necessary conditions for high catalytic metal oxide activity is still not yet achieved. However, it is understood that free coordination sites on the surfaces of metal oxides are important for catalysis. Lower coordination sites, or coordinatively unsaturated sites, are especially imperative, since they have room to bind reactants or adsorbed molecules. Because of the many defects on a crystal, the surface atoms can have a coordination number that ranges between 3 and 5. Since current technology does not yet have a sort of omniscient eye to watch the catalysis occur directly, the coordination sites and their effectiveness can only be measured indirectly. For example, methane adsorption on MgO(100) has been measured with infrared spectroscopy and the data analyzed using density functional theory (DFT), the results of which suggest that only three-coordinated O^{2-} sites will bind the methane. The low coordinated Mg^{2+} sites and the four- and five-coordinated O^{2-} sites were shown to have repulsive interactions with respect to binding methane (*10*). As can be seen, the type of coordination site is important and highly specific for successful catalysis.

Much of the documented studies on the catalytic activity of rocksalt metal oxides (MgO, CaO, NiO, CoO, MnO, SrO and EuO) have been done on the (100) surface since it is the most readily available and stable form for most metal oxides. However, with techniques that allow the exposure of the (110) and (111) surfaces came interest in the catalytic ability of these surfaces. In particular, the (111) surface is expected to be more reactive than the (100) surface because it has alternating layers of reactive anions and cations, as can be seen in Figure 1. In an effort to minimize surface (and structural) energy, these polar (111) surfaces are often hydrogen terminated, making them interesting for hydrogen storage or water splitting (*11*).

In this chapter we focus on work surrounding MgO(111) and NiO(111) nanoparticles as these can be easily synthesized using wet chemical preparations

(*6*, *7*). Although these two materials have the same lattice parameters, MgO is an insulating metal oxide and is expected to have strong Lewis acid and Lewis base sites; NiO, on the other hand, is a transition metal oxide and a semiconductor. Due to these differences between MgO and NiO as metal oxides, it will be instructive to determine which is better for different types of catalysis and why. While this determination has not yet been made, some important experimental and theoretical studies done on the MgO(111) and NiO(111) surfaces are reviewed.

Theoretical Studies of (111) Rock Salt Metal Oxides

For metal oxides with rocksalt (NaCl) structures, the (100) plane is a checkerboard of alternating metal and oxygen atoms. If a cube of rocksalt (with only (100) faces exposed) were cut along three non-adjacent corners, a (111) surface would be exposed. A clean (111) surface is comprised solely of a layer of oxygen atoms (metal atoms) and would have a layer of metal atoms (oxygen atoms) below it, alternating with each layer. This surface would have a large and unfavorable surface energy because these oppositely-charged, alternating layers cause a diverging electric field perpendicular to the surface. As such, a clean (111) surface is relatively thermodynamically unstable, and must somehow reconstruct, facet, adsorb molecules, or in some other way neutralize this diverging electric field (*13*).

Until recently, the surfaces of metal oxide (111) particles were believed to thermally microfacet to more stable (110) or (100) facets (*14*). However, with further theoretical and experimental study, this was later disproved; acid etching was in fact responsible for the faceting of the (111) surfaces to pyramids (*15*). Next, it was proposed that, above 1200 °C, reconstructions of the MgO(111) surface involving groups of oxygen trimers (ozone) and/or single oxygen atoms occur periodically on the surface. The ($\sqrt{3}$ x $\sqrt{3}$)R30° reconstruction is comprised of equilateral oxygen trimers. The (2 x 2) and (2$\sqrt{3}$ x 2$\sqrt{3}$)R30° reconstructions have both oxygen trimers and single oxygen atoms on the surface. Shown by transmission electron diffraction, these reconstructions are the current standing models of the MgO(111) surface above 1200 °C (*16*).

Morphology and Geometry of MgO(111)

Lower temperature morphology of the MgO(111) surface has been debated using experimental and theoretical studies. Whether the low temperature MgO(111) surface was reconstructed or terminated by hydroxylation was not conclusive. It has been shown that the surface oxygen layer of the (111) crystals are stabilized by hydroxylation, and is actually more stable than the clean (100) surface. Gajdardziska-Josifovska et al. combined experimental and theoretical methods to study the MgO(111) surface (*11*).

The experimental methods used disks of MgO single crystal slabs that were cut along the (111) plane, then characterized by reflection high-energy electron diffraction (RHEED), low-energy electron diffraction (LEED), and x-ray photoelectron diffraction (XPD) and Auger electron spectroscopy (AES). The

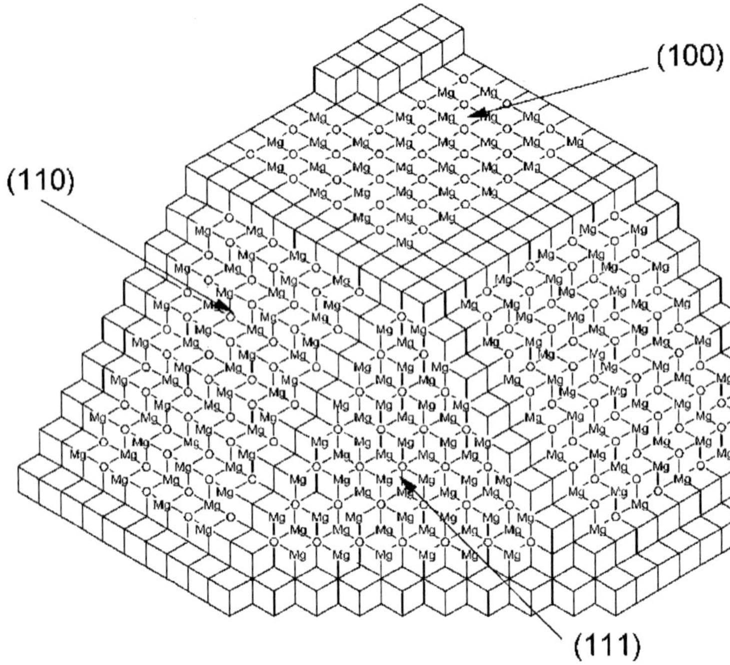

Figure 1. The (100), (110), and (111) facets of a prototypical rocksalt structure metal oxide, MgO. Note that the (100) and (110) facets are composed of alternating cations and anions, while the (111) facet is made up of a single type of ion (either all cations or anions, depending on how the face is cut). (Reference (12) - Reproduced by permission of The Royal Society of Chemistry. Copyright 2008, The Royal Society of Chemistry.)

transmission high-energy electron diffraction (THEED) pattern shows bright (1 x 1) spots which are not present in bulk MgO, implying the MgO(111) crystals kept the (1 x 1) structure. TEM images show terraces of (111) surfaces and steps. Also, the LEED and RHEED on the MgO(111) slabs showed (1 x 1) patterns. Results from XPD/AED analysis favored the models with O- and OH- terminated surfaces instead of Mg-terminated surfaces. Also, the high-energy-resolution O 1s spectra at both normal and grazing emission show a shifted peak that can be delegated to a OH-terminated surface. These XPD data were further analyzed using theoretical methods.

Three theoretical methods were used to analyze the experimental XPD data: single scattering XPD in cluster geometry, multiple scattering XPD in cluster geometry, and multiple scattering XPD in slab geometry. For the XPD calculations, the models compared were the clean Mg-terminated surface, the clean O-terminated surface, and the OH-terminated surface. The distances between the layers were optimized by density functional theory (DFT). Total energy calculations on the OH-terminated model indicate that having the H directly over the O instead of over the hollow site is more stable by 0.16 eV, so that model was used for the XPD calculations. The H plane itself was not used for the calculation since it has negligible scattering power in XPD. The theoretical study examines the azimuthal dependence of the calculated ratio of the Mg KLL Auger electron diffraction intensities to the O1s photoelectron diffraction intensities for electrons exiting at an angle of 7° from the surface, then compares the calculated ratio to the experimentally found ratio. The results of this can be seen in Figure 2.

The O-terminated surface and the OH-terminated surface are closest to the experimental ratio, with the OH-terminated surface being slightly closer. The calculated ratio of the Mg-terminated surface does not bear resemblance to the experimental ratio. This is evidence that the MgO(111) surface is OH-terminated instead of neutrally reconstructed with some oxygen atoms terminating the surface.

The electronic structures of the different surfaces were calculated with DFT and compared to the shifts found in the experimental XPS spectra. The OH-terminated surface showed shifts of the same magnitude and direction for the 2s and 1s states as found in the experimental XPS spectra. The O-terminated surface showed shifts of a different magnitude and of the opposite direction for the 2s and 1s states as compared to the experimental XPS spectra. As such, the calculated electronic structure of the OH-terminated surface is closer to the experimentally found data, so it is more likely that the MgO(111) surface is OH-terminated (*11*).

An *ab initio* study was also performed on MgO(111) using density functional theory (B3LYP) and CRYSTAL code, modeling the geometry of the octopolar reconstructions of MgO. An ideal octopolar reconstruction looks like a pyramid with three atoms of one type forming an equilateral triangle as a base, and one atom of another type (which will be called the apical atom) centered over and above the base equilateral triangle. According to the model, partial relaxation of the octopoles causes the apical atom to move inwards towards the bulk and the atoms forming the equilateral triangle to swell outwards. However, full relaxation causes the atoms forming the equilateral triangle (which is the second layer) to move closer together, towards each other, and the third layer (and subsequent layers, to a lesser degree) to buckle downwards. While experimental studies agree with these theoretical results that buckling occurs in the reconstruction, the details of the buckling cannot be reconciled.

Using the full relaxation model, the cleavage energy, an indication of the stability of a surface, of the MgO(111) octopolar surface was compared to the MgO(100) surface. A summary of the findings can be found in Table I.

The octopolar surface creates a higher cleavage energy than the clean (100) surface. This suggests that the octopolar reconstruction is less favorable than the (100) surface, which is in agreement with the experimental finding that MgO octopolar reconstruction has not been observed. The cleavage energy of the MgO(111) surface was also calculated, and can be found in Table I. The hydroxylated surface is much more stable than either the octopolar surface or the (100) surface, so it can be concluded that the hydroxylated surface of MgO(111) will be favored (*13*).

Morphology and Geometry of NiO(111)

Similar to the *ab initio* study performed on MgO(111), a study was performed on NiO(111) using density functional theory (B3LYP) and CRYSTAL code. Since all (111) rocksalt metal oxides have the same ideal octopolar reconstruction, the

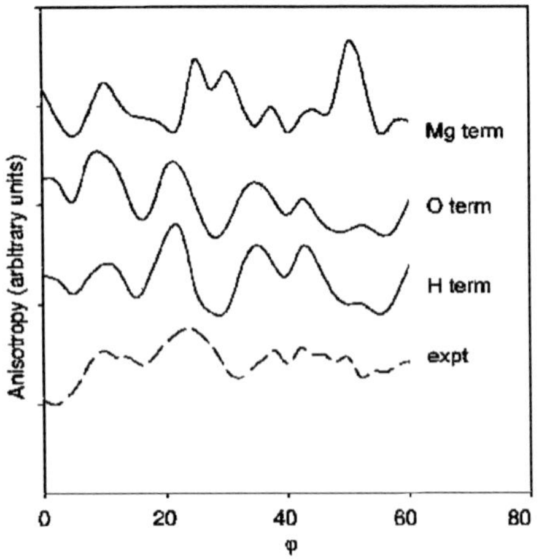

Figure 2. The calculated and experimental ratios of the the Mg KLL Auger electron diffraction intensities to the O1s photoelectron diffraction intensities for electrons exiting at an angle of 7° from the surface. (Reprinted figure with permission from (11). Copyright 2005 by the American Physical Society.)

Table I. Cleavage energies of different MgO surfaces[a]

Surface Type	*Cleavage energy*
MgO(100)	2.38 J/m^2
MgO(111)-octopolar surface	3.08 J/m^2
MgO(111) hydroxylated surface	1.13 J/m^2

[a] Data are from reference (*13*)

same model was used, and the geometry of the octopolar reconstruction of NiO was simulated. The results of the ideal octopolar reconstruction can be found in the section "*Morphology and Geometry of MgO(111)*".

Using the full relaxation model, the cleavage energies of the NiO(111) octopolar surface were compared to those of the NiO(100) surface. A summary of the findings can be found in Table II.

Note that the NiO octopolar surface creates a higher cleavage energy than the clean NiO(100) surface. This again might suggest that the octopolar reconstruction is less favorable than the (100) surface, as in the MgO case. However, the cleavage energy ratio of MgO(111)-octopolar to MgO(100) is 1.29, and is greater than the cleavage energy ratio of 1.12 for NiO(111)-octopolar to NiO(100). This may explain why the octopolar reconstruction for NiO occurs experimentally while the MgO octopolar reconstruction has not been observed. Like MgO(111), for NiO(111) the hydroxylated surface is much more stable than either the octopolar surface or the NiO(100) surface, so it can be surmised that the hydroxylated surfaces will be favored (*13*).

Table II. Cleavage energies of different NiO surfaces[a]

Surface Type	*Cleavage energy*
NiO(100)	5.34 J/m^2
NiO(111)-octopolar surface	6.02 J/m^2
NiO(111) hydroxylated surface	0.99 J/m^2

[a] Data are from reference (*13*)

DFT Study of the Adsorption of CO on NiO(111)

A study on the adsorption of CO on NiO(111) using periodic density functional theory (DFT) was performed (*12*). It used a slab model and UB3LYP in Gaussian 98 to investigate and compare the interactions of the clean and CO-adsorbed surfaces of NiO(100) and NiO(111). The density of states, energies of reconstruction, and energies of adsorption were calculated for the different NiO systems. For the NiO(100) system, the oxygen surface site only has a small difference in the density of states for p_z when compared to bulk oxygen, meaning that it is not a good adsorption site for CO (or NO) and the adsorption energy is small. It is actually the Ni surface site of NiO(100) that adsorbs CO, using a d_{z2} orbital (see Figure 3). The adsorption occurs with a C-Ni interaction in which none of the carbon monoxide's pi character is used.

When looking at the electron population of all orbitals involved in the adsorption, it was found that the Ni site gains some electron population while the CO loses electron population. This implies that the CO is giving electrons to the surface.

When modeling the NiO(111) surface, the DFT calculations predicted what was expected, which is that a surface of only Ni atoms or a surface of only O atoms is unstable. To study CO adsorption on NiO(111), Wang, Li, and Zhang assumed that CO adsorbed perpendicularly on an Ni site, with the carbon-end of the CO facing the Ni (*17*). The calculations then showed that the bond length of C-Ni of the (111) system was shorter compared to the bond length of C-Ni of the (100) system. Also, the bond length of C-O of the (111) system was longer compared to the bond length of C-O of the (100) system. This means that there is a stronger interaction between the C and Ni (and a weakened interaction between the C and O) in the (111) system, implying a stronger adsorption. From looking at the electron population of all orbitals involved in the adsorption, it was found that Ni in the (111) system loses electron density while CO gains electron density, which is opposite of what was found in the (100) system. A four-electron interaction is proposed to exist between the CO and NiO(111).

Adsorption of CO on the alleged reconstructions of NiO(111) was also investigated. The three reconstructions considered were the "1+3" type, the O-ended "vacancy type", and the Ni-ended "vacancy type". The "1+3" type describes a reconstruction in which the surface layer has ¾ of its atoms missing (i.e., ¼ of its atoms remaining) and the second layer to the surface has ¼ of its atoms missing (¾ of its atom remaining). Wang, Li, and Zhang assume that, in the "1+3" type, the surface layer is comprised of oxygen atoms, the second layer is comprised of nitrogen atoms, and CO adsorption occurs with the nitrogen

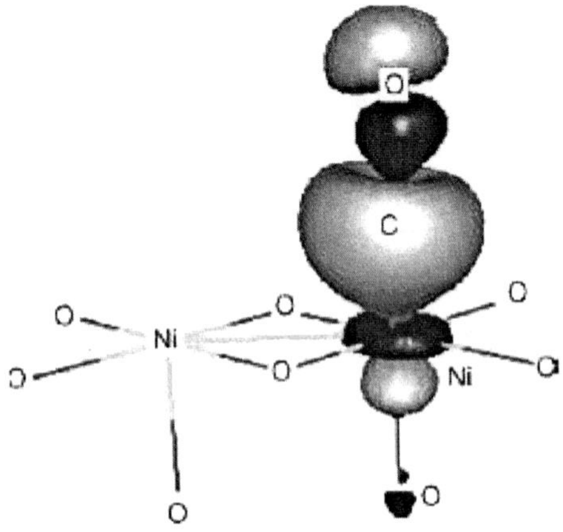

Figure 3. The Gaussian 98 procedure was used to create this figure of the CO orbital interaction with NiO(100). (Reprinted from reference (17), Copyright 2006, with permission from Elsevier.)

atoms on the second layer. The O-ended "vacancy type" has a surface with ¾ of its atoms missing (¼ remaining) and a third layer with ¼ atoms missing (¾ remaining). Note that the surface and third layer are comprised of the same kind of atom, so, in total, a complete layer of one type of atom is missing. For the O-ended "vacancy type", the surface and third layer are oxygen atoms. Likewise, for the Ni-ended "vacancy type", the surface and third layer are nitrogen atoms. In both "vacancy types", only a Ni-CO adsorption is considered; O-CO adsorption is not considered. When comparing the unit cell energies of the three reconstruction types and the regular (1x1) structure of NiO(111), the "1+3" type reconstruction is most stable. However, when comparing the unit cell energies of the reconstruction types and regular (1x1) structure after CO adsorption on Ni, the CO adsorbed on the (1x1) structure has such a relatively large adsorption energy that it becomes more stable than the "1+3" type without CO adsorption, and is slightly more stable than any of the reconstructions with CO adsorbed. This means when CO is adsorbed on the NiO(111) surface in a (1x1) structure, it is unlikely to reconstruct to other morphologies (*17*).

DFT Study of the Adsorption of H_2 on MgO(111)

Chemisorption of H_2 on a MgO(111) surface was modeled using 2D periodic ab initio calculations. With the Crystal 92 program, energy maps of the H_2-MgO(111) interaction were created, using the calculation

$$\Delta E_{ads} = E(H_2/MgO) - E(MgO) - E[H_2(gas)]$$

to determine all energies. $E(H_2/MgO)$ is the energy of the system with the sorbed H_2, E(MgO) is the energy of the MgO(111) slab with the same geometry but no H_2, and $E(H_2(gas))$ is the energy of a free hydrogen gas molecule. The simulation measured the interactions between a lowered H_2 gas molecule over the MgO(111) surface with special attention to five sites:

1. a three-coordinated magnesium ion, Mg(cn3),
2. a four-coordinated oxygen ion in the second layer, O(cn4),
3. a six-coordinated magnesium ion in the third layer, Mg(cn6),
4. a five-coordinated oxygen ion in the second layer, O(cn5),
5. a site in between two five-coordinated oxygen ions in the second layer.

Mg(cn3), O(cn5), O(cn4), and Mg(cn6) are native to the (1x1) structure of MgO(111). O(cn6) and O(cn3) can only be found in a reconstructed MgO(111) surface. Mg(cn3) is also present in the reconstructed surface. These sites are depicted in Figures 4 and 5.

The simulation first studied the magnitude of the interactions between the ideal MgO(111) surface at zero Kelvin, forced into a rigid crystal structure, and a hydrogen molecule lowered vertically (90°) onto the surface, the results of which are found in Figure 6.

For O(cn4), a relatively deep potential well occurs along with a lengthening of the H-H bond (from 0.719 Å to 1.18 Å), meaning that, for an ideal MgO(111) surface at 0 K, O(cn4) is chemisorption site. For this configuration, at the optimized H-H bond length (1.15 Å), the adsorption energy is -199 kJ/ mol. The other sites, O(cn5), Mg(cn3), Mg(cn6), and between the O(cn5)'s, are only physisorption sites. When the simulation was done in the same manner except the hydrogen molecule was lowered horizontally (0°) onto the surface, all sites showed only weak physisorption. The simulation was then run for tilted H_2 adsorption in steps of 10°. The optimized geometry was found with the H_2 molecule tilted below 90°, with the lower hydrogen chemisorbed to a O(cn4) site and the higher hydrogen near a Mg(cn3) site (see Figure 7). The resulting optimized adsorption energy of -477 kJ/mol is more than twice that of the adsorption energy for the vertically-lowered H_2 on the O(cn4) site.

The next simulation studied the magnitude of the interactions between a hydrogen molecule and a relaxed MgO(111) surface at 10 Kelvin. In comparison to the ideal MgO(111) surface at zero Kelvin, the relaxed surface has a 40% smaller interplanar distance between the top two layers. When the hydrogen molecule was lowered vertically onto the surface, no chemisorption sites were found. For all of the sites, the H-H bond length virtually stayed the same. The simulation was then run for tilted H_2 adsorption in steps of 10°. The optimized geometry was again found with the H_2 molecule tilted below 90°, with the lower hydrogen chemisorbed to a O(cn4) site and the higher hydrogen near a Mg(cn3) site (see Figure 8).

The resulting optimized adsorption energy is -124 kJ/mol and is approximately ¼ of the adsorption energy found for the rigid MgO(111) surface. The simulation was done again for a reconstructed MgO(111) surface at 300 K, with the optimal geometry found with the H_2 molecule tilted below 90°, with the lower hydrogen chemisorbed to a O(cn3) site and the higher hydrogen near a Mg(cn3) site. Under these conditions, the optimized adsorbed energy is -63 kJ/mol. From these results, it is shown that a low coordination number is necessary but not sufficient for chemisorption of H_2, and, in the models closest to a (1x1) structure of MgO(111), H_2 absorbs best on a low coordination O(cn4)-Mg(cn3) pair (*18*).

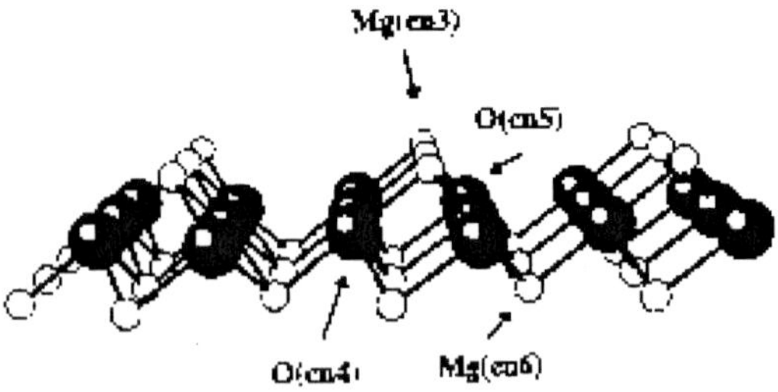

Figure 4. The surface of MgO(111) for both the ideal case at zero Kelvin and the relaxed case at ten Kelvin. (Reprinted with permission from reference (18). Copyright 1998, American Institute of Physics.)

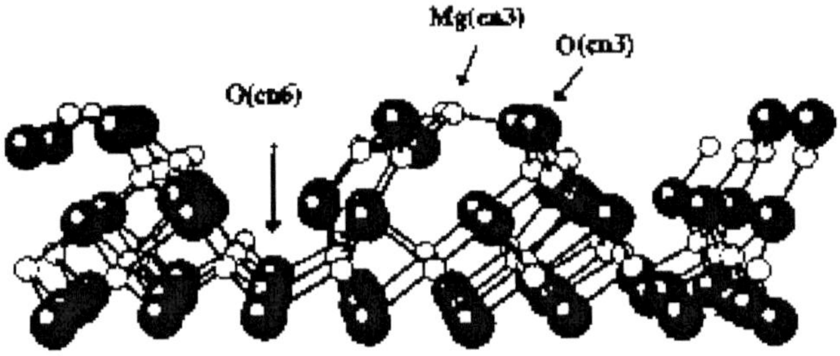

Figure 5. The reconstructed surface of MgO(111) used for the 300 K model. (Reprinted with permission from reference (18). Copyright 1998, American Institute of Physics.)

Wet Chemical Preparation of (111) Rock Salt Metal Oxides

The (111) surface of rock salt metal oxides is of particular interest since it is composed of either all oxygen or all metal ions. The (111) polar surface would be useful for reactions that are catalyzed by oxygen anions or hydroxyl groups. A notable method for synthesizing high surface area, unsupported, (111) rock salt metal oxides uses sol-gel chemistry and supercritical drying to produce aerogels. MgO(111) and NiO(111) are two such examples of this chemistry.

MgO(111) nanosheets are synthesized by a modified aerogel preparation. A magnesium belt is prepared by sandpapering and washing with acetone, then cut into small pieces and dissolved in absolute methanol under argon, forming $Mg(OCH_3)_2$. After dissolution, 4-methoxybenzyl alcohol is added in a molar ratio of Mg:alcohol 2:1. Benzyl alcohol has been experimentally found to be a good template for tailoring some metal oxides, but the reason for this is not concrete. After the mixture was stirred for 5 hours, a solution of water and alcohol is then added dropwise, with the water:Mg molar ratio being 2:1. The water and methanol is used for hydrolysis of the solution, producing the nanosheets. After stirring for 12 hours, the solution is autoclaved. This is then purged with argon for 10 minutes, and then the chamber is pressurized to 10 bar with argon before heating. The autoclave is then heated to 265 °C for 15 hours. After heating, the vapor inside is vented while hot, inducing the supercritical drying. In particular,

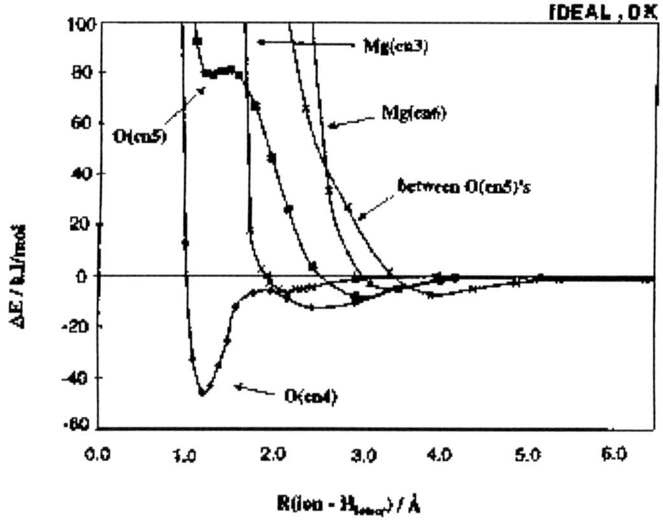

Figure 6. This graphs the interaction energy between H_2 and MgO(111) as the H_2 is lowered vertically onto the ideal surface at zero Kelvin. For this particular graph, the H-H bond length was fixed at 0.719 A. (Reprinted with permission from reference (18). Copyright 1998, American Institute of Physics.)

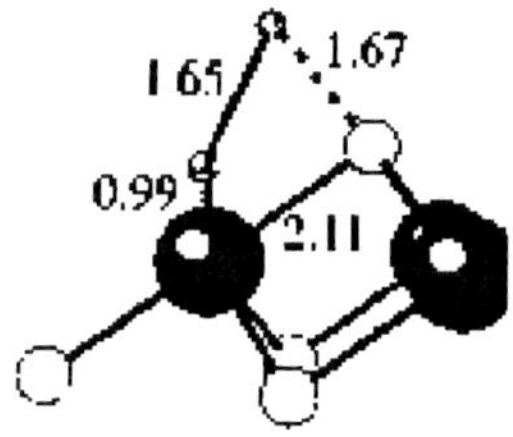

Figure 7. This is the optimized geometry found for the ideal MgO(111) surface at 0 K. (Reprinted with permission from reference (18). Copyright 1998, American Institute of Physics.)

4-methoxybenzyl alcohol, being a stronger acid than methanol, can more strongly interact with the hydroxyl group of the intermediate species Mg(OH)-(OCH_3), and can substitute methanol during mixing. The resulting white, powdery Mg(OH)(OCH_3) precursor is then calcined in air (typically 500°C) to remove any carbon species, resulting in MgO(111) nanosheets with diameter between 50-200 nm and thickness between 3-5 nm. High resolution TEM pictures of the resulting MgO(111) nanosheets can be found in Figure 9 (*6*, *19*).

NiO(111) nanosheets with hexagonal holes are also synthesized by a modified aerogel preparation. $Ni(NO_3)_2{\bullet}6H_2O$ is dissolved in absolute methanol. After complete dissolution, urea and benzyl alcohol are added in the molar ratio Ni: urea : benzyl alcohol of 1: 0.5 : 2. The role of benzyl alcohol is that of a structure-

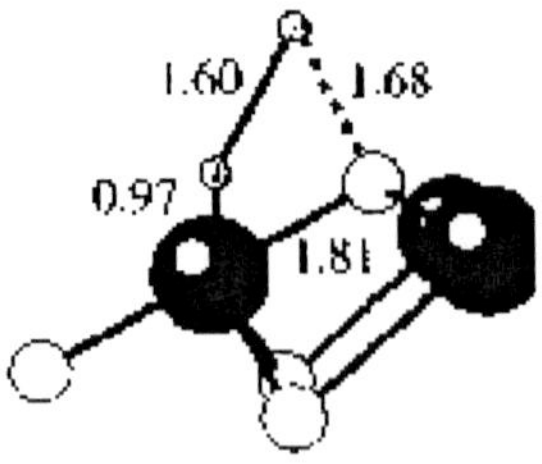

Figure 8. This is the optimized geometry found for the relaxed MgO(111) surface at 10 K. (Reprinted with permission from reference (18). Copyright 1998, American Institute of Physics.)

directing agent. While the exact mechanism of its role is unknown, without benzyl alcohol, NiO nanosheets do not form. Urea is used to control the size of the nanosheets; more urea created smaller-sized nanosheets. It is important in the synthesis, since its hydrolysis neutralizes acid by-product that is made from the reaction of nickel nitrate with methanol and water. Too much acid byproduct will slow the formation of nickel oxide precursor. After one hour stirring, the solution is autoclaved and purged with argon five times, at 7500 torr. Before heating, the autoclave is pressurized to 7500 torr, also with argon. The pressurized chamber is then heated to 200 °C for 5 hours, then increased to 265 °C for 1.5 hours. After heating, the vapor inside is vented while hot, inducing the supercritical drying. The resulting green, powdery NiO precursor is calcined in air (typically 500°C) to remove any carbon species. The ramp rate to 500 °C does not seem to affect the formation of the nanosheets. For NiO precursor calcined in air at 250 °C, very few holes are formed. For NiO precursor calcined in air at 350 °C, many small holes (<10nm) are observed. These observations led to the conclusion that the hexagonal holes are likely due to heat-induced etching at defect sites, with higher calcination temperatures forming larger hexagonal holes. From thermogravimetric and differential scanning calorimetric analysis, it was found that the carbon species on the NiO precursor decompose at 335 °C, and at 360 °C there is a small weight increase, ascribed to oxygen from air filling oxygen vacancies on the surface. A TEM image of the resulting NiO(111) nanosheet with hexagonal holes can be found in Figure 10 (*7*).

(111) Metal Oxides

MgO(111)

MgO, with its strong ionic character and insulating properties (wide band gap), has been the subject of many theoretical and experimental studies. Like all rocksalt metal oxides, its (100) facet is the most thermodynamically stable. However, MgO(111) can be found naturally as the mineral periclase, fueling scientific investigation on arrangement of ions on the surface (*3*). Much of the experimental work done involves thin films, with MgO(111) either as the substrate or the coating film (*19–23*). Usually, the film growth is done in high vacuum or ultra-high vacuum.

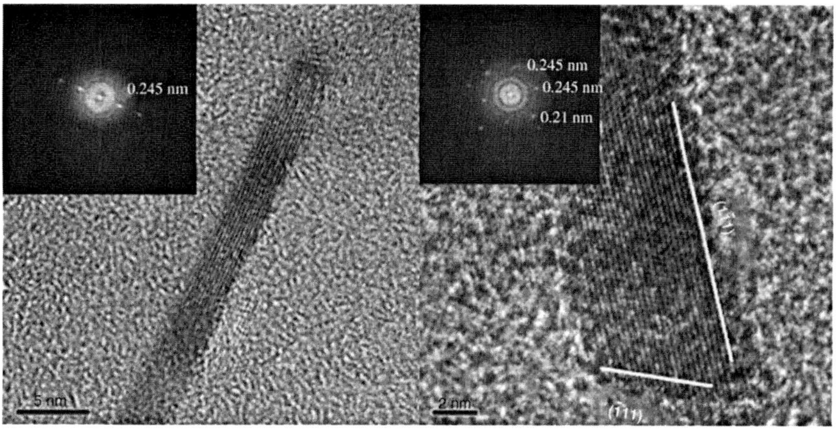

Figure 9. High resolution TEM images and local Fourier transforms of two MgO(111) nanosheets. (Reproduced from (6). Copyright Wiley-VCH Verlag GmbH & Co. KGaA. Reproduced with permission.)

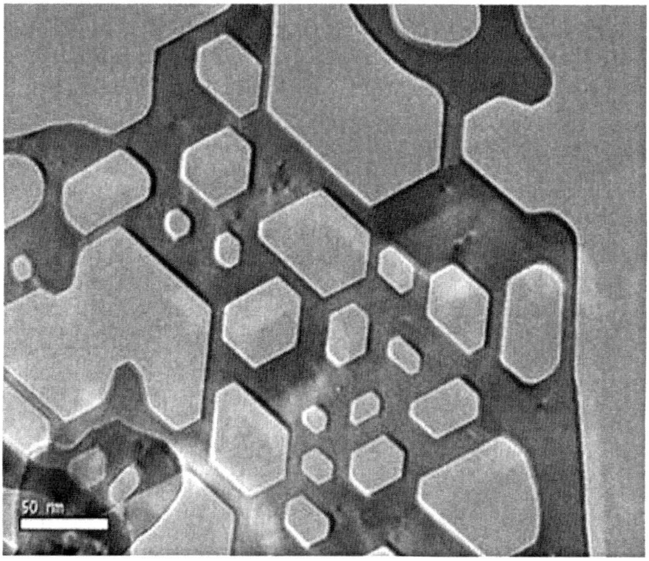

Figure 10. TEM image of an NiO(111) nanosheet calcined at 500 °C. The darker part of the image is the sheet, and the lighter hexagonal portions are the hexagonal holes. (Reproduced from (7). Copyright Wiley-VCH Verlag GmbH & Co. KGaA. Reproduced with permission.)

Based on the given theoretical studies performed on MgO(111), as well as previous work done by the Klabunde group on aerogel preparation and MgO(100) catalysis, the Richards group discovered a wet chemical synthesis of MgO(111) nanosheets (see preparation section). Evidence for the (111) structure is found in high resolution TEM (HRTEM) images; when imaging the edges of the nanosheets, lattice fringes can be seen and show lattice spacings of 0.24 – 0.25

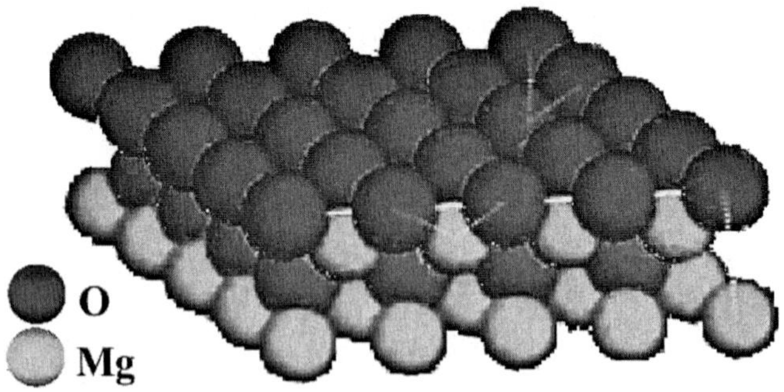

Figure 11. Theoretical (111) surface of MgO. The perfect surface would be composed of either entirely oxygen anions or entirely magnesium cations. (Adapted from (6). Copyright Wiley-VCH Verlag GmbH & Co. KGaA. Reproduced with permission.)

nm. Knowing that MgO packs in the rock salt structure, it can be seen that the experimental lattice spacings match the theoretical (111) lattice spacings. Also, when imaging nanosheets parallel to the TEM grid, the electron diffraction (ED) pattern displays lattice fringes of the [111] zone, showing again that the surfaces of the nanosheets are (111) facets. Figure 1 shows the theoretical, idealized (111), (110), and (100) facets of MgO.

In-situ diffuse reflectance infrared Fourier Transform spectroscopy (DRIFTS), a surface-sensitive technique that measures molecular vibrations, was used to study the surface of the MgO(111) nanosheets. Studying just the nanosheets themselves revealed many of the surface species to be 3-coordinated hydroxyl groups, oxygen anions, and oxygen defects. Also in the DRIFTS, CO_2 was used as a probe molecule to study the surface. Adsorption of CO_2 on the MgO(111) surface resulted in two different types of monodentate carbonate species, a bidentate carbonate species, a bicarbonate species (suggesting there exist hydroxyl groups on the surface), and bulklike carbonate species (in which the peaks are very similar to bulk $MgCO_3$, suggesting both strong base, O^{2-}, and Lewis acid, Mg^{2+}, sites on the surface). Weak adsorptions around 2400 – 2200 cm^{-1} were attributed to physisorbed and end-on chemisorbed CO_2. When heating the adsorbed CO_2 on the MgO(111) to induce desorption, it was found that the most thermally stable species were the monodentate carbonate and bulklike carbonate species, with the bidentate and bicarbonate species adsorbed on the nanosheets being less stable.

Temperature programmed desorption (TPD) of CO_2 on the MgO(111) was also studied. The profile displayed several CO_2 desorption peaks; the higher the CO_2 desorption temperature, the more basic the site of CO_2 adsorption. It was found that low coordinated oxygen anions were the most basic CO_2 adsorption sites, oxygens in Mg^{2+} and O^{2-} pairs were medium basic, and hydroxyl groups were the least basic adsorption sites. The largest peak area was that correlated with the Mg^{2+} and O^{2-} pairs, meaning the majority of the surface basic sites are Mg^{2+} and O^{2-} pairs, corresponding well with the fact that MgO(111) is composed of alternating layers of oxygen and magnesium anions, as can be seen in Figure 11. The results of the CO_2-TPD can be found in Figure 14.

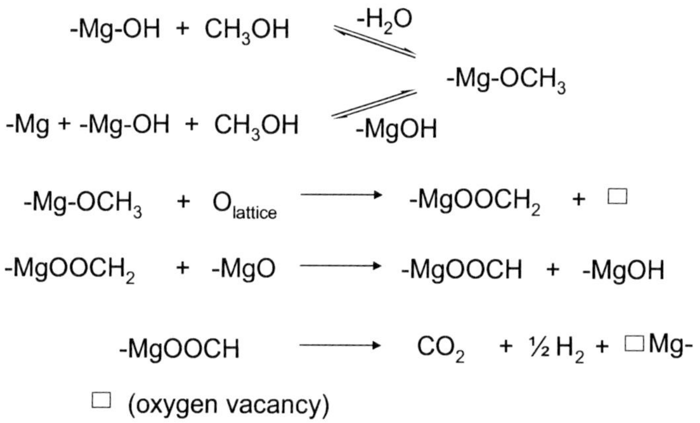

Figure 12. Proposed mechanism of methanol decomposition over MgO(111). (Reproduced from reference (24). Copyright 2007, American Chemical Society.)

In-situ DRIFTS was also used to study the adsorption of methanol on the MgO(111) nanosheets. Studies on conventional MgO(100) have shown that methanol does not adsorb on smooth surfaces of (100), but adsorbs only in the presence of defects on the (100) surface. When MgO(111) was exposed to methanol, the DRIFTS showed both gas phase methanol and weakly adsorbed methanol present. Methoxyl groups and formate species are also seen in the DRIFTS spectra at room temperature. Slowly vacuuming out some of the pressure after methanol had been introduced produced a decrease in the amount of methanol seen in the infrared, implying that some of the methanol adsorption on the MgO(111) is reversible. When the methanol on MgO(111) was heated to 70 °C, a large amount of CO_2 was formed, and within 25 minutes almost all the methanol had been oxidized. During this time period, the amount of CO_2 formed increases while the area of C-H peaks decreases. Comparison of the room temperature and 70°C spectra yields little change in the O-H region, suggesting that adsorbed methanol species interact primarily with oxygen anions and oxygen defects (as opposed to hydroxyl groups). These DRIFTS studies were also carried out on commercial MgO(100) (CM-MgO), and it was found that only trace amounts of methanol were adsorbed with mostly reversible adsorption; a small amount of irreversible methanol adsorption occurred at, presumably, defect sites. At 70 °C, methanol over MgO(100) forms some carbonyl species that increase with time, but with no formation of CO_2. With methanol over MgO(111), at 70 °C, some carbonyl species are formed, but the peak intensity is weak at all times, suggesting that the carbonyl species decompose into CO_2.

Temperature programmed desorption of methanol on MgO(111) found that there exist physisorbed and chemisorbed methanol on the surface, and that the products of methanol decomposition were CO_2, CO, H_2, H_2O, HCOH, and HCOOH. Methoxy groups also desorb, and thus, at low temperature, it is likely that methoxy desorption and surface methanol oxidation (for which adsorbed methoxy species are intermediates) are competing reactions. From the DRIFTS and TPD information gathered, we hypothesized on a methanol decomposition mechanism involving a formate intermediate and creation of an oxygen vacancy on the MgO(111) surface, with main products H_2 and CO_2. Figure 12 shows the proposed mechanism (*24*).

Different types of MgO catalysts (conventional microcrystalline MgO, nanostructured MgO(111), nanoparticle MgO(110), and nanoparticle MgO(100) aerogels) were used to study transesterification of vegetable oil and methanol to biodiesel, see Figure 13. Transesterification of vegetable oil to biodiesel is typically acid or base catalyzed, with base catalysis being the commercial standard since it is ~4000 times faster. Homogeneous catalysts are often used due to their increased performance over heterogeneous catalysts, however the homogeneous catalysts used are considered "non-green" because they cannot be recycled. Thus, MgO, a basic insulator that could be used as a recyclable heterogeneous catalyst, was studied for the transesterification to biodiesel.

The transesterification was studied with sunflower oil and rapeseed oil, under autoclave conditions, microwave conditions, and ultrasound conditions. In the study of transesterification of sunflower oil in autoclave conditions, the results of which are shown in Table III, MgO(111) had the highest conversion and yield of methyl esters. In the study of transesterification of sunflower oil in microwave conditions, all of the types of MgO performed well, with conversions and yield of methyl esters above 95%. When optimizing the calcination temperature for each of the MgO catalysts, it was found that MgO(111) performed best when calcined at 773K, and MgO(110) and (100) both performed best when calcined at 583K. For the microwave conditions, recycling of the catalysts was studied, and it was found that, for MgO(111), recycling led to increased performance in conversion and selectivity over 7 runs, both being around 99% at the last run. For MgO(110) and (100), performance did not decline over the 7 runs. On the study of transesterification of sunflower oil under ultrasound conditions, all MgO catalysts performed with high conversions and selectivity. Analysis of conversion and selectivity with rapeseed oil did not bring many new conclusions, except that conversions and yields were consistently lower with rapeseed oil than with sunflower oil.

Magnesium leaching was studied under the transesterification conditions. For sunflower oil under autoclave and microwave conditions, all MgO catalysts leached less than 10 ppm of Mg in the biodiesel, measured by ICP. This amount can be decreased by filtration with an ion-exchange resin to less than 0.5 ppm. For rapeseed oil under microwave conditions, all MgO catalysts leached a few ppm of Mg in the biodiesel. For sunflower oil under ultrasound conditions, a relatively large amount of magnesium was leached by saponification, in the trend MgO(111) leaching of 2-3% > MgO(110) leaching of 1-2% ≈ MgO(100) leaching of 2% > microcrystalline MgO leaching of 0.5%. For rapeseed oil, the amount of magnesium leached by saponification under ultrasound conditions was less than that for sunflower oil, but followed the same trend. The proposed reason for the large amount of leaching is ultrasound tending to violently break apart particles. Under ultrasound conditions, the amount of Mg leached into the biodiesel (saponification leaching not included) was MgO(111) (128 ppm) > MgO(110) (102 ppm) ≈ MgO(100) (97 ppm) for sunflower oil and MgO(111) (289 ppm) > MgO(110) (251 ppm) ≈ MgO(100) (242 ppm) for rapeseed oil.

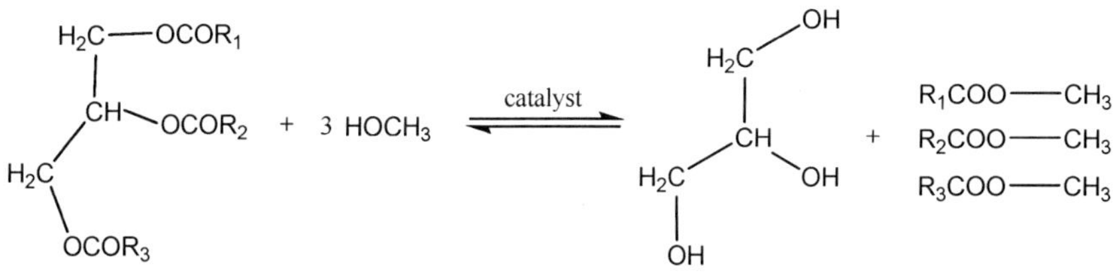

Figure 13. Transesterification of a triglyceride and methanol to biodiesel and glycerol.

Table III. Results of esterification of sunflower oil under autoclave conditions[a]

Catalyst	*Calcination Temperature*	*Reaction Temperature*	*Conversion of sunflower oil*	*Yield in methyl esters*
microcrystalline MgO	N/A	493 K	80%	55%
		373 K	25%	32%
MgO(111)	773 K	343 K	86%	86%
MgO(110)	583 K	343 K	82%	80%
MgO(100)	583 K	343 K	75%	76%
	773 K	343 K	70%	70%

[a] Data are from reference (*12*)

The nanoscale MgO materials were characterized in several ways and compared. MgO(111), the synthesis details of which can be found in the preparation section, has a surface area of 198 m^2/g (measured by BET) and is morphologically nanosheets of 500-200 nm in diameter and 3-5 nm thick. MgO(110), prepared by boiling and calcining commercial MgO from Aldrich, has a surface are of 80 m^2/g and is composed of ~ 8.8 nm nanoparticles. The prepared MgO(100) aerogel has a surface are of 435 m^2/g and crystalline sizes of 1 – 4.5 nm. To compare the difference in basic (as opposed to acidic) oxygen surface sites of the different nanoscale MgOs, CO_2 temperature programmed desorption (CO_2-TPD) was used, the results of which are shown in Figure 14. From the CO_2-TPD results, MgO(111) has the highest amount of basic surface sites (as can be determined by comparing the areas under the curves). This is expected, since MgO(110) and MgO(100) expose magnesium sites on the surface, as well as oxygen sites. For all of the MgO surfaces, the medium-basic oxygen in Mg^{2+} and O^{2-} pairs were the most prevalent, followed in amount by the weakly-basic hydroxyls, followed by a small amount of strongly-basic low coordinated oxygen anion defects.

DRIFTS was used to study the difference in activation (calcination) temperatures between the MgO species, the results of which are summarized in Table IV. The shift of the basic –OH group to lower wavenumbers with the loss of bound water denotes a less energetic, weaker bond, and this is presumably due to the loss of hydrogen bonding on the surface. The temperature at which the maximum intensity of the peak corresponding to basic –OH groups was found varied between MgO species; MgO(111) had a maximum intensity at 773K while MgO(110) and MgO(100) had maximums at 583K. DRIFTS spectra were also taken of the catalysts after transesterification, and the spectra showed many chemisorbed species on the substrates. However, heating above 900K would remove any chemisorbed species without change in the catalyst, supporting high recyclability of the catalysts. Higher temperature was needed (as compared to removing chemisorbed water) to remove the transesterification species because acid species are more strongly chemisorbed than water (*12*).

The MgO(111) nanosheets show good activity for Claisen-Schmidt condensation. The particular reaction studied was the condensation of benzaldehyde and acetophenone. Conversions of benzaldehyde over time were compared using MgO(111), aerogel-prepared MgO (probably (100)), conventionally prepared MgO(100), and commercial MgO. As can be seen in Figure 15, MgO(111) reached 100% conversion in the shortest amount of time. This activity has been ascribed to both the amount of basic sites on MgO(111) and the morphology of the nanosheets; this particular Claisen-Schmidt condensation has been shown to be promoted by O^{2-} sites and the nanosheets have a lot of edges and corners which have highly reactive sites (*6*).

NiO(111)

Nickel oxide, a rock salt structured transition metal oxide, has unique and interesting electronic properties. Like most transition metal oxides, the redox state of nickel oxide can be electrochemically switched. Nickel oxide, in particular, exhibits electrochromism, meaning an applied electric field will reversibly change its optical properties, making it an interesting material for solar energy (*25*). Nickel oxide also has interesting magnetic properties, and is being studied for use in giant magnetoresistive sensors (*26*), magnetic recording, and magnetic memories (*27*). It has also been studied for use in various catalytic reactions (*7*, *28*).

Many efforts have been made to study the growth of NiO(111), especially under ultra high vacuum (UHV) conditions. Studies include growth of ultra-thin films of NiO(111) on Ni(111) (*29*), epitaxial growth of NiO(111) on Cu(111) (*30*), NiO(111) grown on Ni tapes (*31*), and low temperature epitaxial growth of NiO(111) films on MgO(100) and α-Al_2O_3(00l) (*26*). In a particular study by Kitakatsu, Maurice, Hinnen, and Marcus (*32*), NiO(111) thin films were grown on Ni(111) by oxidation of the nickel surface. They found films grown at room temperature were 3 to 4 monolayers thick with NiO(111) and hydroxyl terminated by 0.85 + 0.1 monolayers, yielding an unreconstructed oxide surface structure. However, under UHV, oxidizing Ni(111) at 500K yielded a two to three monolayer thick surface of 93 + 3% NiO(100) and 7 + 3% NiO(111). The small amount of NiO(111) grew in triangular patches and were fully hydroxylated by a monolayer of OH^- groups. Exposing this film to water increased the lateral size of the NiO(111) patches at the expense of NiO(100) on the surface. Under the given circumstances, NiO(100) was not at all hydroxylated. In another study by Barbier et al, they sythesized single crystal NiO(111) as well as a thin NiO(111) film of 5 monolayers on Au(111) and observed the octopolar reconstruction on both, with both being nonreactive to hydroxylation (also studied under UHV). They

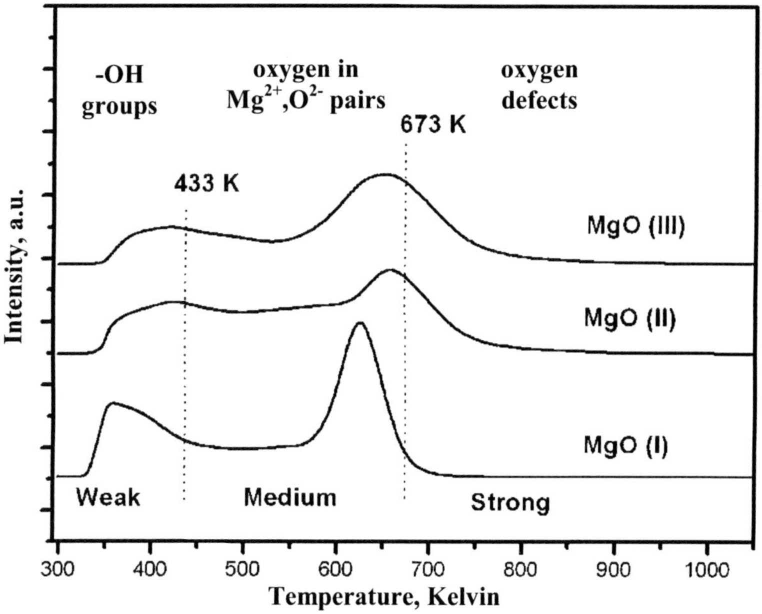

Figure 14. CO_2-TPD for MgO(111), denoted MgO (I), MgO(110), denoted MgO (II), and MgO(100), denoted MgO (III). (Reference (12) - Reproduced by permission of The Royal Society of Chemistry. Copyright 2008, The Royal Society of Chemistry.)

calculated that the single crystal surface was Ni terminated with double steps, while the thin film surface showed both O terminated and Ni terminated domains, separated by single steps (*27*). These studies are representative of the importance of synthesis methods in making a (111) surface, as different procedures can yield different reconstructions of (111).

The wet chemical synthesis of NiO(111) discovered by the Richards group actually came about as an effort to better understand MgO(111). Although nickel oxide and magnesium oxide have different chemical properties, the former is a semiconductor and the latter a basic metal oxide, the two oxides have the same lattice parameters. With this in mind, the Richards group investigated whether the same or similar wet chemical synthesis of MgO(111) would produce NiO(111) with the same morphology. What was found was that, while the wet chemical synthesis of MgO(111) produced smaller platelets, a similar wet chemical synthesis of NiO(111) produced larger sheets with nano-thickness but with larger, micrometer-range lateral size.

To prove the (111) structure of the NiO aerogel (see preparation section), electron diffraction (ED) and XRD were used. During the ED analysis, the incident electron beam was focused along the [111] zone axis, and the resulting hexagonal pattern demonstrates the 111 plane is parallel to the main surface of the nanosheets. Along with proving the (111) structure, the ED pattern showed the NiO(111) to be single crystalline. As further proof of the (111) faceting, the nickel oxide was grown on a silicon wafer, and the XRD spectra showed only the (111) diffraction

Table IV. Summary of DRIFTS results studying calcination temperature effect on active sites[a]

Catalyst	*DRIFTS temperature*	*IR Peaks Observed*
MgO(111)	298 K	basic –OH groups and bound H_2O
	773 K [b]	basic –OH groups, shifted to lower wavenumbers
	973 K	basic –OH groups, shifted to lower wavenumbers
MgO(110)	298 K	basic –OH groups and bound H_2O
	583 K [b]	basic –OH groups, shifted to lower wavenumbers
	773 K	basic –OH groups, shifted to lower wavenumbers
	973 K	basic –OH groups, shifted to lower wavenumbers
MgO(100)	298 K	basic –OH groups and bound H_2O
	583 K [b]	basic –OH groups, shifted to lower wavenumbers
	773 K	basic –OH groups, shifted to lower wavenumbers
	973 K	basic –OH groups, shifted to lower wavenumbers

[a] Data are from reference (*12*). [b] The temperature at which the maxiumum intensity of the peak corresponding to basic –OH groups was found.

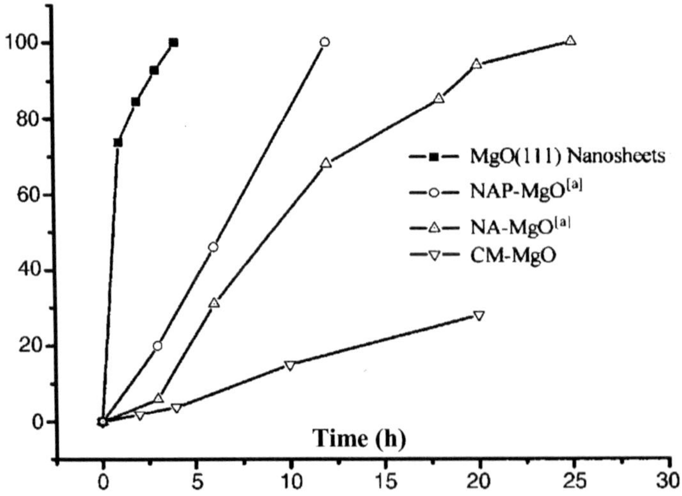

Figure 15. Results of Claisen-Schmidt condensation over different types of MgO. (Reproduced from reference (6). Copyright Wiley-VCH Verlag GmbH & Co. KGaA. Reproduced with permission.)

peak. TEM after calcination showed the NiO(111) nanostructured particles have sheetlike structures with numerous hexagonal holes. The particle diameter ranged up to 1 micrometer, with typical thickness between 1-5 nm, and hexagonal hole size between 20-100 nm. Using BET, the surface area was measured to be 88 m^2/g.

Methanol is used as a common probe molecule for studying catalyst surfaces, and as such the adsorption of methanol on the NiO(111) aerogel was studied by using *in-situ* DRIFTS. After exposing the NiO(111) surface to methanol vapor pressures of 1 and 0.005 torr, formate species were identified in the IR. After exposure to methanol at 70 °C for 5 minutes, a large amount of CO_2 was measured on the NiO(111) surface, and this amount increased with time. From this information, the mechanism involves dissociation of methanol on the NiO(111) surface, formation of a formate intermediate species, and resulting in CO_2 and H_2 production, with the loss of a surface oxygen on the NiO(111) surface. Methanol adsorption was also studied on conventionally-prepared NiO (CP-NiO), made by nickel acetate decomposition at 500°C for 5 hours, producing NiO(100) with a surface area of 32 m^2/g. After exposure of the CP-NiO surface to methanol vapor pressures of 1 and 0.005 torr, no dissociative adsorption or oxidation of methanol occurred. After exposure to methanol at 70°C for 5 minutes, no CO_2 was formed and a small amount of CO was formed. From these studies, it was concluded that the (111) facet of the NiO aerogel was responsible for the dissociation and oxidation of methanol (*7*). The proposed mechanism can be found in Figure 16.

CeO_2 (111)

As an example of a (111) non-rock salt metal oxide, ceria, or CeO_2, has many catalytic applications. For ceria, the (111) termination, shown in Figure 17, is the most thermodynamically stable, and is a non-polar surface made up of equal numbers of Ce^{4+} and O^{2-} sites (*33*). CeO_2 packs in a cubic fluorite structure. Ceria is a good catalyst for CO oxidation, especially when doped with copper, a small amount greatly magnifying its CO oxidation catalytic activity. In addition, thin films of ceria are useful for miniaturized capacitors, oxygen sensors, buffer layers at high temperatures, and optical coatings (*34*). Ceria is also used as an auto exhaust catalysts, due to its redox properties and its being a good oxygen ion storage and transportation medium (*35, 36*). It is also excellent as a water gas shift catalyst (*37*) and is being considered in use for solid oxide fuel cells (*35, 36*).

500Å films of epitaxial CeO_2(111) have been grown on YSZ(111). The ceria films are atomically flat and unreconstructed, as confirmed by RHEED, LEED, and SEM. The films were studied in both oxidized and reduced conditions. For the study, reduced CeO_2 was prepared by annealing between 773K and 973K under ultra high vacuum conditions. Based on XPS data, the reduction of Ce^{4+} to Ce^{3+} is presumed to occur by desorption of oxygen as O_2. At 773K, most of the reduced cerium was in the first monolayer of the ceria. At higher annealing temperatures, there was some subsurface reduction. Oxidized CeO_2 was prepared by exposing the CeO_2 film to O_2 gas at 830K, then cooling to below room temperature in the presence of O_2. It is hypothesized that reduced CeO_2 (111) is less susceptible to water oxidation than (110) or (100) surfaces, since water oxidation has been seen for other reduced ceria powders and not for reduced CeO_2 (111).

On an oxidized CeO_2 (111) surface, low water coverages (the fraction of active sites taken up by a molecule compared to the total number of active sites, both empty and taken) are more stable than high water coverages, as determined by water TPD. This trend is explained by the highest energy adsorption site being

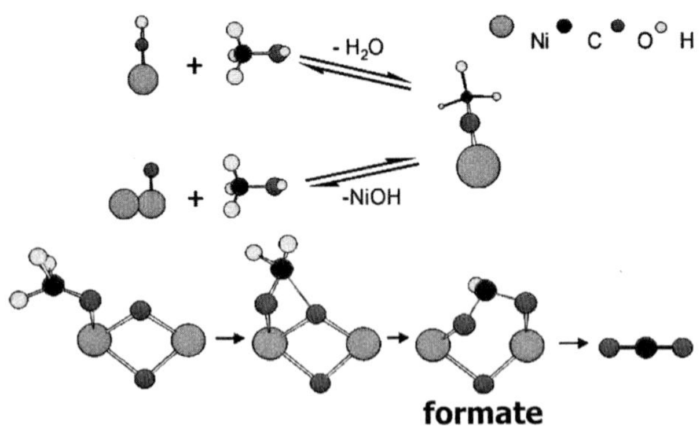

Figure 16. Proposed mechanism for methanol decomposition to H_2 and CO_2 over NiO(111). (Reproduced from (7). Copyright Wiley-VCH Verlag GmbH & Co. KGaA. Reproduced with permission.)

a surface Ce^{4+} ion (with the water molecule hydrogen bonding to surrounding oxygens). These Ce^{4+} ions are relatively far away from each other, such that any further adsorbed water would have difficulty making a hydrogen bonding network with any water molecules in a Ce^{4+} adsorption site. Furthermore, competition for the highest energy adsorption site (Ce^{4+}) would cause high water coverages to be less stable. Also, water molecules in adjacent Ce^{4+} adsorption sites would experience dipole-dipole repulsions (assuming the water molecules adsorbed in the same C_{2v} geometry), again making high water coverages less stable (*33*).

Thin film CeO_2 (111) has been grown epitaxially on many different substrates, including Pt(111), Rh(111), Re(0001) ((*34*) *and references therein*), Ru(0001) (*36*), alumina (*38*), YSZ (*33, 38*), and Cu(111) (*34, 37*). Ultra thin films have also been epitaxially grown on Cu (111) by vacuum deposition. From XPS and LEEDS, 2.5 monolayers of ceria were deposited on the copper. Initial ceria growth occurs by islands of ceria on the copper. The ceria was thermally stable above 970 K, but at 1070 K the (111) surface lost surface oxygens and formed Ce_2O_3 (0001) (*34*).

The adsorption and decomposition of methanol has been studied on CeO_2(111) surfaces. One of the first studies on CeO_2(111) thin films was carried out by Siokou and Nix, who found that methanol adsorbs dissociatively at room temperature (300 K) on not-fully-oxidized CeO_2(111) epitaxially grown on Cu(111). Using TPD and reflection and absorption FTIR (Fourier Transform Infrared) spectroscopy, the intermediates of methanol decomposition were studied. Mainly methoxy species can be seen on the catalyst surface after methanol adsorption. These methoxy species are stable up to 550K, but decompose after 585 K, yielding mostly H_2 and CO, with some formaldehyde and water formation. Some of the methoxy species can be oxidized to formate, occurring mostly at the edges of oxide islands (where there exist sites of higher coordination unsaturation). Compared to the methoxy species, decomposition of the surface formates yields more formaldehyde, and can also decompose into CO_2 (*37*). As an interesting note, Matolín et al. did not observe methanol dissociation after adsorbing methanol on epitaxially grown CeO_2(111) on Cu(111) at 300K, and were unsure of a concrete reason for this discrepancy but did point out the preparation process was different (*39*).

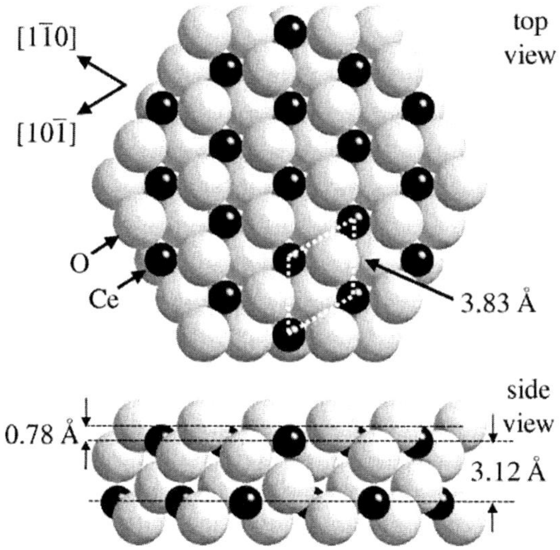

Figure 17. Ideal geometry of CeO_2 (111). (Reprinted from reference (33), Copyright 2003, with permission from Elsevier.)

Furthering studies on methanol, Mullins, Robbins, and Zhou (*36*) looked at methanol adsorption on thin films of CeO_2 (111) vapor deposited on Ru (0001). When the ceria was fully oxidized, methanol produced water at 200K and formaldehyde at 560 K. On reduced ceria, adsorbed methanol produced CO, H_2, formaldehyde, and water. Increasing ceria reduction produced more H_2 and less water. This effect was explained by isotopic exchange experiments; CH_3OD formed hydroxyls on the reduced ceria surface, and would be desorbed as HD, while fully oxidized ceria (they used isotopically labeled $Ce^{18}O_2$) would lose an oxygen to the alcoholic deuterium to produce deuterated water ($H^{18}OD$). This is in agreement with the study by Henderson et al. (*33*), which hypothesized that reduced CeO_2(111) was less susceptible to water oxidation. Matolin et al, by using XPS and soft X-ray synchrotron radiation PES, also found that adsorbed methanol on CeO_2(111)/Cu(111) formed hydroxyl groups, desorbing water, and creating oxygen vacancies, reducing the ceria, but evidenced that the dehydration of water on CeO_2(111)/Cu(111) followed a more complex mechanism than the dehydration of water on CeO_2(111)/Ru(0001) (*39*). Ferrizz et al implied that oxygen vacancies were necessary for methanol adsorption (and thus 100% oxidized ceria would not adsorb methanol) (*38*), however, Beste et al concluded in a DFT study that vacancies are not a requirement for adsorption of methanol on ceria, since methanol adsorption is still exothermic on fully oxidized ceria and decomposition is thermoneutral (*35*).

Other chemistry studies of molecules over CeO_2(111) surfaces have been carried out recently, including formic acid decomposition (*40*) and adsorption and reaction of acetone (*41*). Doping of CeO_2 thin films has also been studied, including samaria doped silica (*42*) and ceria doped with gold (*43*).

Summary and Outlook

From combined experimental and theoretical studies on the surfaces of MgO(111) and NiO(111), it is expected for both that the hydroxylated (111) surface is the most stable surface, more stable even than the (100) surface. Also, for both clean MgO(111) and clean NiO(111) reconstruction to (100) microfacets is energetically more favorable than the octopolar reconstruction, however the ratio between NiO(111) octopolar reconstruction and NiO(100) is close enough to unity to favor a NiO reconstruction to a (2x2) structure (also known as an octopolar reconstruction). In addition, reconstruction of the NiO(111) surface was found to be disfavored after absorption of CO.

For theoretical studies, which usually study slabs of atoms, and thin film studies in vacuum or ultra high vacuum, either a hydroxylated surface prevails or some reconstruction of the (111) structure is necessary. However, in our wet chemical preparation of MgO(111) and NiO(111) aerogels, we found more medium-basic oxygen sites in Mg^{2+} and O^{2-} pairs than weakly-basic hydroxyl groups. These ideas are not conflicting as crystals often need more than one type of crystallographic facet to achieve minimum total surface energy, depending on the morphology (*44*). Since films and aerogels have very different morphology, it is not surprising their (111) surfaces differ in termination.

(111) rocksalt metal oxides prepared by wet chemical syntheses exhibit new and interesting catalysis. The adsorption of CO_2 and decomposition of methanol has been studied on MgO(111) and NiO(111), both showing low temperature decomposition of methanol to H_2 and CO_2. The decomposition of methanol has also been studied on CeO_2 (111), a cubic fluorite metal oxide, and resulted in a different product distribution than the (111) rocksalt metal oxides. Transesterification of vegetable oil to biodiesel was also studied over different types of MgO, with MgO(111) outperforming MgO(110) and MgO(100) under autoclave conditions. In short, the wet chemically synthesized (111) rocksalt metal oxides show great potential as catalysts for reactions promoted by surfaces with many basic sites.

References

1. Goniakowski, J.; Finocchi, F.; Noguera, C. *Rep. Prog. Phys.* **2008**, *71*, 016501.
2. *Metal Oxide Catalysis*; Jackson, S. D., Hargreaves, J. S. J., Eds.; Wiley-VCH: Weinheim, Germany, 2009; Vol. 1.
3. Henrich, V. E.; Cox, P. A. *The Surface Science of Metal Oxides*; Press Syndicate of the University of Cambridge: New York, 1996.
4. *Nanoscale Materials in Chemistry*; Klabunde, K. J., Ed.; John Wiley & Sons, Inc.: New York, 2001.
5. Berg, M.; Jaras, S. *Appl. Catal. Gen.* **1994**, *114*, 227–241.
6. Zhu, K.; Hu, J.; Kuebel, C.; Richards, R. *Angew. Chem., Int. Ed.* **2006**, *45*, 7277–7281.
7. Hu, J.; Zhu, K.; Chen, L.; Yang, H.; Li, Z.; Suchopar, A.; Richards, R. *Adv. Mater.* **2008**, *2*, 267–271.

8. Zion, B. D.; Sibener, S. J. *J. Chem. Phys.* **2007**, *127*, 154720.
9. Bender, M.; Seiferth, O.; Carley, A. F.; Chambers, A.; Freund, H.-J.; Roberts, M. W. *Surf. Sci.* **2002**, *513*, 221–232.
10. Knozinger, H. *Science* **2000**, *287*, 1407.
11. Lazarov, V. K.; Plass, R.; Poon, H-C.; Saldin, D. K.; Weinert, M.; Chambers, A.; Gajdardziska-Josifovska, M. *Phys. Rev. B* **2005**, *71*, 115434.
12. Verziu, M.; Cojocaru, B.; Hu, J.; Richards, R.; Ciuculescu, C.; Filip, P.; Parvulescu, V. I. *Green Chem.* **2008**, *10*, 373–381.
13. Wander, A.; Bush, I. J.; Harrison, N. M. *Phys. Rev. B* **2003**, *68*, 233405.
14. Henrich, V. *Surf. Sci.* **1976**, *57*, 385–392.
15. Plass, R.; Feller, J.; Gajdardziska-Josifovska, M. *Surf. Sci.* **1998**, *414*, 26–37.
16. Plass, R.; Egan, K.; Collazo-Davila, C.; Grozea, D.; Landree, E.; Marks, L. D.; Gajdardziska-Josifovska, M. *Phys. Rev.Lett.* **1998**, *81*, 4891–4894.
17. Wang, W.; Li, J.; Zhang, Y. *Appl. Surf. Sci.* **2006**, *252*, 2673.
18. Hermansson, K.; Baudin, M.; Ensing, B.; Alfredsson, M.; Wocik, M. *J. Chem. Phys.* **1998**, *109*, 7515.
19. Goniakowski, J.; Noguera, C. *Phys. Rev. B* **1999**, *60*, 16120–16128.
20. Montengro, M.; Döbeli, M.; Lippert, T.; Müller, S.; Weidenkaff, A.; Willmott, P.; Wokaun, A. *Appl. Surf. Sci.* **2005**, *247*, 197–203.
21. Goodrich, T.; Cai, Z.; Ziemer, K. *Appl. Surf. Sci.* **2008**, *254*, 3191–3199.
22. Willmann, H.; Beckers, M.; Birch, J.; Mayrhofer, P.; Mitterer, C.; Hultman, L. *Thin Solid Films* **2008**, *517*, 598–602.
23. Warot, B.; Snoeck, E.; Ousset, J. C.; Casanove, M. J.; Dubourg, S.; Bobo, J. F. *Appl. Surf. Sci.* **2002**, *188*, 151–155.
24. Hu, J.; Zhu, K.; Chen, L.; Kubel, C.; Richards, R. *J. Phys. Chem. C* **2007**, *111*, 12038–12044.
25. Purushothaman, K. K.; Muralidharan, G. *Sol. Energy Mater. Sol. Cell* **2009**, *93*, 1195–1201.
26. Lindahl, E.; Lu, J.; Ottosson, M.; Carlsson, J. *J. Cryst. Growth* **2009**, *311*, 4082–4088.
27. Barbier, A.; Mocuta, C.; Kuhlenbeck, H.; Peters, K. F.; Richter, B.; Renaud, G. *Phys. Rev. Lett.* **2000**, *84*, 2897–2900.
28. Carnes, C.; Klabunde, K. J. *J. Mol. Catal. Chem.* **2003**, *194*, 227–236.
29. Kitakatsu, N.; Maurice, V; Marcus, P. *Surf. Sci.* **1998**, *411*, 215–230.
30. Stanescu, S.; Boeglin, C.; Barbier, A.; Deville, J.-P. *Surf. Sci.* **2004**, *549*, 172–182.
31. Woodcock, T. G.; Abell, J. S.; Eickemeyer, J.; Holzapfel, B. *J. Microsc.* **2004**, *216*, 123–130.
32. Kitakatsu, N.; Maurice, V.; Hinnen, C.; Marcus, P. *Surf. Sci.* **1998**, *407*, 36–58.
33. Henderson, M. A.; Perkins, C. L.; Engelhard, M. H.; Thevuthasan, S.; Peden, C. H. F. *Surf. Sci.* **2003**, *526*, 1–18.
34. Matolín, V.; Libra, J.; Matolínová, I.; Nehasil, V.; Sedláček, L.; Šutara, F. *Appl. Surf. Sci.* **2007**, *254*, 153–155.
35. Beste, A.; Mullins, D. R.; Overbury, S.; Harrison, R. *Surf. Sci.* **2008**, *602*, 162–175.
36. Mullins, D. R.; Robbins, M. D.; Zhou, J. *Surf. Sci.* **2006**, *600*, 1547–1558.

37. Siokou, A.; Nix, R. M. *J. Phys. Chem. B* **1999**, *103*, 6984–6997.
38. Ferrizz, R. M.; Wong, G. S.; Egami, T.; Vohs, J. M. *Langmuir* **2001**, *17*, 2464–2470.
39. Matolín, V.; Libra, J.; Škoda, M.; Tsud, N.; Prince, K. C.; Skála, T. *Surf. Sci.* **2009**, *603*, 1087–1092.
40. Senanayake, S. D.; Mullins, D. R. *J. Phys. Chem. C* **2008**, *112*, 9744–9752.
41. Senanayake, S. D.; Gordon, W. O.; Overbury, S. H.; Mullins, D. R. *J. Phys. Chem. C* **2009**, *113*, 6208–6214.
42. Gupta, S.; Kuchibhatla, S. V. N. T.; Engelhard, M. H.; Shutthanandan, V.; Nachimuthu, P.; Jiang, W.; Saraf, L. V.; Thevuthasan, S.; Prasad, S. *Sens. Actuators, B* **2009**, *139*, 380–386.
43. Camellone, M. F.; Fabris, S. *J. Am. Chem. Soc.* **2009**, *131*, 10473–10483.
44. Viswanath, B.; Kundu, P.; Halder, A.; Ravishankar, N. *J. Phys. Chem. C* **2009**, *113*, 16866–16883.

Chapter 5

Selected Environmental Applications of Nanocrystalline Metal Oxides

Slawomir Winecki*

Nanoscale Corporation, Inc., Manhattan, Kansas
***swinecki@nanoscalecorp.com**

This chapter is not intended as a comprehensive overview of environmental applications of nanocrystalline metal oxides. Rather, our intent is to present a few pressing environmental challenges and to demonstrate that these will most likely be solved by application of nanocrystalline metal oxides sorbents. Two areas of applications will be discussed: removal of toxic chemicals from air and capture of pollutants released during coal combustion. Both have a long history; however, they remain very active areas of research due to still unsolved environmental issues.

Following the main theme of this book, we will identify the nanocrystalline structure of sorbents using the usual definition of nanocrystalline materials which requires that the crystallite size, at least in one dimension, must be below 100 nm. It turns out that according to this definition most of the metal oxide-based sorbents with high specific surface area are in fact nanocrystalline metal oxides. This includes many well known sorbents like zeolites or molecular sieves; materials that were used for decades, long before the term "nano" was widely used outside of its original "10^{-9}" meaning. We will also discuss the importance of the reactivity of surface, edges, and corner sites that dominate chemical properties of materials with crystallite sizes of 10 nm and below.

Are All Sorbents Nanocrystalline Sorbents?

The majority of solid sorbents are porous materials with high specific surface area (SSA). Activated carbon sorbents usually have the highest SSA with carbons commonly exceeding the 1,000 m^2/g mark. Sorbents based on metal oxides usually have smaller SSAs that are strongly dependent on the specific material. Oxides that are prone to sintering, calcium oxide being a primary example rarely exceed SSA of 20 m^2/g. Materials like magnesium oxide or titanium dioxide are less sensitive to sintering and their SSA can be as high as 500-700 m^2/g.

Chemical manufacturers work diligently to achieve the maximum possible SSA of all sorbents since this translates to better performance. It is commonly accepted that high SSA increases sorbent capacity for both physical adsorption and chemical interactions. In a typical air filtration application where a solid sorbent removes a chemical pollutant from air, only sorbents with SSA above 300-500 m^2/g are regarded as effective and sorbents with even higher SSA are preferred. These very high SSA and related high porosities are required to compensate for very short contact time between sorbent and polluted stream inside the sorbent bed, as well as to minimize a pressure drop across a filter. For example, filter cartridges for personal protection respirators cannot be thicker than 1-2 inches and the total allowed pressure drop must be below 1-2 inches of water. Even in industrial applications that require smaller SSAs, due to longer contact times and larger pressure drop allowed, SSAs in 20-100 m^2/g range are considered a minimum.

How do these high values of SSA translate to crystalline size? In a simplest approximation, the SSA of spherical particles with diameter d and true density ρ can be expressed as:

$$SSA=6/d\rho$$

Assuming a typical solid density of 3 g/cm^3, the "nano" threshold of 100 nm translates to the mere 20 m^2/g! Does it mean that all high surface area sorbents are in fact nano-sorbents? In our opinion, in many cases the answer is yes; however, it all depends on the specific material. Activated carbons are difficult to classify as nanocrystalline since there is no clear crystal structure. Activated carbons, while examined by microscopic techniques or by X-ray diffraction methods, do not show a crystal structure, instead their structure appears amorphous. In contrast, metal oxide-based sorbents usually produce well defined crystal diffraction patterns while examined by a powder X-ray diffraction (XRD) technique and clear crystallites are visible under a high resolution tunneling electron microscope (TEM). Figures 1 and 2 present both the TEM image and the XRD spectra of NanoActive® MgO Plus produced by NanoScale Corporation. Both techniques, at least in principle, allow for a determination of crystallite size. The TEM image allows for direct observation of crystallites so their size and habit (shape) can be determined. However, this is only possible if the resolution of the microscope allows for it. Specifically, TEM observation of the smallest nanocrystallites is quite challenging. In the high resolution image presented in Figure 1 not only individual crystallites with 2-4 nm sizes can be seen but also individual rows of atoms are visible.

Small nanocrystallites, in 2-20 nm range, can be relatively simply detected from broadening of XRD diffraction peaks. The spectrum shown in Figure 2 illustrates this effect. The width of XRD peaks is dramatically increased for nanocrystalline MgO as compared to the regular form of this material. The broadening is caused by diffraction generated during scattering from small nanocrystallites with size comparable to the wavelength of X-rays. The Scherrer equation relates the mean crystalline size with the broadening of XRD peaks: (*1*, *2*)

$$\beta = \frac{K\lambda}{L\cos\theta}$$

In the above equation, β is the full width half maximum of peaks observed (measured in radians), L is the mean crystallite size (dimension of the crystallite perpendicular to the diffracting net planes), K is a constant shape factor (often taken as 0.9) and λ is the wavelength of the radiation employed (typically 0.15418 nm for an XRD spectrometer using Cu_{α} X-rays). Application of the Scherrer equation to the XRD spectrum of NanoActive MgO indicates the mean crystalline size of approximately 2.5 nm, a value consistent with the TEM image.

Reactivity of Surfaces, Edges and Corners

Several mechanisms are effective in the removal of chemical pollutants by solid sorbents. Mechanisms like physical adsorption, capillary condensation, and chemisorption are widely recognized and described in the literature (*3*, *4*). Chemical reactivity of nanocrystalline materials is based on a different mechanism which is related to very large fractions of atoms present on the surface, edges and corners of these materials. Figure 1 makes it clear that the surface features dominate morphology and chemical reactivity of nanocrystalline materials. Atoms and ions located on the surface are not only accessible to chemical species in contact with the sorbents but also are highly reactive due to their partial saturation and increased surface energy. This high reactivity will include both stoichiometric and catalytic reactions.

The importance of surface, edge and corner effects on the chemical reactivity of nanocrystalline materials strongly depends on the crystallite size. Figure 3 shows results of a model which assumes uniformly seized cubic crystallites and typical values of density and crystal spacing in solids. Assumption of the simple geometry and omission of agglomeration effects allows for estimations of fractions of chemically reactive atoms and ions, as well as for the SSA of the material. It is apparent that a material with crystallite size of 2 nm has SSA in excess of 1,000 m^2/g and approximately 50% of its atoms are located on a surface and are readily available for chemical reactions. The chemical reactivity of the material is further increased by approximately 10% of atoms located on edges and about 1% of atoms located on corners.

The importance of surface effects diminishes for materials with larger crystallite size. For instance, sorbents with 20 nm crystallites have smaller

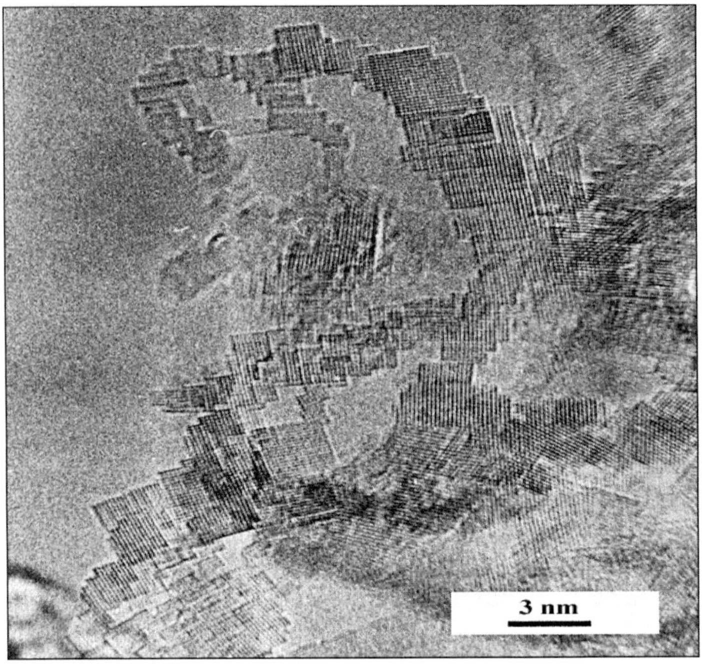

Figure 1. High resolution TEM image of NanoActive MgO Plus.

fractions of reactive atoms and the SSA on the order of 100 m^2/g. This illustrates that the traditionally applied 100 nm size threshold for "nano" materials should be replaced with a somewhat lower value if chemical applications are considered.

The surface-, edge- and corner-reach morphology of nanocrystalline materials, including metal oxides, defines chemical reactivity of these materials. The following general characteristics are present:

- High chemical rates due to a large number of surface sites readily accessible to chemical reactions.
- Chemical reactivity with a very broad range of chemical species as compared to the regular form of the same materials. Reactivity towards new pollutants is generated by highly unsaturated edge and corner sites with enhanced surface energy.
- Polar surface with the highest affinity for polar pollutants and water. Sorbent interaction with humidity must be always considered in applications of these materials.

The following sections describe two specific potential applications of nanocrystalline metal oxides. Arguments for utilization of these materials, as well as technical challenges, are outlined.

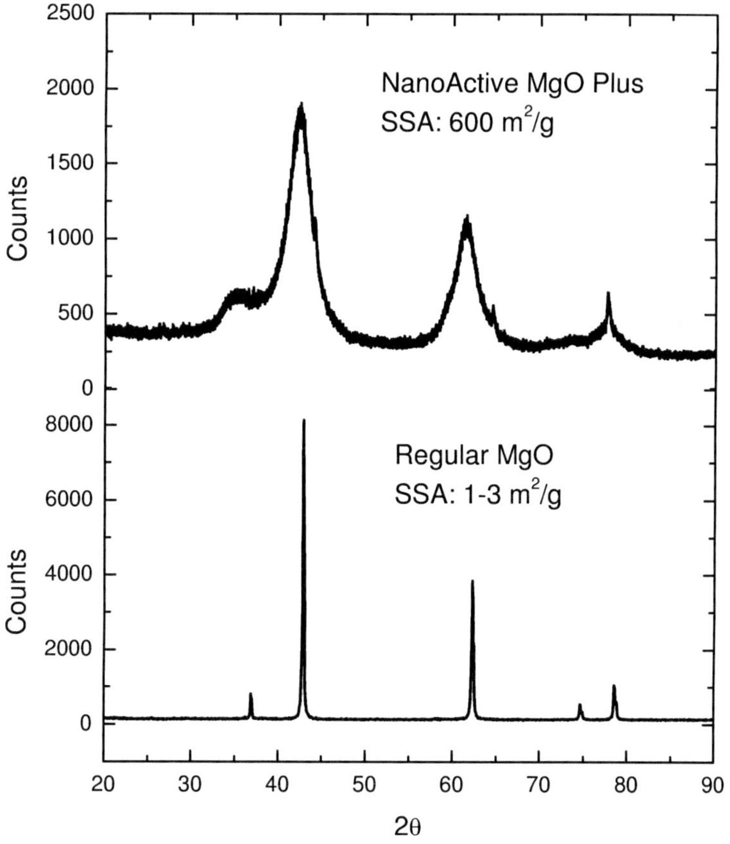

Figure 2. XRD spectrum of high surface area NanoActive MgO Plus. Spectrum of low surface area MgO is shown for comparison.

Applications in Air Filtration

Removal of toxic industrial chemicals (TICs) from air streams is critical in many industrial, laboratory or military/security settings, but at the same time still poses a significant engineering challenge. TICs can contaminate air in industrial activities or accidents, during research and development efforts, or during transportation or storage of these materials. An effective response to such situations typically requires personal protection respirators and, in some cases, larger scale air filtration systems removing pollutants from large spaces or entire buildings. Removal of a single and known air pollutant is quite straightforward since the air filtration equipment can be selected and prepared in advance. Sorbents targeting a specific pollutant tend to have large removal capacities. A much more challenging task arises when multiple and/or unknown pollutants need to be removed from air. These applications require "universal" air filtration sorbents targeting pollutants with a broad range of properties and chemical characteristics. Arguably one of the most challenging aspects of these efforts is an air filtration system targeting a broad range of TICs (preferably all

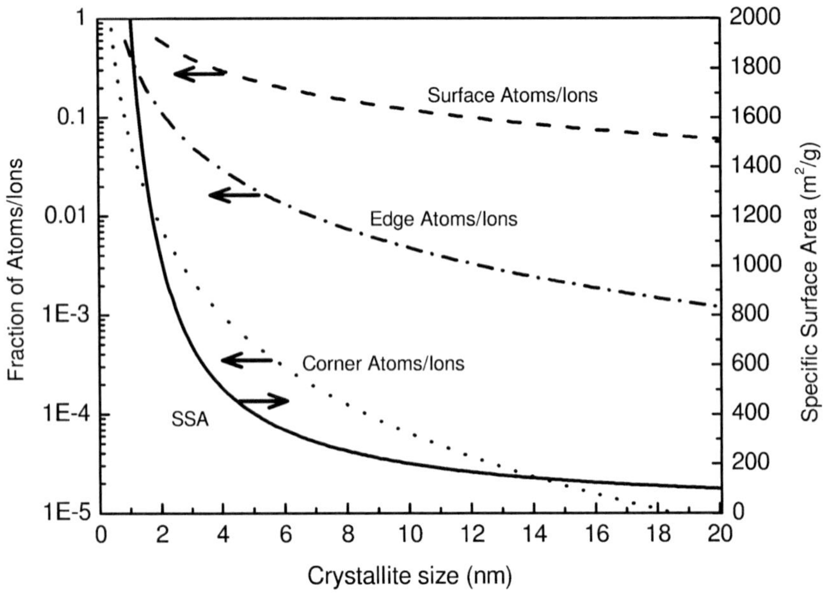

Figure 3. SSA and fractions of chemically reactive atoms/ions for nanocrystalline metal oxides.

of the approximately 150 most common TICs). Development of air filtration sorbents with such broad capabilities is an active area of research pursued by both university groups and private companies.

Currently, filters containing activated carbons are the most widely used air filtration systems. Activated carbons can be easily manufactured with SSA in 1,000-2,000 m^2/g, higher that most other sorbents. In addition, carbon does not react with water chemically and its non-polar surface does not attract water. These properties make carbons resistant to humid air exposure, one of the key requirements for practical air filtration devices. However, carbons are only effective for removal of low volatility TICs that readily adsorb on solid surfaces or inside of narrow pores that provide capillary condensation. Since these physical processes are reversible, pollutants with higher volatility have a tendency to desorb from the sorbents' surface and reenter air. Out of 150 common TICs only approximately 50 TICs can be effectively removed by physical adsorption and condensation mechanisms offered by activated carbons. Limited applicability and the possibility of pollutant desorption are serious drawbacks of activated carbon sorbents.

Sorbents targeting the 100 high volatility TICs must rely on irreversible chemical reactions that immobilize these pollutants on the sorbent. Traditionally, this is accomplished by chemical impregnates distributed over a sorbent's surface. Impregnates react chemically with pollutants adsorbed on sorbents' surface and convert them into less toxic and less volatile products. Table 1 lists the commonly used impregnates together with pollutants being retained (*5*).

Table 1. Common impregnates used to enhance chemical reactivity of solid sorbents

Impregnate	*Chemical contaminant(s) removed by the impregnate*
Copper/silver salts	Phosgene, chlorine, arsine
Iron oxide	Hydrogen sulfide, mercaptans
Manganese IV oxide	Aldehydes
Phosphoric acid	Ammonia
Potassium carbonate	Acid gases, carbon disulfide
Potassium iodide	Hydrogen sulfide, phosphine, mercury, arsine, radioactive methyl iodide
Potassium permanganate	Hydrogen sulfide
Silver	Arsine, phosphine
Sulfur	Mercury
Sulfuric acid	Ammonia, amine, mercury
Triethylenediamine (TEDA)	Radioactive methyl iodide
Zinc oxide	Hydrogen cyanide

Sorbents containing impregnates are great for removal of single pollutants or several chemicals with similar chemical properties. For instance, sorbents impregnated with potassium carbonate can be used for removal of acid gases or carbon disulfide while materials containing sulfur can capture mercury vapors. The problem arises when a universal sorbent is required to treat a large number of pollutants. Table 1 demonstrates that such a task requires a cocktail of many impregnates simultaneously applied on one sorbent. This approach works; however, in order to maintain high surface area of the support only limited amounts of each impregnate can be applied. This results in a universal sorbent with very limited removal capacity for individual pollutants and short useful lifetime in real applications.

One type of a universal sorbent with multiple impregnates is the ASZM-TEDA carbon which has been used in U.S. military nuclear, biological, and chemical filters since 1993. This material is a coal-based activated carbon that has been impregnated with copper, zinc, silver, and molybdenum compounds, in addition to triethylenediamine. ASZM-TEDA carbon provides a high level of protection against a wide range of chemical warfare agents and other pollutants but requires frequent filter replacements (*5*).

Nanocrystalline metal oxides offer an alternative means to combine the universal capability of sorbents, including the ability to capture highly volatile pollutants, with high removal capacities. The needed reactivity is provided by numerous surface sites, not by chemical impregnates. This allows the entire sorbent surface to be reactive toward a great number of different pollutants at large removal capacities (*6–8*). The capability of nanocrystalline metal oxides to react with a multitude of pollutants is well demonstrated by the capabilities of

NanoScale's FAST-ACT® decontamination product (*9*). FAST-ACT is a mixture of nanocrystalline MgO and TiO_2 which was demonstrated to chemically react with numerous air pollutants and other toxic compounds. Table 2 provides a partial list of compounds removed by the nanocrystalline metal oxides contained in FAST-ACT.

The polar surface of metal oxides will attract water molecules and it is critical that this effect does not interfere with capture of pollutants. Prolonged exposure of air filtration sorbents to humid air eliminates application of metal oxides that react chemically with water. For instance, alkaline CaO is not suitable since it easily forms hydroxide that converts to carbonate. Oxides like TiO_2 and Al_2O_3 are more appropriate since they will retain water on their surface without conversion to hydroxides.

Performance of air filtration sorbents is usually evaluated using a breakthrough method. This technique closely resembles packed bed air filters; in particular, it provides a realistic representation of mass transfer and diffusion processes taking place in an air filter. Use of this method gives a high degree of confidence that different filtration media are realistically evaluated and compared. A schematic of a typical breakthrough setup is shown in Figure 4. In this setup, a stream of compressed air is conditioned in a dryer and a humidity controller to achieve the desired relative humidity and flow. Then, the air is mixed with a controlled stream of pollutant to reach the needed concentration. A fixed bed of tested material in a granulate form, is placed in a specially designed test tube that allows for uniform flow distribution that minimizes wall channeling effects. The pressure drop across the bed is measured using a differential pressure gauge. Figure 5 presents the test tube used by NanoScale. The test tube is 30 mm in diameter, which for granulated adsorbents used (mesh 16-35, 0.5-1.2 mm) is sufficient to minimize wall effects. Prior to the breakthrough test the test tube is filled with tested adsorbent that formed a 10 mm thick bed. This thickness was found to be sufficient to prevent bed channeling and premature pollutant breakthrough for superficial air velocity of 6 cm/s. A thicker bed is required for larger air velocities or larger granule sizes. Air flow through the bed is in the downward direction in order to eliminate bed fluidization.

Adsorption performance is evaluated by monitoring toxic pollutant concentrations before and after passing through a fixed bed of tested material. Pollutant concentrations are typically measured by gas-phase gas chromatography or by gas-phase infrared spectroscopy. NanoScale's breakthrough setup uses a GC-MS spectrometer (HP 5890 Series II gas chromatograph and HP 972 mass spectrometer) equipped with an automatic valve system and PC data acquisition workstation. During the analysis, air upstream and downstream of the adsorbent bed is sampled and periodically (5-10 minutes intervals) analyzed by the GC-MS spectrometer. The combination of unique retention time and MS mass fragmentation patterns allow for unambiguous elimination of all background signals and reliable determination of pollutant concentration based on observed signal strength.

A typical breakthrough test is conducted over a period of several hours until the toxic pollutant appears downstream of the filter. Performance of tested materials was evaluated based on the overall shape of breakthrough

Table 2. Chemical contaminates removed by mixture of nanocrystalline MgO and TiO_2 (NanoScale's FAST-ACT product)

Class of Contaminants	*Examples*
Acids	Hydrochloric, hydrofluoric, nitric, phosphoric, sulfuric, sulfur dioxide, hydrogen cyanide
Caustics	Anhydrous ammonia, sodium hydroxide (aq.)
Alcohols	Ethanol, methanol
Chlorinated organics	Acetyl chloride, chloroacetyl chloride, chloroform, methylene chloride
Ketones	Acetone
Oxidizers	Chlorine, ethylene oxide, nitrogen dioxide
Phosphorus/sulfur/nitrogen compounds	Acetonitrile, dimethylmethyl phosphate, methyl mercaptan, paraoxon, parathion, sodium cyanide (aq), 2-chloroethylethyl sulfide, 4-vinylpyridine
Other organics	Ephedrine, p-cresol, pseudoephedrine, and others
Chemical warfare agents	GA, GD, VX, HD, Lewisite

curves (downstream pollutant concentration plotted as a function of time) and breakthrough times. The breakthrough apparatus assembled at NanoScale allows for simultaneous analysis of two samples, one with a dry air stream and another with relative humidity (typically in the 25% to 75% range). Figure 6 shows a set of breakthrough curves obtained for granulated NanoActive ZnO sorbent exposed to the hydrogen sulfide (H_2S). Pollutant concentrations upstream of the sorbent beds remained relatively constant, although fluctuations up to 20% were frequently observed. These fluctuations were due to the difficulties in controlling relatively small flows of pollutants and were random in nature. However, this effect did not affect comparisons between tested sorbents since the applied method of data analysis focused on filtration capacities and used time-averaged concentrations. The filtration capacities were calculated based on the shape of the breakthrough curves, the time averaged agent concentrations upstream of the sorbent beds, and the amount of sorbent used. The capacities were expressed in milligrams of agent removed per gram of sorbent used, units frequently used to characterize air filtration sorbents. In order to obtain reliable values for the removal capacities, as well as their statistical uncertainties, breakthrough tests should be repeated several times.

As it was discussed earlier, the presence of humidity affects performance of nanocrystalline metal oxides. Depending on the specific pollutant-sorbent combination, humidity exposure may shorten or prolong breakthrough times and removal capacities. These effects were apparent during NanoScale's testing of sorbents targeting acetaldehyde and hydrogen chloride pollutants. Breakthrough tests with acetaldehyde (CH_3CHO) were conducted with an air stream containing 540 mg/m^3 (300 ppm) of this pollutant. A superficial gas velocity was 12 ft/min (6 cm/s), and the bed thickness was 10 mm. Tested adsorbents were granulated, with granule size 12-30 mesh (activated carbons) or 16-35 mesh (metal oxides). Figure

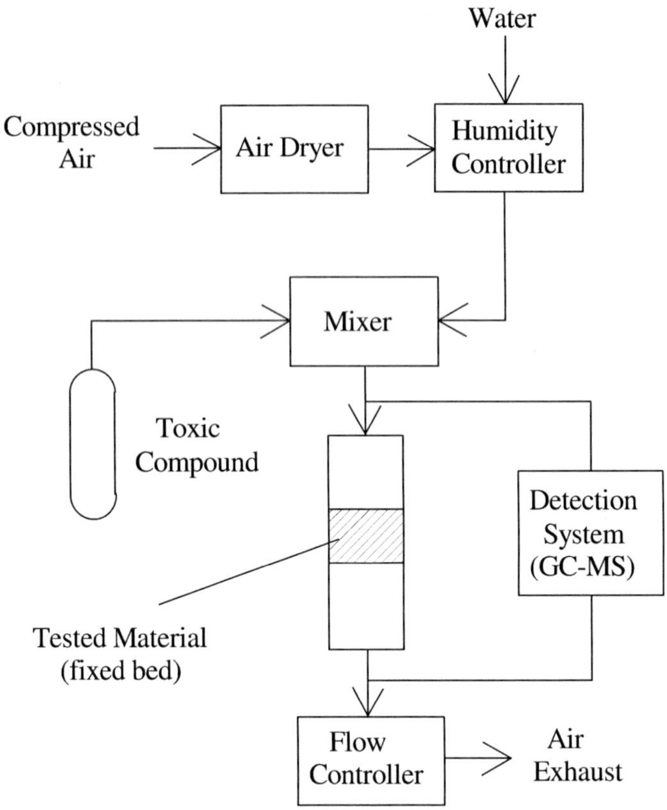

Figure 4. Schematic of the air filtration breakthrough apparatus.

7 presents the acetaldehyde air filtration removal capacities for two carbons and several nanocrystalline metal oxide sorbents.

Under dry conditions, all tested metal oxide formulations outperformed both activated carbons, in some cases by as much as a factor of 2-3. However, under humidified conditions only NanoActive MgO sorbent outperformed carbons while other metal oxides showed limited removal capacities. The anticipated source of this effect is a blocking of the sorbent's active sites by water molecules adsorbed on polar metal oxide surfaces.

Humidity had an enhancing effect on the performance of nanocrystalline metal oxides against hydrogen chloride (HCl). Figure 8 presents the hydrogen chloride air filtration removal capacities for carbon and metal oxide materials. In most cases, the removal capacity measured with humidified air was larger than the capacity determined under dry conditions. Several metal oxide sorbents outperformed activated carbons. Specifically, NanoActive Al_2O_3 Plus tested under humidified conditions outperformed activated carbons by at least a factor of 4, and reached an exceptional removal capacity of (1340 +/- 270) mg/g. Essentially, this sorbent more than doubled its mass before a breakthrough was detected. This enhancing effect of humidity can be easily explained by a large affinity the HCl pollutant has towards water and formation of a highly reactive acid which readily reacts with metal oxides to form chlorides.

Application of nanocrystalline metal oxides as air filtration media should be evaluated not only in terms of removal capacities but also in terms of reduced desorption. Desorption of toxic agents and their release into air streams passing through air filtration devices is a well known phenomenon for activated carbon sorbents (*10*). In order to demonstrate such an effect, specially designed desorption experiments with acetaldehyde were carried out for ASZM-TEDA carbon and NanoActive MgO sorbents. In these tests, sorbents were exposed to an air stream containing acetaldehyde vapor for a four-hour period at a controlled temperature of 10 °C. After this elapsed time, the upstream acetaldehyde concentration was set to zero by closing the valve controlling the pollutant flow, while agent concentrations continued to be monitored upstream and downstream of the filter bed. After an additional four hours (8 hours total since the beginning of the test), the temperature was rapidly increased from 10 °C to 30 °C. Figure 9 presents the results of the experiment for ASZM-TEDA carbon, with acetaldehyde concentrations plotted as a function of time. By not allowing acetaldehyde vapor to enter the apparatus upstream, through by-passing the agent source, the upstream agent concentrations were dramatically reduced (a small agent signal detectable upstream of the sorbent bed after the four-hour mark was caused by agent adsorption on internal parts of the experimental system). In contrast, the downstream acetaldehyde concentrations remained at relatively high levels for several hours. This indicates desorption of the acetaldehyde from the surface of ASZM-TEDA carbon and release of the agent into an air stream emerging from the filter bed, an effect triggered by the reduction of upstream acetaldehyde concentration. Further desorption of the acetaldehyde agent occurred after increasing the bed temperature from 10 °C to 30 °C, conducted after 7.5 hours from the beginning of the experiment. This effect was manifested by the increase in downstream agent concentration detected between the 8th and 10th hours of the experiment.

Desorption of an agent from a sorbent bed, either caused by a change in the upstream agent concentration or by a change in temperature, is highly undesirable and even dangerous in real filtration systems. The possibility of an unexpected release of a concentrated and toxic agent into an air stream leaving a filter seriously limits usefulness of such systems and calls for the development of alternative sorbents that do not have this disadvantage.

An identical desorption experiment, as described above, was carried out for the NanoActive MgO sorbent. The results are presented in Figure 10, and indicate a different desorption of this sorbent when compared to the ASZM-TEDA carbon. Both the change in upstream agent concentration, as well as the change in temperature, did not cause significant desorption or release of acetaldehyde into the air stream leaving the filter bed for the NanoActive MgO sorbent. This result demonstrates a significant advantage of the NanoActive MgO and its destructive adsorption capability over the simple physical adsorption offered by most other sorbents. Not only do the nanocrystalline sorbents promise effective capture of numerous high volatility toxic pollutants but they also prevent their desorption back to an air stream with associated hazards.

Figure 5. Test tube used in the adsorption breakthrough apparatus.

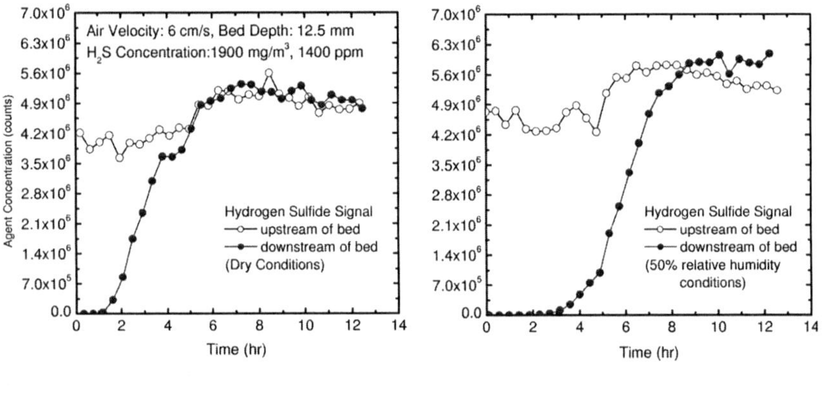

Figure 6. Typical air filtration breakthrough curves obtained for the NanoActive ZnO sorbent exposed to the hydrogen sulfide agent. Tests conducted under dry and 50% relative humidity conditions.

It can be anticipated that in coming years some nanocrystalline metal oxide-based air filtration sorbents will gain more wide spread application and, at least partially, replace activated carbons and other currently used media. Nanocrystalline air filtration sorbents are likely to be mixtures of 2-3 single metal oxides or mixed metal oxides that intimately combine several metals into one material. Materials with very high SSA, preferably above 1,000 m^2/g, and large resistance to water exposure are expected to be the most effective sorbents.

Application in Clean Coal Technologies

Coal was the first fossil fuel used on a massive scale and still represents 23% of the primary energy supply worldwide (*11*). Coal is relatively low-priced and its vast reserves are quite evenly distributed on all continents. In most industrialized countries coal use as transportation fuel or as household cooking or heating fuel has been replaced by petroleum and natural gas products. However, use of coal for electricity generation is still widespread. Worldwide, almost 40% of electric power is produced using 60% of the global coal production. U.S. possesses 27% of proven coal reserves that at the current rate of consumption will last at least 170 years. Currently, 51% of the nation's electricity is produced in 1,100 conventional coal-fired power plants. U.S. Department of Energy (DOE) forecast predicts that the use of coal will increase in the coming years (*12*).

The problem with coal is the significant environmental pollution associated with its use. Emissions of pollutants like SO_2, NO_X, and particulates have long been recognized as major drawbacks of coal combustion. A series of Clean Air Act legislations has forced adoption of pollution controls in all conventional power plants in the nation. The biggest reductions were achieved in particulate emissions, and some progress was achieved in reduction of acid gas releases. Releases of mercury and CO_2, the main cause of global warming, were recognized as major health and environmental risks connected with coal combustion. There is an enormous worldwide research and development effort to develop new clean coal technologies that will allow for environmentally responsible utilization of this fuel.

Mercury pollution control in conventional coal-fired power plants is the major clean coal application where nanocrystalline metal oxides are expected to provide an effective solution. Coal-fired utilities in the U.S. are responsible for an annual emission of nearly 50 tons of mercury. Mercury emitted into the air has a tendency to preferentially pollute rivers, lakes and sea water, eventually accumulating in fish. Because of the growing concern about the toxic effects on a fish-eating population, on March 15, 2005, the United States Environmental Protection Agency (EPA) issued the Clean Air Mercury Rule to permanently cap and reduce mercury emissions from coal-fired power plants. When fully implemented, these rules will reduce utility emissions of mercury from 48 tons a year to 15 tons, a reduction of nearly 70 percent. As a result of this legislation, it is expected that the nation's coal-fired utilities will be required to radically reduce mercury emissions in the next decade. The announcement of these new regulations sparked vigorous development of mercury control technologies and renewed commercial interest in mercury removal.

The industry's response to the EPA mercury rule was the development of a dry sorbent injection method (*13*). In this method, sorbent powder is injected into hot (typically 350 °F) combustion gases emerging from coal combustion boilers. The sorbent intermixes with these gasses and preferentially captures mercury vapors and compounds and subsequently gets captured in an electrostatic precipitator or fabric filter installed in all coal-fired utilities to control particulate releases. This prevents mercury emission into air and retains it in a solid fly ash product collected by all of the coal-fired power plants. The fly ash ends up in

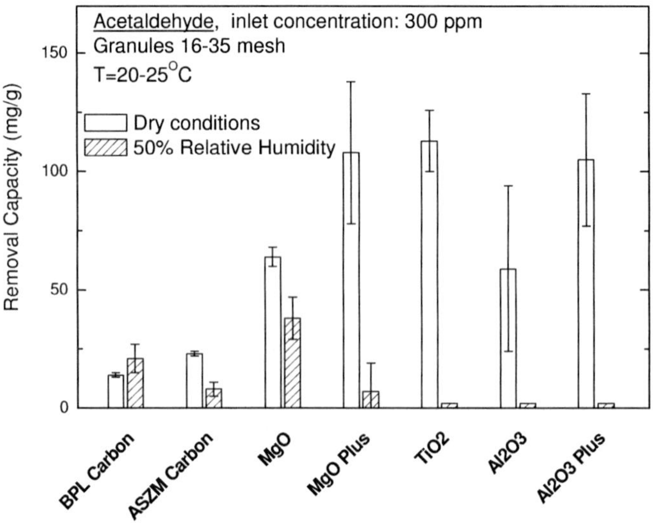

Figure 7. Air filtration removal capacities for various sorbents and the acetaldehyde.

various construction materials including Portland cement, gypsum dry wall or road construction components where mercury species are permanently fixed and do not cause health or environmental concerns. The short contact time between the sorbent particles and the combustion gases, usually only 1-2 seconds, poses severe mass transfer and kinetic limitations on the sorbent. These require a high surface area sorbent, particle size in 1-2 m range, and high surface reactivity and selectivity towards mercury.

The main advantage of the dry injection method is the low cost of equipment since sorbent injectors can be easily added to the existing boiler exhaust systems while sorbent capture is realized by existing particulate control devices. The disadvantage of this method is the high cost of the currently used brominated activated carbon sorbent and its poor compatibility with fly ash applications. The application of carbon injection in coal-fired electric utilities will increase the cost of electricity by as much as 10% (*14*). Dry ash-based Portland cement is weakened by air trapped in activated carbon particles while the appearance of black carbon in white dry wall products is problematic. As a result, fly ash containing carbon frequently cannot be sold as a useful product, and instead requires costly disposal. Due to these factors, non-carbon based mercury sorbents have been studied including manganese dioxide, hopcalite, trona, vanadium oxides, alumina, silica and others (*15, 16*).

Requirements of the dry sorbent injection mercury control method make it ideally suited for nanocrystalline metal oxides. Some low cost and low toxicity oxides such as CaO or MgO may be good porous supports for a chemically impregnated mercury sorbent. In contrast to carbons, the porous structure of alkaline metal oxides will collapse in contact with water eliminating the air entrapment problem. The white color of most oxides makes them appropriate for dry wall applications. Furthermore, alkaline metal oxides may be effective

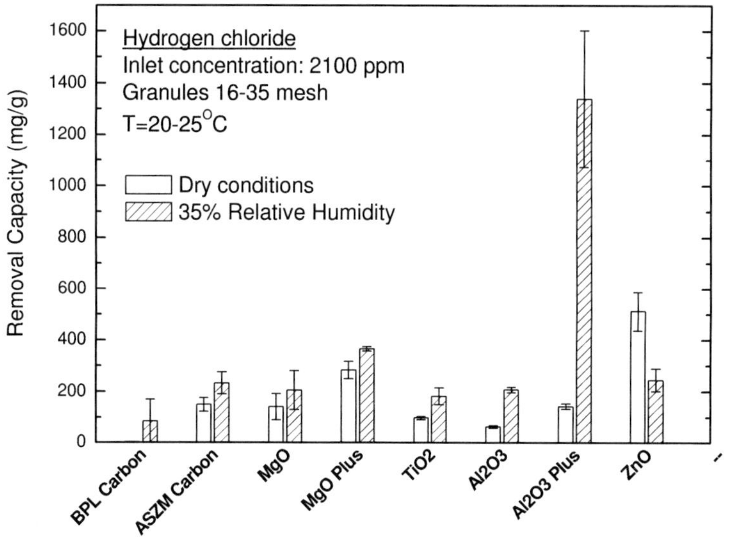

Figure 8. Air filtration removal capacities for various sorbents and the hydrogen chloride agent.

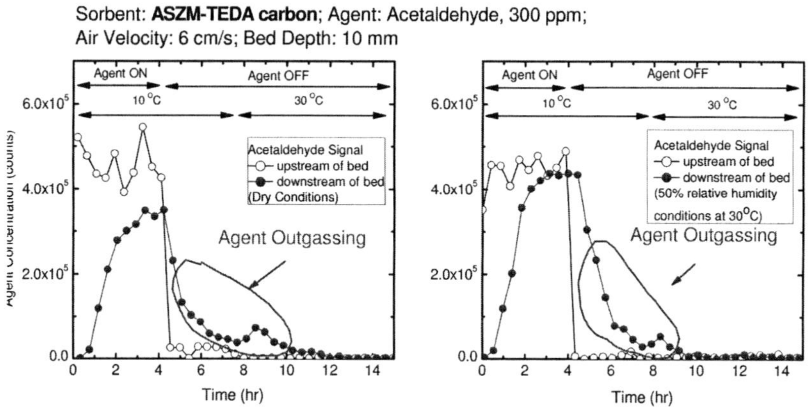

Figure 9. Results of the desorption experiment conducted for ASZM-TEDA carbon and the acetaldehyde agent.

in removal of acid gases SO2 and NO_X that escape the currently used control devices. This will allow the electric power utilities to use a single pollution control to satisfy the requirements of the EPA Clean Air Mercury Rule and to simultaneously obtain credits for reduced releases of SO_2 and NO_X pollutants. Finally, nanocrystalline metal oxides can be manufactured on a large scale as powders in the desired particle size and with high surface area and reactivity. The mercury sorbent application will be a very sizable market opportunity for manufacturers of these sorbents. The market size in the U.S. alone can be estimated as a few hundred million dollars per year, perhaps the biggest commercial opportunity for any sorbent application.

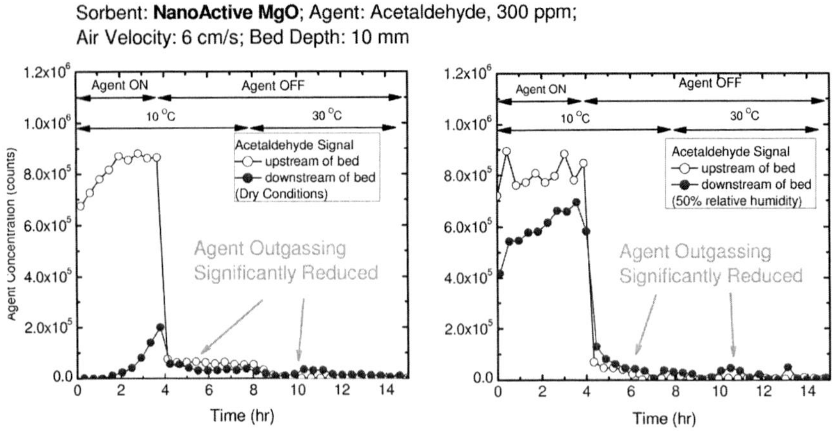

Figure 10. Results of the desorption experiment conducted for NanoActive MgO sorbent and the acetaldehyde agent.

Table 3. Capacity for hydrogen sulfide removal from coal based synthesis gas by a nanocrystalline ZnO based sorbent

Experimental conditions	*Removal capacity (lb H_2S/lb ZnO)*
25 °C, 14.7 psig	0.13
150 °C, 14.7 psig	0.44

The next generation of coal-fired power plants will likely be based on coal gasification systems rather than on coal combustion used currently. Coal gasification has numerous advantages including improved overall efficiency, dramatically reduced emissions of air pollutants, and economical carbon dioxide sequestration options. Specifically the improved efficiency of Integrated Gasification Combined Cycle (IGCC) power plants makes them attractive candidates for future power plants utilizing coal.

Coal-gasification uses only a fraction, typically one third, of the oxygen needed for full combustion of coal. In addition, water is added to the gasification chamber to produce hydrogen. Partial combustion conditions and water addition produces a synthesis gas mixture of CO, CO_2, H_2O, H_2 and some reduced forms of sulfur and nitrogen pollutant gases like H_2S, COS, NH_3, and HCN. Heavy metals like mercury or arsenic are present in the synthesis gas as vapors or high volatility compounds. The gasification is typically carried out using pure oxygen instead of air to avoid nitrogen dilution. Industrial scale gasification is usually carried out at high pressures, typically in the 100-500 psi range and at temperatures 800-1500 °C (*17*). Removal of pollutants in coal gasification systems is easier and cheaper than in conventional combustion boilers because pollutants are more concentrated and smaller equipment is required. The main disadvantage if gasification systems like the IGCC plants is their complexity. Conventional coal-fired boilers have been used in power plant for decades and

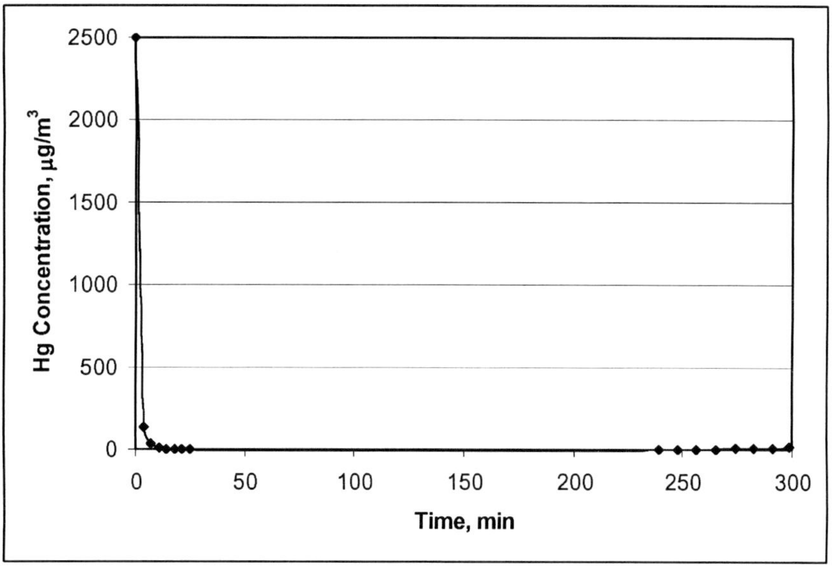

Figure 11. Mercury breakthrough plot for copper oxide based sorbent at 150 °C in simulated synthesis gas.

are relatively simple to operate. Gasification plants are as complex as chemical processing plants and need to be operated as such.

Applications of nanocrystalline metal oxides in coal gasification systems involve high temperature pollutant control methods. Current coal gasification technologies utilize low temperature cleanup methods for removal of sulfur, nitrogen and mercury pollutants that require cool-down of the fuel gas and limit overall efficiency of the plant. Development of effective hot gas cleanup technologies is highly desired as it would increase efficiency, lower capital and operational costs of gasification plants, and reduce the environmental impact of these facilities. The DOE actively promotes the coal gasification technology development and specifically warm synthesis gas cleanup (*18*). Many research groups have developed H_2S metal oxide-based sorbents capable to work at high temperatures (*19–21*). Similarly, metal oxide catalysts suitable for high temperature ammonia decomposition into nitrogen and hydrogen are being optimized (*22*). Low temperature mercury capture from synthesis gas can be accomplished by using activated carbon sorbents. The same task can be carried out at high temperatures by using metal oxide sorbents (*14*).

NanoScale has developed sorbents for clean coal applications including materials for gasification systems. In one study, NanoScale tested its proprietary zinc oxide-based sorbent for removal of hydrogen sulfide from fuel gas. The testing was performed at Western Research Institute in Laramie, WY, using a pilot-scale coal-based gasification system. The synthesis gas was composed mainly of hydrogen, water vapor, carbon monoxide, and carbon dioxide. Hydrogen sulfide was present at 1000-2000 ppm together with some other low-level impurities. Table 3 presents the removal capacity for hydrogen sulfide obtained at ambient and elevated temperature.

The synthesis gas environment and presence of H_2O and CO_2 in the stream did not prevent efficient scrubbing of hydrogen sulfide. The removal capacities are substantial, particularly at the elevated temperature. Importantly, nanocrystalline ZnO significantly outperformed the commercial H_2S sorbent that exhibited limited capacity of 0.07 lb H_2S/lb ZnO (tested on the same setup and under the same conditions as nanocrystalline ZnO).

NanoScale developed several metal oxide formulations for capture of mercury from high-temperature (150–370 °C) synthesis gas. Several nanocrystalline sorbents were evaluated in a lab-scale, fixed bed reactor with the outlet mercury concentration monitored by a semi-continuous mercury analyzer. Elemental mercury vapor was introduced into inert gas (nitrogen) and simulated fuel gas streams using permeation tubes. The gas composition used in these tests was: 30% H_2, 30 % CO, 10% CO_2, 20% H_2O, 4000 ppm H_2S, 4000 ppm COS, and balance of N_2. To simulate the effect of high-pressure gasification conditions, the concentration of mercury in the SFG was maintained at 2500 μg/m3. The gas hourly space velocity was maintained at 150,000 hr^{-1} (STP). Several sorbents showed significant affinity for mercury vapor. In particular, sorbents based on copper oxides achieved a high mercury-sorption capacity. Figure 11 shows the mercury breakthrough plot for this sorbent at 150 °C. Clearly, the sorbents capture 100% of mercury for a period of at least 5 hours.

The 5-hr sorption capacity of the sorbent is calculated to be 2 mg/g sorbent or 0.2 wt%. Based on the trend of the breakthrough curve, it appears that the sorbent has a potential to achieve a much higher Hg-sorption capacity while removing >90% mercury from the incoming gas stream.

In summary, nanocrystalline metal oxide-based sorbents have multiple applications in clean coal technologies. Sorbent for mercury and acid gas removal from combustion gases is an example of an application related to conventional coal-fired power plants. Nanocrystalline metal oxides are expected to enter this area during the next few years. High temperature pollutant control in coal-fired gasification systems is a more long term applications of these materials. Materials that combine moderate SSA with low cost are expected to be the most effective sorbents.

References

1. Scherrer, P. *Göttinger Nachrichten Gesell* **1918**, *2*, 98–99.
2. Patterson, A. L. *Phys. Rev.* **1939**, *56*, 978–982.
3. Noll, K. E.; Gounaris, V.; Hou, W. *Adsorption Technology for Air and Water Pollution Control*; Lewis Publishers, Inc.: Chelsea, MI, 1992; pp 21–48.
4. Ruthven, D. M. *Principles of Adsorption and Adsorption Processes*; Wiley-Interscience: New York, NY, 1984; pp 29–61.
5. *Guidance for Filtration and Air-Cleaning Systems to Protect Building Environments from Airborne Chemical, Biological, or Radiological Attacks*; DHHS (NIOSH) Publication No. 2003-136; Department of Health and Human Services, Centers for Disease Control and Prevention, National Institute for Occupational Safety and Health, 2003; p 58.

6. Stark, J. V.; Park, D. G.; Lagadic, I.; Klabunde, K. J. *Chem. Mater.* **1996**, *8*, 1904–1912.
7. Khaleel, A.; Kapoor, P.; Klabunde, K. J. *Nanostruct. Mater.* **1999**, *11*, 459–468.
8. Koper, O.; Klabunde, K. J. U.S. Patent 5,057,488, 2000.
9. http://www.nanoscalecorp.com.
10. Linders, M. J. G.; Baak, P. J.; Van Bokhoven, J. J. G. M. *Ind. Eng. Chem. Res.* **2007**, *46*, 4034–4039.
11. Olah, G. A.; Goeppert, A.; Prakash, G. K. S. *Beyond Oil and Gas: The Methanol Economy*; Wiley-VCH Verlag GmbH & Co. KGaA: Weinheim, Germany, 2006; pp 28−32.
12. http://www.eia.doe.gov/fuelcoal.html.
13. Sondreal, E.; Pavlish, J. *Center for Air Toxic Metals Newsletter* **2000** (May), 1.
14. Stiegel, G. J.; Longanbach, J. R; Klett, M. G.; Maxwell, R. C.; Rutkowski, M. D. *The Cost of Mercury Removal in an IGCC Plant*; U.S. DOE National Energy Technology Laboratory, 2002; pp 1−27.
15. Granite, E. J.; Pennline, H. W.; Hargis, R. A. *Ind. Eng. Chem. Res.* **2000**, *39*, 1020–1029.
16. Granite, E. J.; Pennline, H. W.; Hargis, R. A. *Sorbents for Mercury Removal from Flue Gas*; U.S. DOE Federal Energy Technology Center, 1998; pp 1−50.
17. Rezaiyan, J.; Cheremisinoff, N. P. *Gasification Technologies. A Primer for Engineers and Scientists*; CRC Press Taylor & Francis Group: Boca Raton, FL, 2005; pp 35−86.
18. http://fossil.energy.gov/programs/powersystems/vision21/index.html.
19. Gangwal, S. K.: Turk, B. S.; Gupta, R. P. U.S. Patent 6,951,635, 2005.
20. Gupta, R. P.; Gangwal, S. K. U.S. Patent 5,254,516, 1993.
21. Slimane, R. B.; Abbasian, J. *Ind. Eng. Chem. Res.* **2000**, *39*, 1338–1344.
22. Feitelberg, A. S.; Ayala, R. E.; Hung, S. L.; Najewicz, D. J. U.S. Patent 6,432,368, 2002.

Chapter 6

Mesoporous Titanium Dioxide

Dan Zhao, Sridhar Budhi, and Ranjit T. Koodali*

Department of Chemistry, University of South Dakota, Vermillion, SD 57069
***Ranjit.Koodali@usd.edu**

Mesoporous titanium dioxide is extensively employed as photocatalysts for environmental remediation of pollutants, active charge transport carriers in dye sensitized solar cells and for photosplitting of water. Depending on the Depending on the ordering of the pores, they may be classified as periodic or aperiodic materials. The aperiodic materials consist of aerogels (prepared by supercritical drying) and xerogels (prepared by ambient temperature drying). Periodic mesoporous materials are usually prepared by employing sacrificial templates (such as surfactants) to impart order into them. Sol-gel chemistry forms the basis for the preparation of both ordered and non-ordered mesoporous materials. The sol-chemistry of silica is widely explored and there are available methods to control the particle size and porosity of siliceous materials; however it is not easy to control the porosity and particle size of titanium dioxide. We will discuss the synthesis of mesoporous materials, the mechanism of photocatalysis and selected applications related to environmental remediation in aqueous and gas phase in this chapter.

1. Introduction

Porous materials are finding important applications in the fields of science and technology as adsorbents (*1*), as catalysts in the production of fine chemicals, in pharmaceuticals for drug delivery, in production of renewable energy sources (*2*), and as sensors and nanodevices. Porous materials are classified by the International Union of Pure and Applied Chemistry (IUPAC) as 1) microporous materials with pore size less than 2 nm, 2) mesoporous materials with pore size ranging between 2 to 50 nm, and 3) macroporous materials with pore size larger than 50 nm (*3*). Of these three types of porous materials, microporous and mesoporous materials find extensive applications. Zeolites which come under the category of microporous materials are well known for shape selectivity because of uniform, and regular pore size and channels. Despite numerous advantages, zeolites are found to be efficient catalysts only for molecules with dimensions in the 10-20 Å range; for catalytic reactions involving molecules with dimensions larger than 20 Å, diffusion is very poor and this often limits the efficiency. Thus, there has been intense research in the design of mesoporous materials to alleviate this problem.

One of the most widely studied porous materials is titanium dioxide. It is extensively used as a pigment in paints and cosmetics, sun block screens, lip sticks, and salad dressings. However, its use as a photocatalyst has attracted the interest of scientists for the past several decades. A number of review articles regarding the principles, mechanisms, and applications of semiconductor photocatalysis have appeared (*4–6*). For additional references and relevant literature, readers can go to recent reviews and books (*7–11*).

In this chapter, a review of mesoporous titanium dioxide is presented with insight into the underlying mechanisms of photocatalytic reaction and selected application of mesoporous TiO_2 in environmental remediation, which is the focus of this book. Hopefully, readers will better appreciate Chapters 10, 11 and 12 that relate to photocatalysis after perusing this chapter.

2. Semiconductor Photocatalysis

Semiconductor photocatalysis is a heterogeneous system in which the photoinduced reactions take place on the surface. The electronic structure of semiconductor material is characterized by a filled valence band and an empty conduction band. The void energy region extending from the top of the valence band to the bottom of the conduction band is called band gap, and usually denoted as E_g. When irradiated by light with energy greater than or equal to the band gap, electrons are excited from the valence band to the conduction band and positively charged holes are left in the valence band. The photogenerated electrons and holes react with the electron donors and electron acceptors absorbed on the semiconductor surface or form active oxygen species and induce a series of redox processes (*12*). The system may also include a dye molecule acting as the photosensitizer. The dye molecules (covalently anchored or physically adsorbed) on the semiconductor surface are excited either to their singlet or triplet state; then electrons of the excited dye molecule can be injected into the conduction band of the semiconductor, if the potential of the conduction band is lower than the redox potential of the excited dye. The electrons can be used for the reduction of other molecules and heavy metal ions adsorbed on the semiconductor surface or scavenged by O_2 to form $O_2^{-\cdot}$ radical (*13*).

In 1972, Fujishima and Honda published their landmark discovery of the photoelectrochemical splitting of water by rutile TiO_2 film electrode; this initiated the field of semiconductor heterogeneous photocatalysis (*14*). TiO_2 is considered to be the most promising semiconductor photocatalyst due to its excellent stability and anticorrosion properties. Semiconductor photocatalysis using TiO_2 as photocatalyst has been applied for a variety of reactions, such as photocatalytic water splitting to produce hydrogen and oxygen, photocatalytic reduction of nitrogen dioxide and carbon dioxide, photocatalytic degradation of a wide variety of toxic organic compounds such as phenols, aromatic carboxylic acids, PCB's, pesticides, and dyes, photocatalytic reduction of heavy metal ions from aqueous solutions, and destruction of microorganisms such as bacteria and viruses (*15–17*).

The inherent electronic structure of semiconductor enables it to serve as photocatalyst under light irradiation. The width of the band gap and the position of the conduction band and valence band are important factors that decide the absorption wavelength and the oxidation and reduction ability of the semiconductor material. Semiconductors with narrow band gap attract more attention due to its capability of using short wavelength visible light. For the application of photodegradation of organic pollutants, the valence band needs to be positive enough to oxidize H_2O and produce hydroxyl radical, while the conduction band needs to be negative enough to allow the efficient transfer of photo-generated electrons to O_2.

2.1. Titanium Dioxide

Titanium dioxide is believed to be the most suitable semiconductor for environmental remediation. The band gap and the position of valence and conduction band are appropriate for the photocatalytic oxidation and reduction of a wide variety of environmental pollutants. The good stability towards photocorrosion, chemical corrosion and the low price also make TiO_2 superior to other semiconductors.

Titanium dioxide has eight crystal structures, anatase ($I4_1/amd$), rutile ($P4_2/mnm$), brookite (*Pbca*), TiO_2-B (*C2/m*), TiO_2-R (*Pbnm*), TiO_2-H (*I4/m*), TiO_2-II (*Pbcn*), and TiO_2-III ($P2_1/c$). Rutile is thermodynamically the most stable form. The others are metastable phases and transform to the rutile phase when they are heated to high temperatures. Anatase and rutile are the most common phases and numerous studies have been done on their synthesis and photocatalytic activity. Brookite is the rarest phase, and it is often difficult to synthesize in the laboratory. Anatase and rutile crystallize in the tetragonal system. The structure of anatase and rutile can be described by TiO_6 octahedra chains. Each Ti^{4+} is coordinated by six O^{2-} and the TiO_6 octahedra share edge or corner oxygen with the adjacent octahedra. The difference between the anatase and the rutile crystal structure is the distortion of the octahedron and the assembly pattern of the octahedra chain. In rutile, the octahedron is not regular and is slightly distorted orthorhombically. The octahedron of anatase is more distorted and hence its symmetry is lower than orthorhombic. Further, the Ti-Ti distances in anatase are greater (3.79 Å and 3.04 Å for anatase *vs* 3.57 Å and 2.96 Å for rutile), whereas the Ti-O bond length is shorter (1.934 Å and 1.980 Å for anatase *vs* 1.949 Å and 1.980 Å for rutile) (*18*). Each octahedron in rutile is in contact with 10 neighbor octahedra (two sharing edge oxygen pairs and eight sharing corner oxygen atoms); in anatase each octahedron is in contact with eight neighbors that include four sharing an edge and four sharing a corner. The differences in the crystal structure cause different mass densities and electronic band structures. The band gaps are 3.2 eV for anatase and 3.0 eV for rutile, corresponding to light of 388 nm and 413 nm wavelength respectively. Among the polymorphic forms of titania, anatase is usually reported to show better photocatalytic activity than rutile. However, the high photocatalytic activity of brookite has also been reported (*19*).

Degussa P 25 is a commercial titanium dioxide nanomaterial produced by Degussa Company. The surface area (BET) is about 50 m^2/g, and the crystallite size is 30 nm while the secondary particle aggregate size is around 0.1 μm. It contains mixed phases of anatase and rutile in a ratio of 7:3 and the average size of anatase and rutile elementary particles are about 85 and 25 nm. Because it is easily available, relatively inexpensive and has high photocatalytic activity, Degussa P 25 has been a standard material in the field of photocatalysis. The superior activity of Degussa P 25 compared to that of other titanium dioxide has been attributed to high adsorption affinity for organic compounds, high surface area, and lower recombination rates. It was found that the photo-generated holes are trapped exclusively on the particle surface while the photo-generated electrons are trapped within the lattice (*20*).

3. Synthesis of Titanium Dioxide

As the most promising photocatalyst, the synthesis of TiO_2 nanomaterials has been extensively studied in recent years. Several synthesis methods, including sol-gel (*21*), hydrothermal and solvothermal, micelle and inverse micelle (*22*), chemical and physical vapor deposition (*23*), electrodeposition (*24*), and direct oxidation method (*25*) have been used for the preparation of TiO_2. Nearly all morphologies of TiO_2, nanoparticles, nanorods, nanowires, nanobelts, nanotubes, nanosheets, have been realized by the various synthetic methods mentioned above.

3.1. *Aperiodic* Titanium Dioxide

Sol-gel method is the most common and versatile method for the preparation of TiO_2 nanoparticles. Titanium(IV) alkoxide is the most frequently used titanium precursor in sol-gel process. Titanium alkoxides are highly reactive and in order to avoid uncontrolled hydrolysis and/or condensation and moderate its reactivity, it is often reacted with ligands such as acetyl acetonate or acetate (*26*). Acids and bases, such as nitric acid, hydrochloric acid and tetramethylammonium hydroxide, are used to catalyze the hydrolysis of titanium precursor. A TiO_2 sol is formed by the hydrolysis and polymerization of the titanium precursor (eqn. 1). In the next step (eqn. 2), the condensation step, further polymerization causes the gelation of the parent sol. The nature of the inorganic network formed depends on the competition between the hydrolysis and condensation reactions. Gels with small pores are formed from small hydroxylated oligomers if hydrolysis is fast and condensation is slow. If condensation is fast and hydrolysis is slow it will result in the formation of a gel consisting of interconnected network. The competition between hydrolysis and condensation is primarily determined by the ability of the metal precursor to adapt to the change in its coordination number and its electrophilic character.

$$M(OR)_n + H_2O \rightarrow \quad [M(OR)_{n-1}(OH)] + ROH \qquad (1)$$

$$M\text{-}OH + M\text{-}OX \rightarrow M\text{-}O\text{-}M + X\text{-}OH\ (X= R \text{ or } H,\ M= Ti) \qquad (2)$$

The gel obtained contains solvent and depending on the method of removal of the solvent, a highly porous aerogel or a dense xerogel will be formed.

Aerogels are *composites of being and nothingness* (*27*), and consists of networks of nanoscale domains containing large volume fraction of mesoporosity (*28*). These are formed when the wet gels are dried under supercritical conditions. The supercritical drying preserves the "porosity" of the gel by preventing the appearance of a liquid-vapor interface which is responsible for the collapse of the pores. If the solvent was removed by evaporation at ambient conditions, then the resulting material is termed as a xerogel. Drying of the solvent under ambient conditions results in the formation of liquid-vapor interface. The appearance of a liquid-vapor interface creates capillary pressure in the pores due to the high surface tension of the solvents. This causes the pores to collapse thereby resulting in some microporosity and lowering of the pore volumes. Ambigels are produced when the polar solvents (alcohols) in the wet gels are exchanged by non-polar solvents such as alkanes which have lower interfacial energy than polar solvents. The pore collapse is less for ambigels when compared with xerogels.

Teichner *et al.* (*29*) prepared titania aerogels by supercritical conditions. The aerogels crystallized in the anatase phase. The Brunauer-Emmett-Teller (BET) specific surface area and total pore volume of the as-synthesized (uncalcined) titania was found to be 168 m^2/g and 0.45 cm^3/g respectively. The porosity of aerogels increased with increase in weight percentage of titanium precursor in the sol. Baiker and coworkers prepared high surface area titania aerogels by acid catalyzed alkoxide sol-gel route followed by high temperature supercritical drying (*30*). The surface area of the aerogels were found to be in the range of 195 to 200 m^2/g which was relatively high compared with the aerogels prepared by other groups earlier. The pore diameter was between 14 nm and 17 nm and pore volume between 0.69 and 0.82 cm^3/g. Ko and coworkers (*31*) synthesized titania aerogels using semi-continuous extraction of CO_2. A systematic study of hydrolysis ratio, amount of acid and calcination temperature was conducted. At low hydrolysis ratio either a soft gel or no gel formed in 24 h. At higher ratios, colloidal gel which contains precipitates were formed and the hydrolysis ratio of four was found to be ideal which produced titania with a surface area of 227 m^2/g after calcination at 773 K. An important conclusion from this study was that supercritical drying temperature affected the textural properties of the titania aerogel. Ayen and Iacobucci (*32*) reported the preparation of titania aerogels (as-synthesized) with surface area of around 600-750 m^2/g prepared by low temperature supercritical extraction using carbon dioxide, but the aerogels prepared were found to be highly amorphous. Ko and coworkers further studied the effect of changing supercritical drying temperature from 373 K to 473 K (*33*). Changing the drying temperature did not impart any change in the surface area of the aerogel; instead, it affected the pore volume significantly. A subsequent study by Baiker and coworkers focused on the influence of extraction duration and state of CO_2 (liquid or supercritical) (*34*). Samples dried using supercritical CO_2 possessed higher surface areas and larger pore volumes than samples dried using liquid CO_2. This study showed that textural properties could be varied by suitably altering the extraction conditions.

Tomkiewicz and his group developed titania aerogels for photocatalytic decontamination of aquatic environments (*35*, *36*) and compared its efficiency with commercial Degussa P-25. The gels were dried by two methods, supercritical drying and freeze drying. Powder X-ray diffractometric (XRD) studies indicated that all annealed and non-annealed aerogels and cryogels were in the anatase phase. Non-annealed aerogels exhibited higher adsorption of salicylic acid than annealed aerogels which was in turn higher than commercial Degussa P-25. The aerogels were characterized using techniques such as electron microscopy (*37*), small-angle neutron scattering (*38*) and Raman microscopy (*39*). The electron microscopic studies suggested that the aerogel consisted of closely packed crystalline nanoparticles of size 5 nm in diameter, which forms loosely packed aggregates of about 50 nm in diameter. A detailed XRD and HRTEM study of the titania aerogels was done by Dauger (*40*).

Supercritical drying, particularly high temperature treatment causes shrinkage and cracks in the titania monoliths. An approach to alleviate this problem was reported by Ayers research group (*41*). Titania aerogel was prepared by dissolving Ti sponge in H_2O_2. The gel was further subjected to CO_2 supercritical drying. Although the as-synthesized aerogel possessed surface area of 350 m^2/g, it drastically reduced to 20 m^2/g on heat treatment at 973 K.

Park and coworkers prepared a series of titania aerogels with varying amounts of nitric acid (*42*). The amount of nitric acid in the initial sol-gel composition was found to be crucial in determining the morphology of the final material. However the amount of acid content had no significant influence on the specific surface area. In a separate study, this research group had studied the influence of aging temperature and time on pore structure properties (*43*). Maliknowska's research group prepared titania aerogel (*44*) with different weight percentages of titanium precursor in alcohol and compared its photocatalytic activity with titania-silica aerogel (*45*). The efficiency of the photocatalysts were found to depend on the nature of the pollutant and not on the solvent employed in the preparation or the weight percentage of the titanium precursor.

It is often very difficult to obtain crystalline titania with monodisperse particle sizes by hydrolytic methods. This is due to lack of precise control of hydrolysis of titanium isopropoxide. To overcome these difficulties Fox's research group prepared titania aerogels by non-hydrolytic acylation/deacylation using titanium isopropoxide and acetic anhydride/trifluoroacetic anhydride (*46*). Titania prepared using trifluoroacetic anhydride was found photocatalytically more active for the oxidation of 1-octanol to octanoic acid than titania prepared using acetic anhydride.

The discussion so far relates to the various preparative methods for obtaining mesoporous Titania that have a random ordering of pores, *i.e. aperiodic* pores. We shall now review, the progress made in the synthesis of periodic titanium dioxide materials.

3.2. Periodic Titanium Dioxide

The discovery of M41S materials by Mobil scientists has opened new directions in the field of porous materials (*47*). The synthesis of these materials is based on the supramolecular assembly of surfactant molecules. In order to prepare periodic (ordered) mesoporous materials, surfactants are normally employed as the sacrificial templates. Of the various types of surfactants, quaternary cationic surfactants are the most commonly used ones. A liquid crystal templating mechanism was originally proposed by Beck. According to one mechanistic pathway proposed by Beck, micellar rods are formed around the surfactant micelles initially and the micellar rods arrange themselves in a hexagonal array. The inorganic precursors undergo hydrolysis and condensation around the hexagonal array to generate the mesoporous material. *In situ* ^{14}N NMR studies, conducted by Davis *et al.* (*48*) concluded that there was no liquid crystal phase present in the medium during the formation of MCM-41 material and hence an alternate mechanism was proposed by Stucky *et al.* (*49*, *50*) called co-operative self-assembly mechanism. The self-assembly mechanism has two important steps. First, the silica species undergoes hydrolysis and binds to the surfactant molecules. In the second step, the silicate species undergo further polymerization at the surfactant-silica interface to generate the mesoporous silica material containing the surfactant molecules. The ultimate structure or phase of silica is determined by the charge-density matching between the surfactant and the silicate species. By tuning the charge density at the organic-inorganic interface, mesoporous materials with different pore morphologies and topologies can be prepared (*51*). According to this mechanism, as long as there is an electrostatic complimentarity one can form ordered mesoporous materials (*52*).

The synthesis of periodic mesoporous TiO_2 was reported first by Antonelli and Ying (*53*). TiO_2 was prepared using an anionic surfactant, *i.e.* tetradecylphosphate and the resulting TiO_2 possessed a surface area of 196 m^2/g and pore size of 32 Å. This method of synthesis was found to be applicable only to phosphate surfactants and not to other cationic and anionic surfactants. However, the phosphate surfactants were not completely removed (*54*) by calcination which was confirmed by elemental analysis. Since these materials had considerable amount of phosphorous retained in them, they are better designated as titanium oxo-phosphates rather than pure titania. The materials prepared using these surfactants possessed low photocatalytic activity due to poisoning of catalytically active sites. Even though the material prepared was claimed to be initially hexagonal, a subsequent report by Navrotsky and coworkers (*55*) proved that titania is not in the hexagonal phase; two-dimensional X-ray diffraction patterns and transmission electron microscopic studies suggested the presence of a lamellar phase. Ohtaki and coworkers (*56*) prepared mesoporous TiO_2 using cetyltrimethyl- ammonium bromide, which crystallized in the lamellar phase.

The synthesis of phosphorous free mesoporous titania was reported by Antonelli using amine surfactants (*57*). Nitrogen adsorption studies of an as-synthesized sample gave an isotherm similar to MCM-41 with a surface area of 710 m^2/g with a narrow pore-size distribution centered around 27 Å. When this material was subjected to heat treatment, anatase phase was formed accompanied by a loss of mesoscopic order which was evident from the broadening of X-ray diffraction patterns. Interestingly, when the length of carbon chain in the template was increased from 12 to 18 carbons, no increase in pore size was observed.

In order to control the hydrolysis and condensation rates of titanium precursors and avoid the formation of bulk titania phases, stabilizing agents such as acetyl acetonate, triethanolamine (*58*) or non-hydrolytic pathways are employed. Ozin and coworkers employed hydrolysis of glycolate complex of titanium (*59*) for controlled release of titanyl species. Mesoporous titania prepared by this method lacked thermal stability. In order to improve the thermal stability, it was post-treated with disilane at 373 K.

Stucky and coworkers (*60*) used amphiphilic poly(alkylene oxide) a tri-block co-polymer to prepare mesoporous metal oxides. A thermally stable two-dimensional hexagonal mesoporous TiO_2 with a surface area of 205 m^2/g, pore wall thickness of 51Å and pore size of 65Å was reported. Since this synthetic strategy involves non-hydrolytic pathway the hydrolysis and condensation rates of titanium precursor are very low. Thus, crystallization of bulk oxide mesophases is avoided and an added advantage is that there is no need for use of stabilizing agents. A novel hybrid nano-propping gemini surfactant containing a siloxane moiety (*61*) was also used to prepare mesopore metal oxides by Stucky and co-workers. The siloxane moiety in the surfactant acts as a stabilizer for the metal oxide framework during calcination. The only drawback of this method was that the periodicity of as-synthesized titania was not high when compared with other metal oxides.

On and coworkers synthesized a phosphorous free lamellar and hexagonal titania (*62*) using cationic surfactant cetyltrimethylammonium chloride ($C_{16}TMA^+Cl^-$) and soluble peroxytitantes. Evaporation-induced self-assembly (EISA) mechanism is an alternative approach to tune the inorganic condensation process to allow the formation of liquid crystal template mesophase with pre-determined geometries. In the EISA process, the inorganic material accumulates around the voids of liquid crystalline phase during the removal of the solvent to yield materials that have well-defined mesostructures. EISA technique has been used to prepare hybrid titania-based materials. These hybrid titania-based material are "titaniotropic," (*i.e.*, hybrid liquid-crystalline phases (*63*)). On thermal treatment, the hybrid materials form porous oxides that have large accessible pore surface area and walls (*64*). Periodic mesoporous titania with high surface areas were prepared by Yoshitake *et al.* (*65*) However, post-synthetic chemical vapor deposition (CVD) of titanium isopropoxide was required to stabilize the titania structure. The samples prepared by CVD have increased thermal stability with respect to time at 573K but not to increased temperatures (*66*). A post-synthetic treatment was developed by Cassiers and coworkers (*67*) to enhance the thermal stability of amorphous titania hybrid materials up to 873K.

Conventional synthesis of mesoporous titania takes anywhere between 2-14 days and is time consuming and new methods have been explored to shorten the synthesis time. Recently sonochemistry is finding its way in the synthesis of porous materials (*68*, *69*). The phenomenon of acoustic cavitation (*i.e.*, the formation, growth and implosive collapse of bubbles in a liquid) produces large amounts of energy thereby generating short-lived localized hot-spots. These hot-spots have transient temperatures as high as 5000 K, pressure of 1800 atm. and cooling rates in excess of 10^8 K/s. Gadenken *et al.* used ultrasonic waves to prepare mesoporous titania in very short time, *i.e.* as little as few hours (*70*). Powder XRD and TEM pictures confirmed that the mesoporous titania had a wormhole-like structure. The surface areas of these materials were in the range 600-850 m^2/g depending on the length of the carbons in the template. But these materials too lost its surface area drastically after calcination at high temperatures. Later this group reported that by appropriate choice of the titanium precursor (titanium isopropoxide or titanium tetrachloride) and reaction temperature it was possible to synthesize titania phase selectively without using any surfactant (*71*). Although this method was able to produce mesoporous titania, the pore-size distribution was found to be very large in the absence of pre- or post synthetic treatments.

Highly photoactive nanosized titania with brookite and anatase phase was prepared (*72*) using simple ultrasonication of a mixture of titanium isopropoxide-ethanol-water and then subjecting the resulting material to heat treatment at 773 K. Titania produced this way was more photoactive than the commercial Degussa P25. Anatase and brookite phases of titania with narrow pore size distribution was prepared using high intensity ultrasound probe in the presence and absence of triblock copolymer ($EO_{20}PO_{70}EO_{20}$) (*73*). The photocatalyst prepared either way was more photoactive than Degussa P25. The presence of triblock copolymer was found to enhance the crystallite size of anatase and brookite. Relatively shorter times and mild synthetic conditions of ultrasound was exploited to prepare titania in a subsequent study (*74*, *75*).

The fabrication of mesoporous Titania with crystalline framework in a short time is a challenge. Even though mesoporous Titania can be prepared by sonochemical methods in a relatively short time, post synthetic treatments are required to impart crystallinity. Wang and coworkers reported the synthesis of crystalline, uniform pore sized, high surface area titania nanopowders by micro-wave assisted sol-gel process in a short time and at relatively lower temperatures (*76*). The use of microwave provides rapid and uniform heating rate, reduces the reaction time and helps in producing products that are uniform and homogeneous. Based on the microwave hydrothermal temperature and method of template removal, the surface area obtained for anatase was found to be in the range 250 – 400 m^2/g. The photocatalytic degradation of Methylene blue was found to be higher for titania prepared by microwave hydrothermal than conventional hydrothermally prepared titania. He and coworkers prepared quasi-one dimensional nanorods that contains mesopore anatase using microwaves (*77*).

So far, we have discussed some of the challenges in preparing mesoporous titania that possess crystalline walls, high surface area and high thermal stability. Another challenge is to prepare a material with the above properties in a relatively short time without involving any post-synthetic treatments. This difficulty arises due to the thinning of the walls that occurs during the transformation from amorphous to crystalline phase by heat treatment or ultraviolet light irradiation. If the crystallization and wall formation rates are balanced, then the direct synthesis of mesoporous material with crystalline wall is possible without the need of any post-modification steps (*78*). This was accomplished by Sakai and coworkers when they reported low temperature direct synthesis of mesoporous titania having crystalline walls through sol-gel reaction using the cationic surfactant CTAB and titanium oxysulfate sulfuric acid hydrate ($TiOSO_4.xH_2SO_4.xH_2O$) at 333 K (*79*). $TiO(OH)_2$ the intermediate species formed during the reaction is thought to play a crucial role in the formation of this material (*80*). The UV-Vis absorption spectra of the material prepared by this method shows a drastic red shift towards the visible light region suggesting these materials have photocatalytic activity in visible light (*81*) which is similar to visible light responsive nitrogen doped titania (*82*).

Hyeon and coworkers reported the room temperature synthesis (*83*) of TiO_2 nanocrystals in aqueous media obviating the need for hydrothermal treament. Based on the salt used titania with various shapes and crystal structures were formed. Room temperature synthesis of mesoporous titania using gemini surfactant was also reported by Sakai and coworkers (*84*).

4. Mechanism of Photocatalysis

4.1. Behavior of Photogenerated Electron and Hole

Numerous studies have been done to probe the mechanism and process of semiconductor photocatalysis with a purpose to improve the quantum efficiency, extend the absorption wavelength of titanium dioxide catalyst, and to explore the large scale application of photocatalytic technique in industry. Research in photocatalysis is strongly correlated with the development of energy sources. Some applications of photocatalysis include photocatalytic reduction of carbon dioxide, hydrogen production by photocatalytic or photoelectrochemical splitting of water, and conversion of solar energy into to electricity using dye sensitized solar cells (*85*).

The initial step in photocatalysis is the generation of electron-hole pairs in the semiconductor particle by the excitation of light. Under the irradiation of light with appropriate energy, electrons and holes are produced in conduction band and valence band respectively. They migrate within the particle and may recombine with each other in the volume or on the surface and dissipate their energy as heat. Electrons that reach the surface are good reductants (+0.5 to -1.5 V vs NHE) whereas holes are powerful oxidants (+1.0 to +3.5 V vs NHE, depending on the semiconductor and the pH value). The electron-hole recombination, which can occur in the volume of the semiconductor particle or on the surface, competes with the charge transfer processes and thus decreases the efficiency of the photocatalytic reaction. The charge transfer processes are more efficient if the electron acceptor and electron donor species are adsorbed on the semiconductor surface.

The efficiency of the photocatalytic reaction is dependent on the particle size of the TiO_2 semiconductor. With smaller particle sizes, the number of active sites increases and hence higher efficiencies are obtained. However, if the particle size is too small, the recombination process becomes dominant and the efficiency decreases.

Laser flash photolysis studies have been used in determining the characteristic times for reactions concerning charge carrier. The time scale for the generation of electron-hole pairs is 100 fs (*86*, *87*). Thus, the photo-generated electrons and holes are trapped on the surfaces of TiO_2 in 100 ps to 10 ns time scale (*88*). The surface-bound hydroxyl radicals are formed in the 10 ns time scale, whereas the electron-hole recombination is observed to proceed in the 10–100 ns range (*89*). The electron transfer to oxygen molecules is calculated to proceed in the 100 μs to ms region, which is slower than other steps (*90*). The overall quantum efficiency of the interfacial charge transfer reaction is determined by two important competitive reactions. These are (1) the competition between charge carrier recombination (100–10 ns) and trapping (100 ps to 10 ns) and (2) competition between trapped carrier recombination and interfacial charge transfer (100 μs to ms). Interfacial charge transfer processes are very important, especially in nanostructured materials because of the very large surface area, and form the basis of many applications such as dye sensitized solar cells (DSSC). Charge carrier trapping is an important process that can influence the overall quantum efficiency. Charge carrier trapping helps to increase the lifetime of the electron and hole by suppressing electron hole recombination. Defects in materials lead to the formation of traps in the bulk and on the surface. The surface and bulk states are localized and hence charge carriers trapped in such states are considered to be localized. Time resolved laser flash photolysis has been used to experimentally observe trapping of conduction band electrons (*91*). Transient studies have demonstrated lifetimes in the nanosecond range for the trapping of the conduction band electrons, whereas the trapping of valence band holes in general required time in the 250 ns region (*92*). Electron spin resonance (ESR) experiments have indicated the existence of trapped Ti^{3+} defect sites, which prove the existence of trapped photo-generated electrons within the bulk of the semiconductor (*93*). Adsorbed oxygen on the TiO_2 surface scavenges the trapped electrons and inhibits the formation of Ti^{3+} sites. Electron trapping is the critical factor that limits the rate of charge carrier recombination between electrons and the photo-oxidized dye molecules in DSSC and is also thought to be the limiting factor in the performance of such solar cell devices. Regarding the trapping of holes, two types of trapped holes have been observed, deeply trapped holes and shallowly trapped holes. Deeply trapped holes are unreactive and rather long-lived, whereas shallowly trapped holes are highly reactive and in thermal equilibrium with free holes.

Apart from ESR, other techniques such as current-voltage measurements; frequency resolved, time-resolved, photocurrent measurements; microwave conductivity; transient absorption; and photoconductivity have been used to probe charge transport dynamics in nanocrystalline TiO_2 (*94*). Among the various techniques, photoconductivity has become a popular and simple technique to understand the competition between the charge carrier recombination and their trapping. Another promising method that has been reported is surface photovoltage spectra (SPS) and electric-field induced surface photovoltage spectra (EFISPS) (*95*). EFISPS can provide a rapid, non-destructive method to evaluate the surface and bulk carrier recombination of nano-sized semiconductors such as ZnO and TiO_2. For these nanostructured materials, the results indicate that the weaker the surface photovoltage signal, the higher is the photocatalytic activity (*95*).

Doping nanocrystalline TiO_2 by transition metal ions usually helps improve the photoinduced charge separation in nanostructure semiconductors and also improves the interfacial charge transfer process at the semiconductor–solution interface. In this case, the electrons in the conduction band and the holes in the valence band are trapped in the defect sites produced by doping and hence the photocatalytic efficiency is improved (*96*). Further, several transition metal ions also enable visible light absorption by providing defect states in the band gap. However, the formed intraband states may serve as recombination centers and decrease the overall photocatalytic efficiency in certain instances. Co-doping two different transition metal ions in nanocrystalline TiO_2 has also proven to be beneficial in improving the photoinduced charge separation due to the cooperative or synergistic action of the two dopants. For example, it has been reported that Fe^{3+} and Eu^{3+} in nanocrystalline TiO_2, play different roles in improving the photoefficiency: Fe^{3+} serves as a hole trap and Eu^{3+} as an electron trap in nanocrystalline TiO_2, and they speed up the anodic and cathodic processes, respectively (*97*).

4.2. Active Oxygen Species in Photocatalytic Reactions

Photo-generated electrons and holes may also react with oxygen and water molecules to form several kinds of active oxygen species in the reaction system. Although the mechanism varies from one reaction to another, superoxide radical, hydrogen peroxide, and hydroxyl radical, are found to play important roles in the degradation and mineralization of the organic compounds. The formation of active oxygen species are summarized by Eqs. (3)–(9):

$$TiO_2(e) + O_2 \rightarrow TiO_2 + O_2^{\cdot-} \quad (3)$$

$$TiO_2(h^+) + H_2O \rightarrow TiO_2 + \cdot OH + H^+ \quad (4)$$

$$O_2^{\cdot-} + TiO_2(e) + 2H^+ \rightarrow TiO_2 + H_2O_2 \quad (5)$$

$$2O_2^{\cdot-} + 2H^+ \rightarrow O_2 + H_2O_2 \quad (6)$$

$$\cdot OH + \cdot OH \rightarrow H_2O_2 \quad (7)$$

$$H_2O_2 + TiO_2(e) \rightarrow \cdot OH + OH^- + TiO_2 \quad (8)$$

$$H_2O_2 + O_2^{\cdot-} \rightarrow \cdot OH + OH^- + O_2 \quad (9)$$

In most photocatalytic reactions, electron is scavenged by oxygen to form $O_2^{\cdot-}$ radical. As stated earlier, electron transfer to oxygen molecules is much slower than other trapping and recombination processes involving the charge carriers. Gerischer and Heller also suggested that the electron transfer to oxygen may be the rate-limiting step in semiconductor photocatalysis (*98*). Though it is generally accepted that the hydroxyl radical is the major species responsible for the degradation of organic pollutants, some research shows that $O_2^{-\cdot}$ radical could also play a role in the mineralization process of the organic pollutant (*99*).

H_2O_2 has also been detected in several photocatalytic reactions. There may be two pathways for the formation of H_2O_2 in aerated aqueous solution: the reduction of absorbed oxygen by conduction band electrons and oxidation of H_2O by valence band holes. Isotopic labeling experiments performed by Hoffman and coworkers suggested that all the oxygen in H_2O_2 ($H^{18}O^{18}OH$) came from $^{18}O_2$ in the photocatalytic oxidation of carboxylic acid by ZnO (*100*). H_2O_2 may also contribute to the degradation of organic pollutants by acting as a direct electron acceptor or as a source of hydroxyl radical.

Hydroxyl radical is believed to be the major species responsible for the oxidation of organic pollutants. The existence of hydroxyl and hydroperoxy radicals has been verified by ESR studies in aqueous TiO_2 suspension (*101*). Intermediates detected during the photocatalytic degradation of halogenated aromatic compounds have been typically hydroxylated compounds, which gives indirect evidence for the role of hydroxyl radicals as the primary reactive oxidant species, since hydroxylated intermediates are formed when these substrates are reacted with known precursors of hydroxyl radicals (*15*). It is believed the oxidation of chlorinated hydrocarbons with abstractable hydrogen atoms, such as chloroform, is initiated by the abstraction of hydrogen atom by the surface-bound hydroxyl radical (*102*). However, another pathway that involves direct hole transfer to the organics was also proposed by some researchers. In some cases, the substrates were oxidized both by the TiO_2-bound hydroxyl radical oxidation and direct hole oxidation pathways. On the basis of detailed kinetic analysis and the time evolution of the intermediates during the photocatalytic transformation of phenol in the presence of different alcohols, Pelizzetti and co-workers suggested that the oxidation of phenol proceeds 90% through the reaction with TiO_2-bound hydroxyl radicals; the remaining 10% is *via* direct interaction with the hole (*103*). Hoffmann and co-workers have reported that the direct hole oxidation pathway was more important for strongly adsorbing 4-chlorocatechol, whereas hydroxyl radical attack was dominated for the poorly adsorbing 4-chlorophenol (*104*). It seems that the oxidation mechanism of the substrate depends on its structure and adsorption ability on the surface of TiO_2. The substrates with weak adsorption on the surface of TiO_2 and susceptive to hydroxyl radical attack are oxidized mainly through hydroxyl radical pathway. The substrates, which are tightly bound on the surface of TiO_2, is dominated through the direct hole oxidation pathway.

4.3. Visible Light Photocatalysts

The application of TiO_2 photocatalyst is limited by its wide band gap and the short wavelength absorption in UV region. In order to extend the light response of TiO_2 from UV to visible region and increase its quantum efficiency, several ways were developed for the modification of TiO_2 material. Doing TiO_2 nanomaterials with metal and nonmetal elements alters the chemical composition, the electronic properties and, hence the optical properties of TiO_2 nanomaterials. Various metals, such as Cr, Ce, Co, V, Fe, Ni, and W, and nonmetal elements, such as N, C, S, B, F, Cl, and Br, were successfully doped into TiO_2 nanomaterials (*82*, *105–108*). The absorption in visible region was significantly increased by doping TiO_2 with metal ions, which is due to the charge-transfer transition between the *d* electrons of the metal dopant and the conduction band or valence band of TiO_2. It is postulated that doping by transition metal ions decreases the recombination rate (by trapping electrons) and hence increases the overall efficiency of the photocatalytic reactions in several photocatalytic reactions. However, there is an optimum metal content at which the efficiency is the highest. Also The absorption edge of the nonmetal-doped TiO_2 is also shifted to longer wavelength and an obvious absorption in visible region was observed. Though the doping of nonmetal brought TiO_2 nanomaterials with visible light photocatalytic activity, the electronic structure and the excitation mechanism are still unclear. The narrowing of the band gap, nitrogen impurity centers localized in the band gap and the formation of oxygen vacancies and color centers were all proposed to be responsible for the light absorption and photocatalytic activity of nonmetal-doped TiO_2 nanomaterials in visible region (*82*, *109*). Modification of TiO_2 with materials that have narrower band gap is another approach to extend the optical absorption into the visible region. The sensitizer could be narrow band gap inorganic semiconductors, noble metal nanoparticles, or organic dyes. The sensitization of TiO_2 by several narrow band gap semiconductors, such as CdS, CdSe, PbS, Bi_2S_3, Ag_2S and Sb_2S_3, were reported and the charge transfer from the sensitizer and TiO_2 matrix was studied (*110*, *111*). Ag, Au, and Pt nanoparticles, which have plasmon absorption in visible region, were also used to sensitize TiO_2 nanomaterials (*112*, *113*). The absorption of the composite material can be easily tuned by changing of the size of noble metal nanoparticles. Organic dye sensitization of TiO_2 nanomaterials has been used by several researchers in the field of dye-sensitized solar cells (DSSCs) (*85*, *114*, *115*). Transition metal complex, such as polypyridine complexes, phthalocyanine, and metallophorphyrins, have been linked to TiO_2 nanoparticle surface and used to absorb light in both UV and visible region.

5. Applications in Environmental Remediation

As an advanced oxidation process, semiconductor photocatalysis has been extensively studied for the degradation of toxic pollutants in the environment. Among various oxide semiconductor materials, TiO_2 is the most efficient and suitable photocatalyst for environmental application due to its strong oxidizing power and stability against photo- and chemical corrosion. A wide variety of organic and inorganic compounds such as aliphatics, aromatics, halogenated aliphatics and aromatics, polychlorinated biphenyls (PCBs), pesticides, herbicides, dyes, nitrogen oxide and carbon oxide, chromium species, mercury (II) and other non-metal ions have been successfully degraded. The organic compounds were completely oxidized to CO_2, H_2O and associated inorganic ions. In the past decades, numerous works about the photocatalytic degradation using TiO_2 photocatalyst have been reported. The underlying degradation mechanism of organic compounds has also been studied in detail. Several methods were developed to improve the photocatalytic activity and quantum efficiency of TiO_2 photocatalyst, which are the two most important factors required for the practical applications of this technique. It is well known that the photocatalytic activity of TiO_2 is sensitive to its crystal phase, crystallinity, particle shape, surface area and active sites on the surface. Periodic mesoporous TiO_2 prepared by using surfactants are characterized by high specific surface area and large porosity; these provide more adsorption and reactive sites on the surface and facilitate better diffusion of reactants and products during the photocatalytic reaction. TiO_2 nanoparticles with mesopores and crystalline wall have proven to be a better photocatalyst than common TiO_2 nanoparticles. The application of mesoporous TiO_2 for the photocatalytic degradation of toxic pollutants is reviewed in this section.

5.1. Photocatalytic Degradation of Selected Pollutants in Water

5.1.1. Chlorophenols and Phenols

Chlorophenols are widely used as pesticides and wood preservatives and cause pollution to the soil and aquatic environment. Chlorinated compounds are persistent in the environment because of the stable C-Cl bond. They have been classified as toxic or priority pollutants by the U.S. Environmental Protection Agency (EPA) and the European Commission. The reaction pathway of the degradation of chlorophenols catalyzed by TiO_2 under UV irradiation has been studied by many researchers (*116–118*). It has been shown that the first step of the photocatalytic degradation of 4-chlorophenol includes two competitive pathways: substitution to form hydroquinone or hydroxylation to form 4-chlorocatechol. The further photo oxidation steps lead to the formation of 1, 2, 4-benzenetriol, 5-chloro-1, 2, 4-benzenetriol, and ring cleavage products of diacids and acid-aldehydes. Mesoporous TiO_2 was used as photocatalyst and a higher activity was achieved compared to nonporous TiO_2 material (*119–122*).

Chen *et al.* studied the activity of three-dimensional mesoporous TiO_2 for photodegradation of phenol. The mesoporous TiO_2 in their study was prepared by non-hydrolytic evaporation-induced self-assembly (EISA) method. The authors found that the pore texture, phase composition and photocatalytic efficiency could be well-controlled by changing the molar ratio of $TiCl_4/Ti(OBu)_4$ precursors (*119*). Mesoporous TiO_2 spheres with tunable chamber structure synthesized using a template-free solvothermal method was also reported to exhibit high photocatalytic activity for phenol degradation. The authors found that the unique sphere-in-sphere chamber structure allowed multiple reflections of UV light hence resulting in greatly enhanced photocatalytic activity (*120*). A novel hydro-alcohol thermal method was developed by Xu *et al.* to prepare thermally stable mesoporous core-shell structured nanocrystallite TiO_2 microspheres for the photodegradation of phenol in aqueous suspension. The sample calcined at 733K gave the best photocatalytic activity and demonstrated to be far superior to that of the commercial Degussa P25 counterpart (*122*). Dai *et al.* examined the effects of templates on the structure, stability and photocatalytic activity of mesostructured TiO_2 (*123*). The uniformity, stability and the structure of mesoporous TiO_2 synthesized with dodecylphosphate, hexadecylphosphate, or octadecylphosphate as templates is superior to that of mesoporous TiO_2 prepared with dodecylamine. However, it is difficult to remove the organo-phosphates completely by calcination or other methods because of its strong binding to the pore walls.

Visible light absorbing TiO_2 materials were examined by several research groups. N-doped TiO_2 mesoporous microtubes were fabricated by Xu *et al.* and the ultraviolet and visible light photocatalytic activity for phenol and methyl orange degradation was studied (*124*). High photocatalytic activity was also achieved for this mesoporous TiO_2. Hao *et al.* investigated the photocatalytic activity of iron (III) and nitrogen co-doped mesoporous TiO_2 for degradation of 2, 4-dichlorophenol in solution under visible light irradiation. Higher photocatalytic activity of the co-doped mesoporous TiO_2 powders was observed compared to N-doped mesoporous TiO_2 sample and P25 (*121*).

5.1.2. Dyes

Organic synthetic dyes are used during the production of textile, leather goods, painting, food, plastic, and cosmetics. The release of industrial dyes into effluent water has caused serious environmental problems in several parts of the world (*125*). Many physical and chemical techniques, including adsorption by active carbon, ultrafiltration, reverse osmosis, biodegradation, chlorination, ozonation, and Advanced Oxidation Processes (AOPs), were developed for the decolorization and mineralization of dye pollutants (*126*). TiO_2 photocatalysis is an efficient way for the degradation of dyes using solar energy. Hundreds of research papers about the photocatalytic degradation of dye pollutants by TiO_2 based photocatalyst have been published (*16, 127*). The mesoporous structure, high specific surface area and high adsorption ability significantly enhance the photocatalytic activity of TiO_2 for dye degradation (*128–131*).

The effect of pore architecture on reactivity of mesoporous nanocrystalline anatase thin films was studied by Carreon *et al.* (*128*) Methylene blue was chosen as a probe molecule to evaluate the adsorption ability and photocatalytic activity. 3D cubic mesoporous nanocrystalline anatase was comprised of more open cubic framework, larger surface area and less obstructed 3D diffusion paths of guest molecules, and showed superior photocatalytic activity (*128*). Yu *et al.* prepared crystalline bi-phase TiO_2 hollow microspheres with mesoporous shells and these materials showed high activity for the decolorization of dye methyl orange (*132*). Dong *et al.* studied the photocatalytic degradation of dye Rhodamine B (RhB) with thermally stable anatase–silica composites. The mesoporous TiO_2–SiO_2 nanocomposites had a highly ordered hexagonal mesostructure and exhibited excellent photocatalytic activities compared to commercial catalyst Degussa P25 (*133*). Saif *et al.* studied the effect of trivalent lanthanide ions doping on the photocatalytic activity of mesoporous TiO_2 (*134*). The commercially available textile dye Remazol Red RB-133 degradation was used as a probe reaction. Ln(III) doping brought remarkable improvement in the photoactivity over pure TiO_2. Lee *et al.* prepared pristine mesoporous TiO_2 and Tungsten doped mesoporous

TiO_2 using CTAB as a template *via* sol-gel method (*135*). The performance of mesoporous TiO_2 was much better than that of the non-mesoporous P25 TiO_2, and Tungsten doped mesoporous TiO_2 was better than pristine mesoporous TiO_2 for the oxidation of Methyl orange. Li *et al.* synthesized Cobalt doped amorphous mesoporous titania-silica with Ti/Si mass ratio of 0.8 and systematically studied the photocatalytic degradation of six cationic dyes (Gentian Violet, Methyl Violet, Methylene blue, Fuchsin basic, Safranine T, and RhB) under UV and visible light irradiation. They found that the degradation rates of Co-TiO_2-SiO_2 for Gentian Violet, Methyl Violet, Methylene blue, Fuchsin basic and Safranine T were greater in alkaline media than in acid and neutral media, while it did not exhibit any significant activity for the photodegradation of RhB in alkaline media or in acid media under visible light irradiation.

Wang *et al.* prepared N-doped mesoporous titania *via* the hydrolysis of titanium tetraisopropoxide using Tween 80 as pore-directing agent in a mixed aqueous solution of isopropanol and acetic acid in the presence of urea (*136*). Nitrogen doping caused the absorption edge of $TiO_{2-x}N_x$ to shift to the visible region. 10%N-TiO_2 exhibited the highest photocatalytic activity for degradation of Methyl orange solution under solar simulated light irradiation. Visible-light-active mesoporous N-doped TiO_2 was also prepared by the precipitation of titanyl oxalate complex ($[TiO(C_2O_4)_2]_2^-$) by ammonium hydroxide at a low temperature followed by calcination at different temperatures (*137*). The authors found that the sample calcined at 400 °C has Bronsted acid sites arising from covalently bonded dicarboxyl groups, which greatly enhanced the adsorption capacity for Methyl orange. Hierarchically ordered mesoporous-macroporous N-doped TiO_2 showed significant improvement of photocatalytic activity for the photodegradation of Methyl orange and RhB under UV and visible-light irradiation (*138*). The authors ascribed the high activity due to the incorporation of nitrogen into the TiO_2 lattice and the presence of the hierarchical meso-macroporous structure. The drastically enhanced activity of nitrogen doped mesoporous TiO_2 for RhB photodegradation was also observed by Liu *et al.* (*139*) Photoluminescence spectra reveal that nitrogen doped mesoporous TiO_2 possesses abundant surface states, which play a vital role in trapping photoinduced carriers and prolonging the lifetime of the carriers. Xie *et al.* prepared mesoporous rod-like F-N-co-doped TiO_2 powder photocatalysts with anatase phase *via* a sol-gel route using cetyltrimethyl ammonium bromide as surfactant. These photocatalysts show more than 6 times higher visible-light-induced catalytic degradation for Methyl orange than that of Degussa P25 (*140*).

The synergistic effects of B/N doping on the visible-light photocatalytic activity of mesoporous TiO_2 was studied by Liu *et al.* (*141*) The formation of an O-Ti-B-N structure led to the construction of a favorable surface structure that facilitates the separation and transfer of charge carriers, thereby enhancing the photocatalytic activity.

5.2. Photocatalytic Degradation of Selected Pollutants in Gas-Phase

5.2.1. Aldehyde and Acetone

Formaldehyde, acetaldehyde and acetone are common chemicals used in household products and the chemical industry. They are volatile organic compounds (VOCs) and a major constituent of indoor pollutants. Thus, the elimination of gaseous formaldehyde, acetaldehyde and acetone in air from indoor environments is of great interest. Photocatalytic oxidation of formaldehyde, acetaldehyde and acetone with nonporous TiO_2 was investigated by several groups (*142*, *143*). High surface area mesoporous TiO_2 is favorable for the adsorption and photocatalytic degradation of volatile organic compounds in gas phase (*144–146*). The photodecomposition of acetone by zeolite-like mesoporous TiO_2 nanocrystalline thin films was studied by Yu *et al.* (*147*) A comparison with a conventional TiO_2 film showed that the ordered mesoporous TiO_2 nanocrystalline film had over two times the specific photocatalytic activity as the conventional TiO_2 film. The high photocatalytic activity of the mesoporous TiO_2 thin films can be explained by the large specific surface area and the three-dimensionally connected mesoporous architecture. Yu *et al.* studied the photocatalytic activity of hierarchically macro/mesoporous TiO_2 for gaseous photocatalytic oxidation decomposition of acetone (*148*). The hierarchically porous TiO_2 showed photocatalytic activity about three times higher than that of Degussa P-25.

Mesoporous anatase TiO_2 nanofiber prepared by hydrothermal post-treatment of titanate nanotube was used for photocatalytic oxidation of acetone in air (*149*). The photocatalytic activity of the TiO_2 nanofibers prepared by this method exceeded that of Degussa P25 when the hydrothermal post-treatment time was done at 473 K for 3-24 h. This was attributed to the fact that the mesoporous TiO_2 nanofibers had smaller crystallite size, larger specific surface area, and higher pore volume. Zhan *et al.* synthesized long TiO_2 hollow fibers with mesoporous walls using electrospinning technique (*150*). The as-prepared hollow fibers were as long as 30 cm with an outer diameter of 0.1-4 μm and wall thickness of 60-500 nm. The hollow fibers exhibited higher photocatalytic activities toward degradation of Methylene blue and gaseous formaldehyde relative to TiO_2 powders.

Kozlova *et al.* studied the effect of precursors and templates on the photocatalytic activity of mesoporous TiO_2 for dimethyl methylphosphonate photodegradation in liquid phase and for acetone oxidation in gas phase (*151*). Fan *et al.* synthesized visible-light-active mesoporous Cr-doped TiO_2 photocatalyst with worm-like channels using an evaporation-induced self-assembly approach and investigated the performance for gaseous acetaldehyde photodecomposition. The results showed that Cr-doped mesoporous TiO_2 exhibited higher photocatalytic activities than pure mesoporous TiO_2 and nonporous Cr-doped TiO_2 under visible light irradiation (*152*). Yu *et al.* prepared mesoporous Au–TiO_2 and evaluated the photocatalytic oxidation of formaldehyde in air. The activity of Au–TiO_2 nanocomposite microspheres was higher than that of pure TiO_2 and Degussa P25 (*153*).

N,S-codoped TiO_2 was reported to be a much better photocatalyst for the photocatalytic oxidation of acetone and formaldehyde under UV light and daylight irradiation in air (*154*). The daylight-induced photocatalytic activities of the as-prepared N, S-codoped TiO_2 powders were about ten times greater than that of Degussa P25. The high activities of the N,S-codoped TiO_2 was attributed to the results of the synergetic effects of strong absorption in the UV-Vis region, red shift in the absorption edge, good crystallization, large surface area and bi-phasic structure of N,S-codoped TiO_2. Inorganic nonmetal anions, such as F^-, PO_4^{3-}, SO_4^{2-}, and CF_3COOH have been reported to be effective modifiers to enhance the photocatalytic activity of mesoporous TiO_2 for photocatalytic degradation of acetone (*155–158*). Surface-fluorinated mesoporous TiO_2 was found to show much better photocatalytic activity for the oxidative decomposition of acetone in air under UV light illumination compared to Degussa P25 (*155*). The high photocatalytic activity was explained by the strong electron-withdrawing ability of the surface Ti-F groups, which reduces the recombination of photo-generated electrons and holes, and enhances the formation of free OH radicals. Similar effect has been observed for treatment of mesoporous TiO_2 by sulfuric acid (*157*). Sulfuric acid treatment enhanced the photocatalytic activity of mesoporous TiO_2 film by 2-4 times. The increase in activity was ascribed to the increase in absorbed hydroxyl on the surface of TiO_2 catalyst. Enhancement of the photocatalytic activity of mesoporous TiO_2 thin films by treatment with trifluoroacetic acid was also studied (*156*). Trifluoroacetic acid complex bound on the surface of TiO_2 acted as an electron scavenger and, thus, reduced the recombination of photogenerated electrons and holes.

5.2.2. Toluene

Toluene is also classified as a volatile organic compound (VOC). Mesoporous titania nanospheres fabricated by spray-hydrolytic method was used to decompose toluene in air (*159*). The sample calcined at 673 K showed photocatalytic activity about two times higher than that of Degussa P25. UV and visible light photocatalytic removal of concentrated toluene in the gas phase was studied by Bosc *et al.* using mesoporous anatase TiO_2 supported WO_3 photocatalysts (*160*). The coupling of low contents of WO_3 with mesoporous anatase TiO_2 led to high photocatalytic efficiency using UV and visible light activation. Sinha *et al.* reported the removal of titania using mesoporous ceria-titania (*161*). The material had great thermal stability after high-temperature treatment. Toluene removal performance was further enhanced by impregnation with Pt.

5.2.3. Ethylene and Trichloroethylene (TCE)

TCE is used as solvent for extraction of waxes, oils, fats, and tars and in several consumer products such as paint removers/strippers and carpet/rug cleaning fluids. It is among the most common persistent and abundant pollutants in groundwater in the United States. According to the Agency for Toxic Substances and Disease Registry (ATSDR), TCE is the most frequently reported organic contaminant in groundwater. The complete mineralization of TCE to HCl and

CO_2 has been achieved in photocatalytic degradation with TiO_2. Cao *et al.* tested the photocatalytic activity of mesoporous titania-silica aerogels for gas phase photocatalytic oxidation of trichloroethylene in a new photoreactor configuration where the reactant gases flow through the aerogel block (*162*). Wang *et al.* incorporated light-harvesting macroporous channels into a mesoporous TiO_2 framework to increase its photocatalytic activity for ethylene photodegradation in gas-phase (*163*). The macrochannels increased the photo- absorption efficiency and allowed efficient diffusion of the gaseous molecules. The catalyst calcined at 623 K possessed an intact macro/mesoporous structure and showed photocatalytic reactivity about 60% higher than that of commercial Degussa P25.

5.2.4. Nitrogen Oxides

Nitrogen oxides are major contributors to acid rain. They are also common hazardous gaseous pollutants in the indoor environment. Long-term exposure may cause various health problems. Mesoporous TiO_2 is found to be a highly efficient photocatalyst for the photocatalytic degradation of nitrogen oxides in the gas phase. TiO_2 mesoporous films prepared by a template-assisted procedure based on the evaporation-induced self-assembly mechanism were shown to be efficient photocatalysts for the oxidation of nitrogen oxide (*164*, *165*). Because of the strong adsorption of the intermediate products, a high selectivity towards nitric acid was achieved. Yu *et al.* synthesized mesoporous nanocrystalline C-doped TiO_2 photocatalysts through a direct solution-phase carbonization using titanium tetrachloride and diethanolamine as precursors and applied the photocatalyst to the photocatalytic degradation of NO under simulated solar-light irradiation (*166*). The absorption of the as-prepared TiO_2 was extended to the visible light region due to the substitution of oxygen sites by carbon atoms. The samples showed more effective removal efficiency than commercial photocatalyst (P25).

6. Concluding Remarks and Future Outlook

From the examples cited in this chapter, it is clear that mesoporous titania of different morphologies can be prepared and utilized successfully for environmental applications in the aqueous and gas phase. Although success has been achieved in the laboratory for complete removal of aqueous and gas phase pollutants, decontamination technologies using photocatalysis are not widespread. However in recent years, the principle of photocatalysis has been exploited for the development of self-cleaning glass windows, anti-fogging mirrors, and for improving indoor air quality. Future research should focus on efforts to improve the quantum efficiency by designing novel TiO_2 based materials that possess hierarchical ordering of pores and containing visible light absorbing species so that solar light can be harnessed efficiently.

Acknowledgments

RTK thanks the 2010 Center for Research and Development of Light-Activated Materials, and NSF- CHE-0722632.

References

1. Izumi, J. Mitsubishi VOC Recovery Process; Mitsubishi Heavy Industries, Ltd.: 1996.
2. Huang, S. Y.; Schlichthorl, G.; Nozik, A. J.; Gratzel, M.; Frank, A. J. *J. Phys. Chem. B* **1997**, *101*, 2576–2582.
3. Everett, D. H. IUPAC Manual of Symbols and Terminology, Appendix **2**, Part I, Colloid and Surface Chemistry. *Pure Appl. Chem.* **1972**, *31*, 578–638.
4. Bard, A. J.; Fox, M. A. *Acc. Chem. Res.* **1995**, *28*, 141–145.
5. Linsebigler, A. L.; Lu, G.; Yates, J. T. *Chem. Rev.* **1995**, *95*, 735–758.
6. Gaya, U. I.; Abdullah, A. H. *J. Photochem. Photobiol. C: Photochem. Rev.* **2008**, *9*, 1–12.
7. Hoffmann, M.R.; Martin, S.T.; Choi, W.; Bahnemann, D.W. *Chem. Rev.* **1995**, *95*, 69–96.
8. Fujishima, A.; Rao, T.N.; Tryk, D.A. *J. Photochem. Photobiol., C* **2000**, *1*, 1–21.
9. Gratzel, M.; Kalyanasundaram, K., Ed.; *Kinetics and Catalysis of Microheterogeneous Systems*; Marcel Dekker: New York, 1991.
10. Anpo, M., Ed.; *Photofunctional Zeolites: Synthesis, Characterization, Photocatalytic Reaction, Light Harvesting*; Nova Science: New York, 2000.
11. Kaneko, M.; Okura, I., Ed.; *Photocatalysis: Science and Technology* Springer-Verlag: New York, 2002.
12. Linsebigler, A. L.; Lu, G.; Yates, J. T. *Chem. Rev.* **1995**, *95*, 735–758.
13. Chatterjee, D.; Dasgupta, S. *J. Photochem. Photobiol., C* **2005**, *6*, 186–205.
14. Fujishima, A.; Honda, K. *Nature* **1972**, *238*, 37–38.
15. Zhao, D.; Chen, C.; Wang, Y.; Ji, H.; Ma, W.; Zang, L.; Zhao, J. *J. Phys. Chem. C* **2008**, *112*, 5993–6001.
16. Zhao, D.; Chen, C.; Wang, Y.; Ma, W.; Zhao, J.; Rajh, T.; Zang, L. *Environ. Sci. Technol.* **2008**, *42*, 308–314.
17. Sun, B.; Reddy, E. P.; Smirniotis, P. G. *Environ. Sci. Technol.* **2005**, *39*, 6251–6259.
18. Burdett, J. K.; Hughbanks, T.; Miller, G. J.; Richardson, J. W.; Smith, J. V. *J. Am. Chem. Soc.* **2002**, *109*, 3639–3646.
19. Ohtani, B.; Handa, J.-i.; Nishimoto, S.-i.; Kagiya, T. *Chem. Phys. Lett.* **1985**, *120*, 292–294.
20. Hurum, D. C.; Gray, K. A.; Rajh, T.; Thurnauer, M. C. *J. Phys. Chem. B* **2004**, *109*, 977–980.
21. Liu, P.; Bandara, J.; Lin, Y.; Elgin, D.; Allard, L. F.; Sun, Y.-P. *Langmuir* **2002**, *18*, 10398–10401.
22. Zhang, D.; Qi, L.; Ma, J.; Cheng, H. *J. Mater. Chem.* **2002**, *12*, 3677–3680.
23. Wu, J.-M.; Shih, H. C.; Wu, W.-T. *Chem. Phys. Lett.* **2005**, *413*, 490–494.
24. Lei, Y.; Zhang, L. D.; Fan, J. C. *Chem. Phys. Lett.* **2001**, *338*, 231–236.

25. Wu, J.-M. *J. Cryst. Growth* **2004**, *269*, 347–355.
26. Rolison, D. R.; Dunn, B. *J. Mater. Chem.* **2001**, *11*, 963–980.
27. Swider, K. E.; Hagans, P. L.; Merzbacher, C. I.; Rolison, D. R. *Chem. Mater.* **1997**, *9*, 1248–1255.
28. Kistler, S. S. *Nature* **1931**, *127*, 741–741.
29. Teichner, S. J.; Nicolaon, G. A.; Vicarini, M. A.; Gardes, G. E. E. *Adv. Colloid Interface Sci.* **1976**, *5*, 245–273.
30. Schneider, M.; Baiker, A. *J. Mater. Chem.* **1992**, *2*, 587–589.
31. Campbell, L. K.; Na, B. K.; Ko, E. I. *Chem. Mater.* **1992**, *4*, 1329–1333.
32. Ayen, R. J.; Iacobucci, P. A. *Rev. Chem. Eng.* **1988**, *5*, 157–198.
33. Brodsky, C. J.; Ko, E. I. *J. Mater. Chem.* **1994**, *4*, 651–652.
34. Dutoit, D. C. M.; Schneider, M.; Baiker, A. *J. Porous Mater.* **1995**, *1*, 165–174.
35. Dagan, G.; Tomkiewicz, M. *J. Phys. Chem.* **1993**, *97*, 12651–12655.
36. Dagan, G.; Tomkiewicz, M. *J. Non-Cryst. Solids* **1994**, *175*, 294–302.
37. Zhu, Z.; Lin, M.; Dagan, G.; Tomkiewicz, M. *J. Phys. Chem.* **1995**, *99*, 15945–15949.
38. Zhu, Z.; Dagan, G.; Tomkiewicz, M. *J. Phys. Chem.* **1995**, *99*, 15950–15954.
39. Kelly, S.; Pollak, F. H.; Tomkiewicz, M. *J. Phys. Chem. B* **1997**, *101*, 2730–2734.
40. Masson, O.; Rieux, V.; Guinebretiere, R.; Dauger, A. *Nanostruct. Mater.* **1996**, *7*, 725–731.
41. Ayers, M. R.; Hunt, A. J. *Mater. Lett.* **1998**, *34*, 290–293.
42. Suh, D. J.; Park, T. J. *Chem. Mater.* **1996**, *8*, 509–513.
43. Suh, D. J.; Park, T. J. *J. Mater. Sci. Lett.* **1997**, *16*, 490–492.
44. Malinowska, B.; Walendziewski, J.; Robert, D.; Weber, J. V.; Stolarski, M. *Int. J. Photoenergy* **2003**, *5*, 147–152.
45. Malinowska, B.; Walendziewski, J.; Robert, D.; Weber, J. V.; Stolarski, M. *Appl. Catal., B.* **2003**, *46*, 441–451.
46. Guo, G.; Whitesell, J. K.; Fox, M. A. *J. Phys. Chem. B* **2005**, *109*, 18781–18785.
47. Kresge, C. T.; Leonowicz, M. E.; Roth, W. J.; Vartuli, J. C.; Beck, J. S. *Nature* **1992**, *359*, 710–712.
48. Chen, C. Y.; Burkett, S. L.; Li, H. X.; Davis, M. E. *Microporous Mater.* **1993**, *2*, 27–34.
49. Huo, Q. S.; Margolese, D. I.; Ciesla, U.; Feng, P. Y.; Gier, T. E.; Sieger, P.; Leon, R.; Petroff, P. M.; Schüth, F.; Stucky, G. D. *Nature* **1994**, *368*, 317–321.
50. Huo, Q. S.; Margolese, D. I.; Ciesla, U.; Demuth, D. G.; Feng, P. Y.; Gier, T. E.; Sieger, P.; Firouzi, A.; Chmelka, B. F.; Schüth, F.; Stucky, G. D. *Chem. Mater.* **1994**, *6*, 1176–1191.
51. Glinka, C. J.; Nicol, J. M.; Stucky, G. D.; Ramli, E.; Margolese, D.; Huo, Q.; Higgins, J. B.; Leonowicz, M. E. *J. Porous Mater.* **1996**, *3*, 93–98.
52. Corma, A. *Chem. Rev.* **1997**, *97*, 2373–2419.
53. Antonelli, D. M.; Ying, J. Y. *Angew. Chem., Int. Ed. Engl.* **1995**, *34*, 2014–2017.
54. Stone, V. F., Jr.; Davis, R. J. *Chem. Mater.* **1998**, *10*, 1468–1474.

55. Putnam, R. L.; Nakagawa, N.; McGrath, K. M.; Yao, N.; Aksay, I. A.; Grunner, S. M.; Navrotsky, A. *Chem Mater.* **1997**, *9*, 2690–2693.
56. Fujii, H.; Ohtaki, M.; Eguchi, K. *J. Am. Chem. Soc.* **1998**, *120*, 6832–6833.
57. Antonelli, D. M. *Microporous Mesoporous Mater.* **1999**, *30*, 315–319.
58. Carbera, S.; Haskouri, J. E.; Porter, A. B.; Porter, D. B.; Marcos, M. D.; Amoros, P. *Solid State Sci.* **2000**, *2*, 513–518.
59. Khushalani, D.; Ozin, G. A.; Kuperman, A. *J. Mater. Chem.* **1999**, *9*, 1491–1500.
60. Yang, P.; Zhao, D.; Margolese, D. I.; Chmelka, B. F.; Stucky, G. D. *Nature* **1998**, *396*, 152–155.
61. Lyu, Y. Y.; Yi, S. H.; Shon, J. K.; Chang, S.; Pu, L. S.; Lee, S. Y.; Yie, J. E.; Char, K.; Stucky, G. D.; Kim, J. M. *J. Am. Chem. Soc.* **2004**, *126*, 2310–2311.
62. On, D. T. *Langmuir* **1999**, *15*, 8561–8564.
63. Firouzi, A.; Atef, F.; Oertli, A. G.; Stucky, G. D.; Chmelka, B. F. *J. Am. Chem. Soc.* **1997**, *119*, 3596–3610.
64. Soler-Illia, G. J. de A. A.; Louis, A.; Sanchez, C. *Chem. Mater.* **2002**, *14*, 750–759.
65. Yoshitake, H.; Sugihara, T.; Tatsumi, T. *Chem. Mater.* **2002**, *14*, 1023–1029.
66. Cassiers, K.; Linssen, T.; Mathieu, M.; Bai, Y. Q.; Zhu, H. Y.; Cool, P.; Vansant, E. F. *J. Phys. Chem. B* **2004**, *108*, 3713–3721.
67. Cassiers, K.; Linssen, T.; Meynen, V.; Van der Voort, P.; Cool, P.; Vansant, E. F. *Chem. Commun.* **2003**, 1178–1179.
68. *Ultrasound: Its Chemical, Physical, and Biological Effects*; Suslick, K. S., Ed.; VCH: Weinheim, 1988.
69. Suslick, K. S.; Choe, S. B. *Nature* **1991**, *353*, 414–416.
70. Wang, Y.; Tang, X.; Yin, L.; Huang, W.; Hacohen, Y. R.; Gedanken, A. *Adv. Mater.* **2000**, *12*, 1183–1186.
71. Huang, W.; Tang, X.; Wang, Y.; Koltypin, Y.; Gedanken, A. *Chem. Commun.* **2000**, 1415–1416.
72. Yu, J. C.; Yu, J. G.; Ho, W. K.; Zhang, L. Z. *Chem. Commun.* **2001**, 1942–1943.
73. Yu, J. C.; Zhang, L.; Yu, J. *Chem. Mater.* **2002**, *14*, 4647–4653.
74. Latt, K. K.; Kobayashi, T. *Ultrason. Sonochem.* **2008**, *15*, 484–491.
75. Liu, Y.; Li, Y.; Wang, Y.; Xie, L.; Zheng, J.; Li, X. *J. Hazard. Mater.* **2008**, *150*, 153–157.
76. Wang, H. W.; Kuo, C. H.; Lin, H. C.; Kuo, I.; Cheng, C. *J. Am. Ceram. Soc.* **2006**, *89*, 3388–3392.
77. Jia, X.; He, W.; Zhang, X.; Zhao, H.; Li, Z.; Feng, Y. *Nanotechnology* **2007**, *18*, 075602.
78. Katou, T.; Lee, B.; Lu, D.; Kondo, J.N.; Hara, M.; Domen, K. *Angew. Chem., Int. Ed. Engl.* **2003**, *115*, 2484–2487.
79. Shibata, H.; Ogura, T.; Mukai, T.; Ohkubo, T.; Sakai, H.; Abe, M. *J. Am. Chem. Soc.* **2005**, *127*, 16396–16397.
80. Shibata, H.; Mihara, H.; Ogura, T.; Mukai, T.; Kohno, H.; Ohkubo, T.; Sakai, H.; Abe, M. *Chem. Mater.* **2006**, *18*, 2256–2260.

81. Shibata, H.; Ogura, T.; Sakai, H.; Matsumoto, M.; Abe, M. *J. Jpn. Soc. Colour Mater.* **2008**, *81*, 235–239.
82. Asahi, R.; Morikawa, T.; Ohwaki, T.; Aoki, K.; Taga, Y. *Science* **2001**, *293*, 269–270.
83. Han, S.; Choi, S. H.; Kim, S. S.; Cho, M.; Jang, B.; Kim, D. Y.; Yoon, J.; Hyeon, T. *Small* **2005**, *1*, 812–816.
84. Abe, M.; Shibata, H.; Tsubone, K.; Mihara, H.; Sakai, T.; Torigoe, K.; Dai, S.; Okhubo, T.; Utsumi, S.; Sakamoto, K.; Matsumoto, M.; Sakai, H. *J. Jpn. Soc. Colour Mater.* **2007**, *80*, 450–452.
85. Grätzel, M. *Nature* **2001**, *414*, 338–344.
86. Skinner, D. E.; Colombo, D. P.; Cavaleri, J. J.; Bowman, R. M. *J. Phys. Chem.* **1995**, *99*, 7853–7856.
87. Colombo, D. P.; Roussel, K. A.; Saeh, J.; Skinner, D. E.; Cavaleri, J. J.; Bowman, R. M. *Chem. Phys. Lett.* **1995**, *232*, 207–214.
88. Serpone, N.; Lawless, D.; Khairutdinov, R.; Pelizzetti, E. *J. Phys. Chem.* **1995**, *99*, 16655–16661.
89. Martin, S.T.; Herrmann, H.; Choi, W.; Hoffmann, M. R. *J. Chem. Soc. Faraday Trans.* **1994**, *90*, 3315–3322.
90. Sun, L.; Bolton, J. R. *J. Phys. Chem.* **1996**, *100*, 4127–4134.
91. Bahnemann, D. W.; Hilgendorff, M.; Memming, R. *J. Phys. Chem. B* **1997**, *101*, 4265–4275.
92. Rothenberger, G.; Moser, J.; Grätzel, M.; Serpone, N.; Sharma, D. K. *J. Am. Chem. Soc.* **2002**, *107*, 8054–8059.
93. Grätzel, M.; Howe, R. F. *J. Phys. Chem.* **2002**, *94*, 2566–2572.
94. Schlichthorl, G.; Park, N. G.; Frank, A. J. *J. Phys. Chem. B* **1999**, *103*, 782–791.
95. Liqiang, J.; Xiaojun, S.; Jing, S.; Weimin, C.; Zili, X.; Yaoguo, D.; Honggang, F. *Sol. Energy Mater. Sol. Cells* **2003**, *79*, 133–151.
96. Martin, S. T.; Morrison, C. L.; Hoffmann, M. R. *J. Phys. Chem.* **1994**, *98*, 13695–13704.
97. Yang, P.; Lu, C.; Hua, N.; Du, Y. *Mater. Lett.* **2002**, *57*, 794–801.
98. Gerischer, H.; Heller, A. *J. Phys. Chem.* **1991**, *95*, 5261–5267.
99. Yang, J.; Chen, C. C.; Ji, H. W.; Ma, W. H.; Zhao, J. C. *J. Phys. Chem. B* **2005**, *109*, 21900–21907.
100. Hoffman, A. J.; Carraway, E. R.; Hoffmann, M. R. *Environ. Sci. Technol.* **2002**, *28*, 776–785.
101. Chen, C. C.; Lei, P. X.; Ji, H. W.; Ma, W. H.; Zhao, J. C.; Hidaka, H.; Serpone, N. *Environ. Sci. Technol.* **2004**, *38*, 329–337.
102. Kormann, C.; Bahnemann, D. W.; Hoffmann, M. R. *Environ. Sci. Technol.* **1991**, *25*, 494–500.
103. Minero, C.; Mariella, G.; Maurino, V.; Vione, D.; Pelizzetti, E. *Langmuir* **2000**, *16*, 8964–8972.
104. Kesselman, J. M.; Weres, O.; Lewis, N. S.; Hoffmann, M. R. *J. Phys. Chem. B* **1997**, *101*, 2637–2643.
105. Khan, S. U. M.; Al-Shahry, M.; Ingler, W. B., Jr. *Science* **2002**, *297*, 2243–2245.
106. Ohno, T.; Mitsui, T.; Matsumura, M. *Chem. Lett.* **2003**, *32*, 364–365.

107. Yu, J. C.; Yu, J. G.; Ho, W. K.; Jiang, Z. T.; Zhang, L. Z. *Chem. Mater.* **2002**, *14*, 3808–3816.
108. Luo, H.; Takata, T.; Lee, Y.; Zhao, J.; Domen, K.; Yan, Y. S. *Chem. Mater.* **2004**, *16*, 846–849.
109. Irie, H.; Watanabe, Y.; Hashimoto, K. *J. Phys. Chem. B* **2003**, *107*, 5483–5486.
110. Vogel, R.; Hoyer, P.; Weller, H. *J. Phys. Chem.* **1994**, *98*, 3183–3188.
111. Robel, I.; Subramanian, V.; Kuno, M.; Kamat, P.V. *J. Am. Chem. Soc.* **2006**, *128*, 2385–2393.
112. Ohko, Y.; Tatsuma, T.; Fujii, T.; Naoi, K.; Niwa, C.; Kubota, Y.; Fujishima, A. *Nature Mater.* **2003**, *2*, 29–31.
113. Tian, Y.; Tatsuma, T. *Chem. Commun.* **2004**, 1810–1811.
114. Grätzel, M. *J. Photochem. Photobiol., C* **2003**, *4*, 145–153.
115. Grätzel, M. *J. Photochem. Photobiol., A* **2004**, *164*, 3–14.
116. Theurich, J.; Lindner, M.; Bahnemann, D. W. *Langmuir* **1996**, *12*, 6368–6376.
117. Al-Ekabi, H.; Serpone, N.; Pelizzetti, E.; Minero, C.; Fox, M. A.; Draper, R. B. *Langmuir* **2002**, *5*, 250–255.
118. Ollis, D. F. *J. Phys. Chem. B* **2005**, *109*, 2439–2444.
119. Chen, L.; Yao, B.; Cao, Y.; Fan, K. *J. Phys. Chem. C* **2007**, *111*, 11849–11853.
120. Li, H.; Bian, Z.; Zhu, J.; Zhang, D.; Li, G.; Huo, Y.; Li, H.; Lu, Y. *J. Am. Chem. Soc.* **2007**, *129*, 8406–8407.
121. Hao, H.; Zhang, J. *Microporous Mesoporous Mater.* **2009**, *121*, 52–57.
122. Xu, J.-H.; Dai, W.-L.; Li, J.; Cao, Y.; Li, H.; Fan, K. *J. Photochem. Photobiol., A* **2008**, *195*, 284–294.
123. Dai, Q.; Shi, L. Y.; Luo, Y. G.; Blin, J. L.; Li, D. J.; Yuan, C. W.; Su, B. L. *J. Photochem. Photobiol., A* **2002**, *148*, 295–301.
124. Xu, J.-H.; Dai, W.-L.; Li, J.; Cao, Y.; Li, H.; He, H.; Fan, K. *Catal. Commun.* **2008**, *9*, 146–152.
125. Brown, M. A.; De Vito, S. C. *Crit. Rev. Environ. Sci. Technol.* **1993**, *23*, 249–324.
126. Spadaro, J. T.; Isabelle, L.; Renganathan, V. *Environ. Sci. Technol.* **2002**, *28*, 1389–1393.
127. Vinodgopal, K.; Kamat, P. V. *Environ. Sci. Technol.* **2002**, *29*, 841–845.
128. Carreon, M. A.; Choi, S. Y.; Mamak, M.; Chopra, N.; Ozin, G. A. *J. Mater. Chem.* **2007**, *17*, 82–89.
129. Li, J.; Liu, S.; He, Y.; Wang, J. *Microporous Mesoporous Mater.* **2008**, *115*, 416–425.
130. Beyers, E.; Cool, P.; Vansant, E. F. *Microporous Mesoporous Mater.* **2007**, *99*, 112–117.
131. Liu, G.; Zhao, Y.; Sun, C.; Li, F.; Lu, G. Q.; Cheng, H.-M. *Angew. Chem., Int. Ed.* **2008**, *47*, 4516–4520.
132. Yu, J.; Wang, G. *J. Phys. Chem. Solids* **2008**, *69*, 1147–1151.
133. Dong, W.; Sun, Y.; Lee, C. W.; Hua, W.; Lu, X.; Shi, Y.; Zhang, S.; Chen, J.; Zhao, D. *J. Am. Chem. Soc.* **2007**, *129*, 13894–13904.

134. Saif, M.; Abdel-Mottaleb, M. S. A. *Inorg. Chim. Acta* **2007**, *360*, 2863–2874.
135. Lee, A.-C.; Lin, R.-H.; Yang, C.-Y.; Lin, M.-H.; Wang, W.-Y. *Mater. Chem. Phys.* **2008**, *109*, 275–280.
136. Wang, Y.; Zhou, G.; Li, T.; Qiao, W.; Li, Y. *Catal. Commun.* **2009**, *10*, 412–415.
137. Fang, J.; Wang, F.; Qian, K.; Bao, H.; Jiang, Z.; Huang, W. *J. Phys. Chem. C* **2008**, *112*, 18150–18156.
138. Shao, G.-S.; Zhang, X.-J.; Yuan, Z.-Y. *Appl. Catal., B* **2008**, *82*, 208–218.
139. Liu, G.; Wang, X.; Wang, L.; Chen, Z.; Li, F.; Lu, G. Q.; Cheng, H.-M. *J. Colloid Interface Sci.* **2009**, *334*, 171–175.
140. Xie, Y.; Zhao, X.; Li, Y.; Zhao, Q.; Zhou, X.; Yuan, Q. *J. Solid State Chem.* **2008**, *181*, 1936–1942.
141. Liu, G.; Zhao, Y.; Sun, C.; Li, F.; Lu, G. Q.; Cheng, H.-M. *Angew. Chem. Intl. Ed.* **2008**, *47*, 4516–4520.
142. Choi, W.; Ko, J.Y.; Park, H.; Chung, J.S. *Appl. Catal., B* **2001**, *31*, 209–220.
143. Blount, M. C.; Kim, D. H.; Falconer, J. L. *Environ. Sci. Technol.* **2001**, *35*, 2988–2994.
144. Yu, J.; Liu, S.; Yu, H. *J. Catal.* **2007**, *249*, 59–66.
145. Yu, J.; Wang, G.; Cheng, B.; Zhou, M. *Appl. Catal., B* **2007**, *69*, 171–180.
146. Zhou, M.; Yu, J.; Cheng, B. *J. Hazard. Mater.* **2006**, *137*, 1838–1847.
147. Yu, J.C.; Wang, X.; Fu, X. *Chem. Mater.* **2004**, *16*, 1523–1530.
148. Yu, J.; Zhang, L.; Cheng, B.; Su, Y. *J. Phys. Chem. C* **2007**, *111*, 10582–10589.
149. Yu, J.; Yu, H.; Cheng, B.; Zhao, X.; Zhang, Q. *J. Photochem. Photobiol., A* **2006**, *182*, 121–127.
150. Zhan, S.; Chen, D.; Jiao, X.; Tao, C. *J. Phys. Chem. B* **2006**, *110*, 11199–11204.
151. Kozlova, E. A.; Vorontsov, A. V. *Appl. Catal., B* **2007**, *77*, 35–45.
152. Fan, X.; Chen, X.; Zhu, S.; Li, Z.; Yu, T.; Ye, J.; Zou, Z. *J. Mol. Catal. A: Chem.* **2008**, *284*, 155–160.
153. Yu, J.; Yue, L.; Liu, S.; Huang, B.; Zhang, X. *J. Colloid Interface Sci.* **2009**, *334*, 58–64.
154. Yu, J.; Zhou, M.; Cheng, B.; Zhao, X. *J. Mol. Catal. A: Chem.* **2006**, *246*, 176–184.
155. Yu, J.; Wang, W.; Cheng, B.; Su, B.-L. *J. Phys. Chem. C* **2009**, *113*, 6743–6750.
156. Yu, J. C.; Ho, W.; Yu, J.; Hark, S. K.; Iu, K. *Langmuir* **2003**, *19*, 3889–3896.
157. Yu, J. C.; Yu, J.; Zhao, J. *Appl. Catal., B* **2002**, *36*, 31–43.
158. Yu, J. C.; Zhang, L.; Zheng, Z.; Zhao, J. *Chem. Mater.* **2003**, *15*, 2280–2286.
159. Zhou, M.; Yu, J.; Liu, S.; Zhai, P.; Huang, B. *Appl. Catal., B* **2009**, *89*, 160–166.
160. Bosc, F.; Edwards, D.; Keller, N.; Keller, V.; Ayral, A. *Thin Solid Films* **2006**, *495*, 272–279.
161. Sinha, A. K.; Suzuki, K. *J. Phys. Chem. B* **2005**, *109*, 1708–1714.
162. Cao, S.; Yeung, K. L.; Yue, P.-L. *Appl. Catal., B* **2007**, *76*, 64–72.
163. Wang, X.; Yu, J. C.; Ho, C.; Hou, Y.; Fu, X. *Langmuir* **2005**, *21*, 2552–2559.

164. Bannat, I.; Wessels, K.; Oekermann, T.; Rathousky, J.; Bahnemann, D.; Wark, M. *Chem. Mater.* **2009**, *21*, 1645–1653.
165. Kalousek, V.; Tschirch, J.; Bahnemann, D.; Rathousk, J. *Superlattices Microstruct.* **2008**, *44*, 506–513.
166. Huang, Y.; Ho, W.; Lee, S.; Zhang, L.; Li, G.; Yu, J. C. *Langmuir* **2008**, *24*, 3510–3516.

Chapter 7

Decontamination of Chemical Warfare Agents with Nanosize Metal Oxides

George W. Wagner*

U.S. Army Edgewood Chemical Biological Center, Attn: RDCB-DRP-R, 5183 Blackhawk Rd., Aberdeen Proving Ground, MD 21010-5424
*E-mail: george.wagner@us.army.mil

Chemical warfare agents (CWA) VX, GD, and HD react with nanosize metal oxides to form non-toxic product, thus affording their decontamination. Reactions of these CWA with several nanocrystalline reactive sorbents of varied particle size and surface area are examined with regard to both reaction kinetics and their ability to decontaminate Chemical Agent Resistant Coating (CARC) paint, used on military vehicles. The results are compared to the currently-fielded Sorbent Decon System (SDS) A-200 sorbent. For VX the best sorbent was $nTiO_2$ which exhibited a half-life for sorbed VX of less than 2 min. Half-lives of tens-of-minutes were observed for GD on $nTiO_2$, nMgO, and the commercial FAST-ACT® sorbent. Half-lives of a few to many hours were observed for HD on nAl_2O_3, $nTiO_2$, FAST-ACT®, and A-200, but only with sufficient surface hydration. With regard to reactivity only, A-200 did not perform as well as the nanocrystalline sorbents, especially for VX and GD. However, all of the sorbents, A-200 included, provided for the comparable removal/decontamination of HD and GD from CARC panels, ca. 75 % and 87 %, respectively.

Introduction

The decontamination of chemical warfare agents (CWA) (*1*) such as nerve agents VX, GB (Sarin), GD (Soman), and blister agent HD (Mustard) is of ongoing concern owing to the continuing potential for these highly toxic materials to be released into the environment. Although CWA use has historically been relegated to use on soldiers in remote battlefields (*2*), contemporary concern is also focused on terrorist attacks in populated areas (*3*). In perhaps the most infamous example of the latter, the Aum Shinrikyo sect conducted the 1995 Tokyo Subway Attack using Sarin Gas; but it was not widely publicized that the Japanese cult had earlier employed VX to kill a former member in 1994 (*4*). Thus, both GB and VX have already been employed by non-military entities in urban settings on civilians. So the need continues for the development of decontamination material and procedures to restore the environment - both battlefields and urban areas - to safe-use by the populace, soldiers and civilians alike.

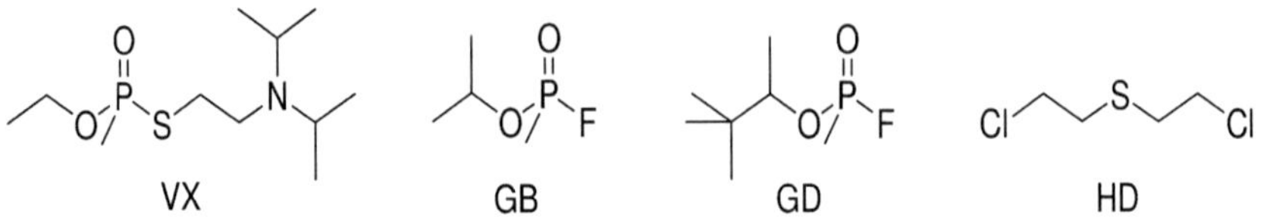

Metal oxides, already known for their versatile uses as adsorbents, catalysts, and catalyst supports, also show general reactivity for CWA (*5–10*). Indeed, the currently-fielded M-100 Sorbent Decontamination System (SDS) utilizes an aluminum oxide-based "reactive sorbent" (A-200) (*11*). But it is the promise of increased reactivity of nanosize metal oxides, owing to their increased surface areas, highly reactive edge and corner defect sites, and unusual, stabilized lattice planes (*12*), which render them particularly attractive for possible use as improved, next-generation reactive sorbents.

The reactions exhibited by nanosize MgO (nMgO) (*6*), CaO (nCaO) (*7*), and Al_2O_3 (nAl_2O_3) (*8*) toward VX, GD/GB, and HD, have previously been reviewed in a monograph by the same title as the current (*13*). These reactions, the mechanisms of which are typical of metal oxide (MO) materials in general, are summarized in Scheme 1.

The reactions observed on metal oxides, including nanosize metal oxides, are analogous to the hydrolysis reactions of these agents in solution (*1*). For example, GB, GD, and VX are all hydrolyzed to their non-toxic phosphonic acids isopropyl methylphosphonate (IMPA), pinacolyl methylphosphonate (PMPA) and ethyl methylphosphonate (EMPA), respectively. Additionally, secondary hydrolysis of these products can occur to yield methylphosphonate (MPA), which has been exhibited by EMPA on Al_2O_3 (*8*).

Besides hydrolysis to thiodiglycol (TG), HD also can undergo elimination of HCl in basic solution to yield divinyl sulfide (DVS) (*1*), and this mechanism is also observed on basic metal oxides such as MgO (*6*), CaO (*7*), and even Al_2O_3 (*8*). Of particular note is the reaction with nCaO which exhibited autocatalytic behavior towards HD elimination whereas conventional CaO did not (*7*). As in aqueous solution, high loadings of HD on wet metal oxides tend to form sulfonium ions such as CH-TG (*7–9*) (Scheme 2) which forms via coupling of the (half-hydrolyzed) chlorohydrin (CH) intermediate ($HOCH_2CH_2SCH_2CH_2Cl$) with TG.

Yet, there are unique aspects to the reactions of these agents with metal oxides - for example, the formation of various surface-bound species. As depicted in Scheme 2, on dry Al_2O_3 HD has been shown to form its TG-alkoxide (*8*), a species which would presumably be hydrolyzed to TG if subjected to moisture (*14*). The

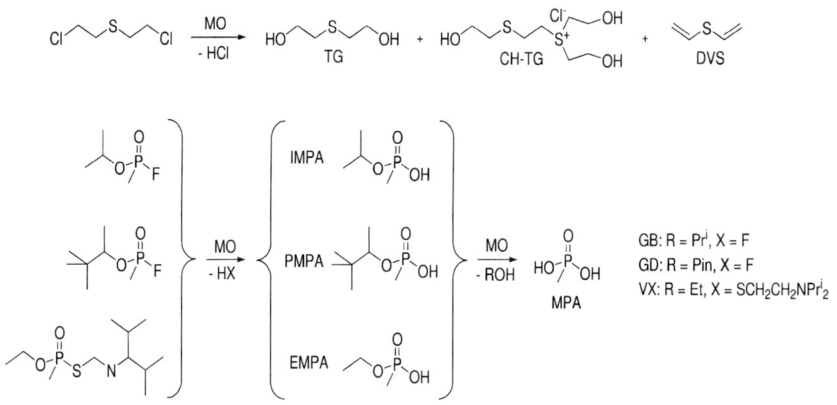

Scheme 1

phosphonic acids also form surface-bound species, both bidentate (IMPA, PMPA, EMPA) and tridentate (MPA). It has also been shown that on Al_2O_3 surface erosion occurs to yield discrete aluminophosphonate compounds (*8*), presumably due to excess acid (*15*).

Observed steady-state half-lives for neat GD, HD, and VX liquid deposited on nMgO were 28 min, 17.8 h and 68 h, respectively, which were attributed to diffusion limitations as rather fast, initial reactions were observed prior the rates slowing (*6*); direct comparison to conventional MgO was not done. For nCaO GD and VX showed the same fast initial/slow steady-state reaction behavior, with a much greater half-life of 4.5 h for GD but a comparable half-life of > 67 h for VX (*7*). But as mentioned above HD exhibited autocatalytic elimination behavior on nCaO, resulting in a shorter half-life of 8.5 h (*7*). Comparison to conventional CaO was done for HD where not catalytic behavior was observed and the steady-state half-life was substantially longer (> 49 days). Also for nCaO, it was found that the HD reaction was very sensitive to the amount of surface-sorbed water present, ranging from >11 days for the freshly-dried material (100 °C); 8.7 h for the as-received sample; 3.5 h for the aged, as-received sample; and > 46 h for the wet material (*7*). Thus, the results showed that an optimum amount of sorbed surface water would yield the fastest reaction – too much or too little water would seriously degrade performance.

For nAl_2O_3 GD, HD, and VX exhibited half-lives of 1.8 h, 6.3 h, and 154 h, respectively (*8*). A comparison done between this material and conventional Al_2O_3 found quite similar half-lives for GD and VX (1.8 h and 163 h) (*9*). But for HD the observed reaction was much faster on conventional Al_2O_3 (half-life only 44 min) than on nAl_2O_3 (half-life 6.3 h); furthermore, the size of the deposited neat HD drop had a substantial impact on the observed reaction rate: simply increasing the HD dropsize from 3.5 μL to 13 μL on conventional Al_2O_3 increased the half-life from 44 min to 70.9 h (no substantial dropsize dependency on the GD and VX reactions were observed) (*9*). Thus, besides the amount of water sorbed on the surface (noted above for nCaO), overall reaction rates are also sensitive to the size of deposited agent drops. Therefore, both deposition amounts and water-content need to be controlled and explored to fully assess the reactivity of reactive sorbents.

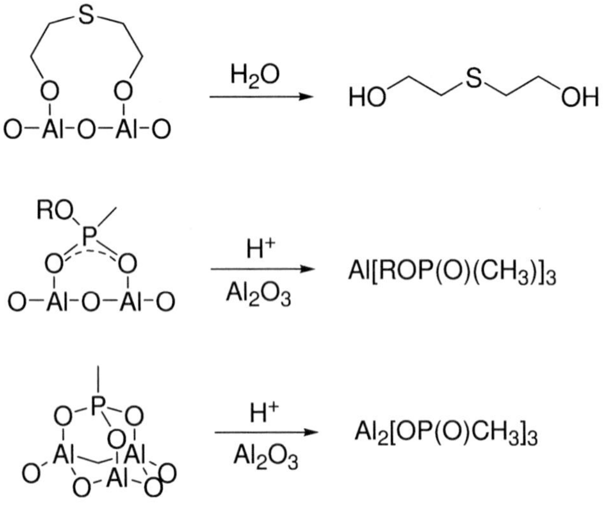

Scheme 2

For the more recent results (*16*) reported in this monograph, a series of candidate nanocrystalline sorbents, nMgO, $nTiO_2$, nAl_2O_3, and FAST-ACT (Nanoscale Corp), as well as A-200 (used as a control), were examined for their reactivity towards VX, GD, and HD, utilizing controlled conditions of air-exposure (water-content) and identical agent dropsize (5 μL). Additionally, the ability of these candidate sorbents and A-200 (control) to decontaminate panels painted with Chemical Agent Resistant Coating (CARC) – the standard, hardened paint ubiquitous on military land vehicles – was also assessed.

Results and Discussion

Candidate Sorbents

The sorbents studied are listed in Table 1. Particle size and surface area was varied for some materials to determine their effect on both reactivity and surface decontamination efficacy.

Reactivity Testing

Utilizing the same solid-state MAS NMR technique used in the prior studies (*7*, *8*) the ability of each sorbent to react with a single, 5-μL drop was examined. Both freshly-opened samples and 24-h air-exposed materials were tested as a measure of the air-sensitivity of the sorbents which, of course, must necessarily be exposed to air during their intended use – to decontaminate surfaces on military vehicles and equipment (*11*). These results are shown in Table 2. Owing to diffusion limitations (see Introduction) some sorbents exhibited initial, fast reactions followed by slower, steady-state reactions; in these instances it is the half-life of the latter that should be considered to determine the persistence of agent on a given sorbent.

Table 1. Candidate Nanocrystalline Sorbent Properties

Sorbent	*Particle Size Distribution (Volume %)*				*Surface Area*
	< 5 μm	*5-20 μm*	*20-50 μm*	*> 50 μm*	*m^2/g*
nMgO #1	7.65	53.51	37.40	1.44	718
nMgO #2	23.84	60.70	14.63	0.84	241
nMgO #3	0.00	33.17	60.54	6.28	771
$nTiO_2$ #1	28.08	28.02	15.55	28.35	489
$nTiO_2$ #2	13.79	28.56	50.60	7.06	492
nAl_2O_3	90.34	8.45	1.21	0.00	304
FAST-ACT	27.93	53.53	10.31	8.23	326
FAST-ACT-2	8.36	29.50	55.81	6.33	659

nMgO Sorbents

For the nMgO sorbents, GD tended to react quickest with half-lives on the order of tens of minutes. That the GD reactions did not exhibit severe diffusion limitations can be ascribed to both its rather high volatility (compared to VX and HD) and good water solubility (GD is soluble in water but not miscible), the latter attribute presumably allowing it to dissolve and diffuse within surface-bound water layers. Note, however, that despite air-exposure and the attendant potential water adsorption, some degradation of the GD reactivity occurred. This is easily understood by the observation that the nMgO sorbents tended to gain the most weight upon air-exposure (upwards of 30 %) and that ^{13}C MAS NMR (not shown) showed the formation of carbonate ($CO_3^{=}$) on this material as a result of reaction with ambient CO_2 in the air. Such a process would tend to neutralize the very basic MgO surface; thus, the deleterious effect on GD hydrolysis. Note that carbonate was not detected on any of the fresh nMgO's consistent with their freshness prior to testing.

VX also tended to react more quickly with fresh nMgO (half-life on the order of hours) and did not appear to be severely diffusion limited on nMgO #1, perhaps as a result of spontaneous spreading on this particular nMgO. However, air-exposure did cause a major loss in observed VX reactivity/diffusion. In its protonated state VX is water soluble. However, with a pK_a of 8.6 (*17*), VX is most likely "free-based" by the basic MgO surface into the unprotonated state; thus, it would suffer limited water solubility and, hence, diffusion (barring spontaneous spreading), within surface water layers.

For HD on nMgO the reaction appears entirely diffusion-limited (in no small part due to its water insolubility) – there is no fast, initial reaction. Yet, consistent with the very basic nMgO surface, the major reaction mechanism for HD is elimination to its vinyl products, varying from 50 % on nMgO #1 to 67 % on nMgO #2 and > 90 % on nMgO #3 (the latter material evidently possesses the highest basicity). Of course, as a result of the aforementioned carbonation/neutralization of surface basicity, elimination is curtailed in the

Table 2. Half-Lives 5- μL VX, GD, and HD Drops Deposited on Candidate Reactive Sorbents

Sorbent	*Treatment*	*VX*	*GD*	*HD*
nMgO #1	fresh	4.2 h	24 min	15 h
	air-exposed	9.3 h[a]/43 h[b]	28 min	24 h
nMgO #2	fresh	1.3 h[a]/11 h[b]	15 min	15 h
	air-exposed	53 h	47 min	35 h
nMgO #3	fresh	not tested	not tested	2.3 h[a]/24 h[b]
$nTiO_2$ #1	fresh	23 min (run 1)	2.8 h	5.4 h[a]/17 h[b]
		23 min (run 2)		
	air-exposed	< 2 min (run 1)	29 min	1.5 h[a]/11 h[b]
		< 2 min (run 2)		
$nTiO_2$ #2	fresh	36 min	17 min[a]/1.3 h[b]	1.5 h[a]/31 h[b]
	air-exposed	8.5 h	9.3 min[a]/23 min[b]	1.5 h[a]/10 h[b]
nAl_2O_3	fresh	2.1 h[a]/32 h[b]	3.0 h	6.3 h[a]/34 h[b]
	air-exposed	2.8 h[a]/31 h[b]	2.0 h	5.7 h
FAST-ACT	fresh	41 min[a]/2.2 h[b]	16 min	4.0 h[a]/38 h[b]
	air-exposed	26 min[a]/3.9 h[b]	34 min	4.3 h[a]/14 h[b]
FAST-ACT-2	fresh	55 min[a]/3.3 h[b]	1.2 h	3.4 h[a]/28 h[b]
	air-exposed	1.5 h	24 min	2.2 h[a]/13 h[b]
A-200	fresh	7.4 h[a]/56 h[b]	1.6 h[a]/13 h[b]	29 h
	air-exposed	14 h[a]/54 h[b]	4.0 h	19 h

[a] Initial Rate. [b] Final steady-state rate.

air-exposed samples to < 10 % in nMgO #1 and 50 % in nMgO #2 and the overall reactivity suffers. It is further interesting to note that air-exposure resulted in greater carbonation of nMgO #2 compared to nMgO #1 (as indicated by their ^{13}C MAS NMR spectra, not shown) which is consistent with the former's higher apparent basicity. Presumably, the highest basicity nMgO #3 would have suffered the most carbonation, but it was not tested.

Regarding the varied particle size and surface area of the nMgO sorbents studied, GD tended to react quickest with the fresh sorbent possessing the smaller particle size distribution and lower surface area, nMgO #2. However, upon air-exposure, nMgO #1 was able to retain its reactivity better. That this is so is perhaps related to its larger particle size which would tend to react slower with ambient CO_2. This is consistent with the somewhat lower weight gain of the nMgO #2 sorbent during 24-h air-exposure, 25-26 wt% vs. 28-30 wt% for nMgO #1. HD tended to react equally well with the two fresh materials, but, again, the larger-particle material tended to retain its reactivity better during 24-h air-exposure. As for VX, it tended to react better with the larger-particle size, higher-surface area nMgO #1, even after 24-h air-exposure.

nTiO_2 Sorbents

Although VX may be unprotonated and, thus, possess limited water solubility within surface water layers on nMgO, this is obviously not the case for nTiO_2 where the half-life for VX is an astoundingly short 23 min. Moreover, upon air-exposure nTiO_2 picks up 13-16 % water, further enhancing VX diffusion and reaction such that the half-life is unbelievably under 2 min! Such fast VX reactions, where an apparently non-diffusion limited half life of < 30 min occurs, has previously been seen on nanotubular titania (NTT) (*10*). It is known that the surface hydroxyls (*18*) of titania (and NTT (*10*)) are more acidic than MgO, and even Al_2O_3 (still basic enough to eliminate HD, as noted above), thus accounting for the presumed VX protonation and facilitated diffusion (especially when sufficient hydration layers are present) on nTiO_2 (and NTT). That the VX half-lives are a bit longer on nTiO_2 #2 could be due to fortuitous differences in water content; for example, upon air-exposure, it picked up only 15.6 % weight compared to 20.8 % for nTiO_2 #1. The effect of sufficient hydration layers on (water-soluble) agent diffusion is also apparent in the significantly shorter half-life of GD (29 min) on air-exposed nTiO_2 #1 compared to the fresh material (2.8 h). For nTiO_2 #2 the behavior of GD is quite different in that, following an initially fast reaction, a diffusion-controlled reaction ensues (albeit still at a faster rate than on nTiO_2 #1). It is not clear if these differences are due to particle morphology, surface characteristics, or hydration effects. However, the HD results suggest that nTiO_2 #1 has the highest initial water content as its hydrolysis is fastest on this as-received material. But after air-exposure both nTiO_2's possess virtually indistinguishable HD-reaction behavior.

Like nMgO, GD tended to react best with the smaller-particle nTiO_2 #2, before and after 24-h air-exposure. Also like nMgO, VX, conversely, reacted best with the larger-particle nTiO_2 #1, before and after 24-h air-exposure. Both materials possessed identical surface areas. Again, HD did not appear to favor one material better than the other.

nAl_2O_3 and A200 Sorbents

For HD on nAl_2O_3, the air-exposed sample only picks up 1-6 % water but this is evidently enough reduce the (diffusion-limited) half-life from 34 h to 5.7 h – the shortest sustained half-life for HD observed on any of the sorbents. Besides hydrolysis, similar but minor amounts of HD elimination products are observed on both fresh and air-exposed material (surface carbonation of the most basic sites does not seem to occur as it does with nMgO). The GD half-life is similarly reduced from 3 h to 2 h (yet still slower than on the more basic nMgO surface). But for VX there is virtually no change; it is extremely persistent on both dry and hydrated materials. The reason for the persistency of VX on alumina, and MgO for that matter, can be attributed to the tight binding of its hydrolysis product EMPA to these basic surfaces (as detected by ^{31}P MAS NMR (*7*, *8*)); thus, EMPA is unable to assume its usual role to autocatalytically hydrolyze VX (*17*). EMPA does not bind tightly to the $nTiO_2$ (nor NTT (*10*)) surface, so it remains free to dismantle VX with aplomb.

These observations for nAl_2O_3 also apply to A-200 which tends to pick up even less water upon air-exposure (1-2 %). In a similar manner, VX also remains persistent on both the fresh and air-exposed materials as EMPA is still sidelined. However, the half-life for GD is considerably reduced on the air-exposed sample from 13 h (diffusion-limited) to 4 h as a result of simply picking up 1.8 % water. Modest gain is also seen for HD reactivity on the slightly wetter material, reducing its half-life from 29 h to 19 h. As with nAl_2O_3 quite similar but minor amounts elimination products are observed on both fresh and wet materials.

FAST-ACT and FAST-ACT-2

VX reacts rather well with both FAST-ACT and FAST-ACT-2, even after air-exposure, as half-lives do not exceed a few hours. GD gives mixed results, having the shortest half-life on fresh FAST-ACT (16 min) which effectively doubles on the air-exposed material (34 min), whereas on FAST-ACT-2 the half-life is shorter on the air-exposed material (24 min) and quite long on the fresh (1.2 h). HD shows its typically slow, water-starved reactivity on the dry, fresh sorbents (unlike nMgO, only minor elimination of HD occurs), whereas enhancement occurs for the air-exposed, hydrated materials. Typical weight gains are 18 – 24 % and 22 – 30 % for air-exposed FAST-ACT and FAST-ACT-2, respectively.

Like nMgO, GD reacted best with the fresh, smaller-particle size, lower-surface area FAST-ACT, whereas the larger-particle, higher-surface area FAST-ACT-2 maintained its reactivity upon air exposure (actually improving). Also like nMgO, HD reacted similarly with both particle sizes. However, unlike nMgO, VX tended to react better with the smaller-particle size, smaller-surface area FAST-ACT.

Reactivity Testing Summary

From the reactivity results it is easy to see that $nTiO_2$ has the most potential to afford quick, simultaneous reactivity for all three agents, provided sufficient water is present. Although current water levels of 16 – 21 % are effective at reducing the VX and GD half-lives to < 2 min and 29 min, respectively, the HD half-life is still 11 h – nearly twice as long as that of air-exposed nAl_2O_3 (5.7 h). However, perhaps even higher water levels could achieve even further reduction in the HD half-life while maintaining, or even further enhancing, its unprecedented VX and GD reactivity. Clearly, hydration levels and their effect on reactivity need to be thoroughly explored and will be the subject of future testing.

With regard to particle size and/or surface area, the smaller particle size versions of nMgO and FAST-ACT tended to react fastest with GD, even

Table 3. Shortest Sustained Half-Lives Observed by Candidate Sorbents (Fresh or Air-Exposed) for VX, GD, and HD

Agent	*nMgO*	*$nTiO_2$*	*nAl_2O_3*	*FAST-ACT*	*A-200*
VX	4.2 h[a]	< 2 min[b]	31 h[c]	1.5 h[d]	54 h[c]
GD	15 min[e]	23 min[f]	2.0 h[c]	16 min[g]	4.0 h[c]
HD	15 h()	11 h[b]	5.7 h[c]	13 h[d]	19 h[c]

[a] Fresh nMgO #1. [b] Air-exposed $nTiO_2$ #1. [c] Air-exposed material. [d] Air-exposed FAST-ACT-2. [e] Fresh nMgO #2. [f] Air-exposed $nTiO_2$ #2. [g] Fresh FAST-ACT.

Table 4. Sorbent Reactivity Ranking Based on Observed Agent Half-Lives

VX	$nTiO_2$ >> FAST-ACT ≈ nMgO > nAl_2O_3 ≥ A-200
GD	nMgO ≈ FAST-ACT ≈ $nTiO_2$ > nAl_2O_3 ≥ A-200
HD	nAl_2O_3 ≥ $nTiO_2$ ≈ FAST-ACT ≈ nMgO ≈ A-200

though they possessed reduced surface areas. Conversely, VX reacted best with higher-surface area nMgO and FAST-ACT. However, for HD, no pronounced particle size and/or surface area effects could be discerned.

Overall, nano-based sorbents were highly superior to A-200, having much shorter half-lives, with $nTiO_2$ being the most effective. The shortest sustained half-lives observed for the agents on the sorbents, either provided by the fresh or air-exposed materials, are collected in Table 3. Again, detailed studies of the effect of hydration level on reactivity need to be completed to adequately determine the optimum reactivity of each sorbent.

From the shortest sustained half-lives given in Table 3, rankings of the sorbents with regard to their propensity to react with VX, GD, and HD in the most expeditious manner can be gleaned. These rankings are shown in Table 4, where half-lives of the same order-of-magnitude are considered equivalent (minutes > tens-of-minutes > hours > tens-of-hours). As noted before, these rankings should be considered tentative until a comprehensive study of the effect of water-content on the reaction kinetics is completed. However, the ranking strongly suggests $nTiO_2$ as the sorbent possessing the most potential to provide simultaneous, fast reaction with all three agents since its efficacy is by far the best for the most toxic (and persistent) agent VX; comparable to nMgO and FAST-ACT for GD; and reasonably close to that of nAl_2O_3 for HD. Moreover, unlike nMgO, $nTiO_2$ is not deactivated by air, rendering it a more robust choice for potential use as a reactive sorbent.

CARC Surface Decontamination Efficacy Testing

Results for the decontamination of HD and GD on CARC panels are given in Table 5. VX testing was not done owing the similarity of the results of the candidate sorbents for these agents (see below). Average values and standard deviations are shown for the six-panel replicates. Following decontamination of

Table 5. CARC Panel Decontamination Test Results[a]

Sorbent	*HD*				*GD*			
	Contact Hazard	*Residual Hazard*	*Total Hazard*	*% Decon[b]*	*Contact Hazard*	*Residual Hazard*	*Total Hazard*	*% Decon[b]*
nMgO #1	1.8 (0.45)	2.4 (0.14)	4.2 (0.59)	74 (3.8)	1.7 (0.52)	0.95 (0.23)	2.6 (0.36)	87 (1.6)
nMgO #2	2.5 (0.21)	2.4 (0.23)	4.9 (0.40)	69 (3.8)	1.4 (0.46)	1.2 (0.17)	2.8 (0.39)	86 (2.0)
$nTiO_2$ #1	2.0 (0.56)	1.9[c] (0.35)	4.0[c] (0.70)	75[c] (4.4)	1.4 (0.098)	1.2 (0.27)	2.6 (0.23)	87 (1.2)
nAl_2O_3	2.5 (0.30)	1.4 (0.16)	3.9 (0.41)	76 (2.4)	1.2 (0.26)	1.4 (0.16)	2.6 (0.15)	87 (0.75)
A-200	2.5 (0.44)	1.3 (0.46)	3.8 (0.47)	77 (2.9)	2.0 (0.24)	0.65 (0.093)	2.6 (0.24)	87 (1.0)
No Decon Control[d]	14 (1.5)	1.6 (0.36)	16 (1.2)	98 (9.3)	12 (1.0)	1.6 (0.50)	13 (1.2)	66 (5.8)

[a] Six replicate 2-in diameter panels contaminated with 10 g/m^2 agent (1000 $\mu g/cm^2$). 16-μg HD or 20-μg GD was evenly spread across the panel and allowed to stand 15 min prior to decontamination. Panel decontaminated with sorbent by rubbing for 1 min. Units in μg (agent recovered). Standard deviation given in parentheses. [b] % Decon = 100 × (total agent applied – total agent recovered)/total agent applied. [c] Average of three replicates. [d] No sorbent applied (to determine agent recovery efficiency).

the panel, with the sorbent powders in this instance, Contact Hazard is the amount of agent that can be picked up from the surface by direct contact with it (measured by placing a film of latex upon it (*16*)) and Residual Hazard is the remaining agent on the surface (measured by total extraction of the panel in chloroform (*16*)). Both values are expressed in μg (weight of agent recovered). These amounts are summed to yield the total amount recovered from the panel. The % Decon is calculated using the total recovered amount and applied starting amount. Control experiments were performed to assess recovery efficiencies (in the absence of applied decontaminating sorbent).

The results show that 69-77 % of HD and 86-87 % GD can be removed from the panels but rubbing for 1 min with the candidate sorbents. As these values are practically within the experimental error, no major difference is evident for the efficacy of the candidate sorbents – each one pretty much performed as well as the next [except, perhaps, for nMgO #2/HD (69 %) which was slightly lower than the rest (74-77%)].

It needs to be pointed out that for the control experiment the excellent recovery of HD from the CARC panels (98 %, in the absence of applied sorbent) shows that the manipulations performed for the application, standing period, and contact hazard assessment result in negligible loss of this agent and that the chloroform extraction was adequate to fully remove it from the CARC surface. However, for GD, unlike the case of HD, some of this agent ran over the sides of the panel during the 15-min standing period (prior to decontamination); thus, the low recovery efficiency of about 66 %.

Conclusions

The reactivity testing revealed that $nTiO_2$, by far, possesses the fastest reaction with VX (half-life < 2 min) of any of the other candidates and its reaction rate with GD was comparable to that of nMgO and FAST-ACT (half-lives tens-of-minutes). However, this fast rate is only observed when $nTiO_2$ is sufficiently hydrated. Hydration of $nTiO_2$ also increases the reaction rates of GD and HD. Yet, the HD half-life on hydrated TiO_2 is still on the order of several hours, comparable, but still nearly twice as long as that of hydrated nAl_2O_3. Thus, the full hydration regime of these two sorbents should be explored to determine which actually provides the best HD reactivity.

The CARC panel test results for HD and GD did not enable a clear distinction of the efficacy of the candidate sorbents as all tended to provide for nearly identical removal of HD and GD from the CARC panels (VX panel testing was not attempted because of this). However, for HD, it did appear that nMgO #1 may have performed slightly less well than the other sorbents. For GD all the sorbents performed virtually identically.

Finally, regarding particle size and surface area, small particles tended to provide for the fastest GD reactions, whereas higher surface area and/or larger particle sizes tended to favor VX reaction. The reaction of HD was quite insensitive to either particle size or surface area. For the decontamination of CARC paint, smaller particle size, even those under 5 μm, did not significantly improve the removal/decontamination of GD or HD.

Acknowledgments

Helpful discussions with Dr. Olga Koper and Mr. David A. Jones regarding the properties of nanoparticles used in this research are gratefully acknowledged. Many thanks go to Mmes. Nicole K. Fletcher and Tara L. Sewell for their assistance with the agent operations and to Messrs. Lawrence R. Procell, Philip W. Bartram, and J. Thomas Lynn for their advice on panel testing procedures. This work was performed under a Cooperative Research and Development Agreement (CRADA), Project No. 0714C, between Edgewood Chemical Biological Center and Nanoscale Corp. Funding was provided by contract #W911SR-07-C-0062.

References

1. Yang, Y.-C.; Baker, J. A.; Ward, J. R. *Chem. Rev.* **1992**, *92*, 1729–1743.
2. Moore, W. *Gas Attack!: Chemical Warfare 1915-1918 and Afterwards*; Leo Cooper: London, 1987.
3. Howard, R. D. *Weapons of Mass Destruction and Terrorism*; McGraw Hill: Dubuque, IA, 2008.
4. Zurer, P. *Chem. Eng. News* **1988**, *76* (Aug 31), 7.

5. Wagner, G. W.; Bartram, P. W. *Langmuir* **1999**, *15*, 8113–8118.
6. Wagner, G. W.; Bartram, P. W.; Koper, O.; Klabunde, K. J. *J. Phys. Chem. B* **1999**, *103*, 3225–3228.
7. Wagner, G. W.; Koper, O. B.; Lucas, E.; Decker, S.; Klabunde, K. J. *J. Phys. Chem. B* **2000**, *104*, 5118–5123.
8. Wagner, G. W.; Procell, L. R.; O'Connor, R. J.; Munavalli, S.; Carnes, C. L.; Kapoor, P. N.; Klabunde, K. J. *J. Am. Chem. Soc.* **2001**, *123*, 1636–1644.
9. Wagner, G. W.; Procell, L. R.; Munavalli, S. *J. Phys. Chem. C* **2007**, *111*, 17564–17569.
10. Wagner, G. W.; Chen, Q.; Wu, Y. *J. Phys. Chem. C* **2008**, *112*, 11901–11906.
11. Olsen, C. T.; Hayes, T.; Babin, M. A.; LeClaire, R. D. *Government Testing of the Sorbent Decontamination System (SDS) in Accordance With the Decision Tree Network*; AD-B307 502; U.S. Army MRMC: Fort Detrick, MD, 2005 (unclassified).
12. (a) Klanbunde, K. J.; Stark, J.; Koper, O.; Mohs, C.; Park, D. G.; Decker, S.; Jiang, Y.; Lagadic, I.; Zhang, D. *J. Phys. Chem.* **1996**, *100*, 12142−12153. (b) Stark, J. V.; Park, D. G.; Lagadic, I.; Klabunde, K. J. *Chem. Mater.* **1996**, *8*, 6182−6188.
13. Wagner, G. W.; Procell, L. R.; Koper, O. B.; Klabunde, K. J. Decontamination of Chemical Warfare Agents with Nanosize Metal Oxides. In *Defense Applications of Nanomaterials*; Miziolek, A. W.; Karna, S. P.; Mauro, J. M; Vaia, R. A., Eds.; ACS Symposium Series 891; American Chemical Society: Washington, DC, 2005; pp 139−152.
14. Pilkenton, S.; Hwang, S.-J.; Raftery, D. *J. Phys. Chem. B* **1999**, *103*, 11152–11160.
15. Wagner, G. W.; Fry, R. A. *J. Phys. Chem. C* **2010**, *113*, 13352–13357.
16. Wagner, G. W. *Decontamination Efficacy of Candidate Nanocrystalline Sorbents with Comparison to SDS A-200 Sorbent: Reactivity and CARC Panel Testing*; U.S. Army ECBC: Aberdeen Proving Ground, MD, in press.
17. Yang, Y.-C.; Szafraniec, L. L.; Beaudry, W. T.; Rohrbaugh, D. K.; Procell, L. R.; Samuel, J. B. *J. Org. Chem.* **1996**, *61*, 8407–8413.
18. Mastikhin, V. M.; Mudrakovsky, I. L.; Nosov, A. V. *Prog. Nucl. Magn. Reson. Spectrosc.* **1991**, *23*, 259–299.

Chapter 8

Advanced Lubricant Additives of Dialkyldithiophosphate (DDP)-Functionalized Molybdenum Sulfide Nanoparticles and Their Tribological Performance for Boundary Lubrication

Dmytro Demydov,[1] Atanu Adhvaryu,[2] Philip McCluskey,[2] and Ajay P. Malshe[1,*]

[1]Mechanical Engineering, University of Arkansas, Fayetteville, Arkansas 72701
[2]Caterpillar Inc., Technical Center, Peoria, Illinois 61552
[*]Contacting author e-mail: apm2@uark.edu

Advanced lubricants can improve productivity through energy savings and reliability of engineered systems, and there is a growing interest in application of nanostructured lubrication materials stimulated by offering valuable properties and exhibiting novel lubrication behaviors. The hybrid chemo-mechanical milling process of molybdenum sulfide with second component based on phosphorous element groups was used as a synthetic approach to prepare multi-component nanoparticle additives. Their addition to base oil could provide advanced lubrication by: (1) supplying nanosized lubricating agents which reduce friction and wear at the asperities contact zone, (2) enabling strong tendency to adsorb on the surfaces, (3) reacting with the surface, forming lubricating tribofilms to sustain high loads and high temperatures, and (4) enabling all these at minimal cost and less environmental impact by reduction or substitution of toxic additives in formulated oils. The authors have successfully synthesized layered inorganic solid nanoparticles based on molybdenum sulfide which were functionalized with dialkyldithiophosphate groups. The structural properties of novel additives were characterized using TEM and particle size analysis. These additives have demonstrated superior tribological properties under harsh friction conditions and their tribological behaviors were compared using Four-ball and Block-on-ring tests. The formed tribofilms in the wear scars were studied using XPS, Auger, and TOF-SIMS techniques. Addition of functionalized MoS_2 nanoparticles significantly improves anti-wear properties of base oil, and there is a friction reduction in comparison with samples of base oil, conventional MoS_2, or non-modified MoS_2 nanoparticles.

1. Introduction

Tribology is the science and technology of interacting surfaces in relative motion. Friction and wear are inherent when two or more surfaces interact (*1*). Wear in various forms is one of the major reasons for failure of engineering components such as gears, piston rings, engines, undercarriages, bearings, camshafts, etc., all of which are critically important for smooth and efficient operations of various industrial systems such as automotive, aerospace, mining, and machining (*2*). At the moment, in most industrialized nations (including US), the annual cost of friction- and wear-related energy and material losses is estimated to be five to seven percent of their gross national products (*3*). Also, in addition to equipment failure, tribological issues can cause significant energy consumption in the above-mentioned industrial systems mainly because of high friction and accelerated wear caused by abrasive third-body particles which may result from wear. For example, total frictional losses alone in a typical diesel engine may account for more than 10% of the total fuel energy (depending on the engine size, driving condition, etc.) (*4*). Thus, addressing such a critical lubrication problem would be significantly beneficial from both economic and environmental standpoints (*5*).

With the increasing level of complexity and severity under which modern engines operate, it is becoming more challenging to conform to environmental regulations and standards. The current trend has moved toward deriving maximum energy from an engine system with minimum input with extended component life and oil change intervals. With efficiency being paramount, friction and wear of dynamic components within the engine system and their performance in a lubricated environment are key. The presence of high temperature, pressure, particulate/wear debris along with acid condensates can have detrimental effect on engine components, such as, corrosion, wear, loss of boundary film, etc., if protection from additive system is not adequate.

Advanced lubricants can improve productivity through energy savings and reliability of engineered systems, and there is a growing interest in application of nanostructured lubrication materials stimulated by offering valuable properties and exhibiting novel lubrication behaviors.

1.1. Boundary Lubrication Regime

In general, three different lubrication stages may occur, where boundary layer lubrication is the most severe condition in terms of temperature and pressure. As stated by Hsu et al. (*6*), boundary lubrication is defined as the lubrication regime where the average oil film thickness is less than composite surface roughness and the surface asperities come into contact with each other under relative motion (*7*). This situation governs the life of critical components and to date has no complete solution.

In boundary layer lubrication, mating parts are exposed to severe contact conditions of high load, low velocity, extreme pressure, and high local temperature. For example, pressures can exceed 1-2 GPa and "flash temperature" can reach 150-300°C. In boundary lubrication regimes there is no space between mating surfaces; they are in direct physical contact. Mating parts (especially those that are typically overcoated with wear resistant coatings and with a supply of un-formulated liquid lubricants) can offer a line of defense for a few thousand to a few tens of thousands of cycles, depending on the operating conditions. However, in a boundary layer situation, at low speeds, high temperatures and/or high loads, the protective film of lubricating oil becomes very thin and ruptures (*8*). Howeer the stochastic behavior of the physical contact between the mating parts results in two problems: (1) abrupt exposure of bare metal mating surfaces, when the coating(s) are scratched and worn away, and (2) formation of abrasive particles, which are produced as a result of friction and wear, aggravating the problem, and thereby significantly affecting the life of critical components.

1.2. Lubrication Additives for Boundary Lubrication

In the state-of-the-art boundary layer lubrication, various measures have been explored to address these challenges, particularly using extreme pressure (EP) additives, e.g. zinc dialkyl dithiophosphate (ZDDP) in formulated / synthetic oils. These EP additive molecules attach to open surfaces and are activated by the heat of the friction at asperity-to-asperity contact due to sliding and rubbing and chemically react with the surface to form a glassy phosphate film, which prevents the surfaces from welding together. However, EPs are typically not uniform in the contact zone due to the stochastic nature of the contact process. Thus, their formation is sporadic, resulting in uneven and rough surfaces. Further, this problem is accelerated due to the fact that it takes finite time and temperature for activation of these molecules, and thus, interim "dry start up period" results into major wear of the mating parts (*9*). Also, the behavior of these molecules is not yet well understood, despite significant research over the last several decades leaving it still as a "black art" (*6*, *10*). Additionally, these EP additives cause harmful emissions, and pose not only engineering challenges but also environmental threats, according to new environmentally acceptable (EA) emission standards.

These critical limitations have led researchers to explore new possible additive based solutions such as solid lubricant particulates mixed with liquid lubricants (*11*). However, most of the solutions are microparticulate-based and are passive (not reactive to pressures and temperatures). Furthermore, the intricate spacing between the mating surfaces is critical (less than half micron), and limits the possibility of using micron to submicron sized solid lubricant particles dispersed in the oil. Micron size particles larger than 0.5 microns (500 nm): (1) cause a lack of effective entry of solid lubricant particles in this boundary layer resulting in clogging, (2) do not offer an effective means to deliver-activate (under extreme pressure and temperature) and bond to the dry open surface area to form transient film, and (3) fail to trap and overcoat the lethal abrasive particles in the fluid. Hence, most of the available micro particles based products have failed to address the boundary layer friction problems adequately.

The unique combination of nanostructure and active chemistry of nanolubricants, when applied as an additive, for example to engine oil, will help reduce friction and wear in engines, and increase their fuel economy. Since the US relies heavily on land based transportation of various forms, even a modest 1.0% increase in efficiency will have a significant impact on the national economy. Strategically, this helps to relieve the US dependence on foreign oil. In addition to engines, these nanolubricants may impact the following areas: (1) industrial gear boxes, sliding tracks or rails, cold and hot forging, and extrusion; (2) application of the nanolubricant as an additive or paste will facilitate friction reduction, contributing to energy saving; (3) machining and related manufacturing processes and (4) nurturing the development of nanotechnology in the US and offering a competitive edge in the international market.

2. Advanced Lubrication Additives

These critical limitations of formulated oils with ZDDP and microparticulate-based additives encouraged chemists to explore innovation based nanoscience and chemistry of active nanostructured lubricants (*12*). This chapter offers a novel nanoscience and chemistry-based approach combining the best attributes of inorganic solid lubricant particulates with those of organic based EP additive solutions (*13*) and the development of active nanostructures of inorganic dry solid lubricant integrated with active organic molecules. The active nanostructures consist of layered pressure sensitive inorganic nanoparticles of the well-known lubricant molybdenum disulfide (MoS_2) integrated with organic temperature sensitive metaloorganic molecules (dialkyl dithiophosphate groups, DDP) that functionally attached to nanoparticles. This solution meets boundary lubrication constraints and can be supplied at the mating surface asperities at micro/nano scales. These nanoparticles deliver sacrificial lubrication transfer films under high pressures and work in symbiosis with organic agents to deliver polyphosphates glassy films on the mating surfaces, thus reducing friction and wear significantly.

Boundary lubrication limitations provide guidance on several considerations necessary for the determined design of an effective and well-understood boundary lubricant system in the application zone and thus, what kind of additive nanoparticle design is necessary. In the boundary layer application region, the reaction product pathway that will provide the lines of defense against severe boundary layer situations must include both (1) hard, durable, load carrying components such as a phosphate layer, generated from metaloorganic polymeric complexes which can react with mating parts at high temperatures and (2) a softer, more easily mechanically shearable, pressure sensitive deformable (exfoliable) component such as MoS_2 nanoparticles. Depending on the severity of environmental conditions at the contact zone, suspended active nanoparticles can continuously replenish the hard and soft components, assuring a faster rate of formation and deposition than depletion/removal, thereby creating a self-repairing situation. This is an important consideration for durability and reliability of critical mechanical components over time.

2.1. Review of Nanoparticulate Lubrication Additives

Currently, state-of-the-art technologies offer passive nanostructured particulate-based dry, non-active, solid lubricants that have been synthesized primarily from MoS_2, WS_2, hexagonal boron nitride (hBN), and other related materials (*14*, *15*). These materials have been synthesized in the form of fullerene-like nanoparticles and nanotubes of these inorganic lubricants using tedious parametric windows. The synthesis is primarily performed using chemical processing routes, e.g. gas phase chemical processing, a bottom-up nanomanufacturing process. Typically, these processes use multiple steps, and are performed using environmentally and otherwise hazardous chemicals such as CS_2, P_2S_5, H_2 and H_2S with high synthesis temperatures (about 850°C) which also generate toxic by-product waste. In addition, capital costs are high and it can easily cost several millions of dollars to start a small pilot plant (*16*).

It is important to note that available technologies suffer from (1) inability to produce nanoparticle lubricants that can respond to pressure and temperature conditions during boundary layer severe friction and wear, (2) insolubility in hydrocarbons, (3) a requirement to be used as a solid lubricant component providing a low friction coefficient, (4) unpredictable synthesis processes due to requirements for high levels of controls on the process parametric window, (5) high cost and low manufacturing production volume and (6) environmentally unfriendly processing.

2.2. Development of Novel Nanoparticulate Lubrication Additives

A large amount of work has been done on nanoparticles and published by groups from Israel (Tennne et al., Weizmann Institute, Rehovot), China (Hu et al., Hefei Institute of Tribology, Hefei) and Russia (Bakunin et al., Topchiev Institute of Petrochemical Synthesis, Moscow) and others. It is evident that this research is emerging and demands timely attention due to major impact of this research in the area of boundary layer lubrication that aids energy efficiency and durability of heavy machinery and equipment and related industries (Table I).

Table I. Comparative analysis of nanoparticles synthesis and their tribological properties

Synthesis Technique	*Average Size (nm), shape, and de-agglomeration of synthesized nanoparticles*	*Their chemical character: active or passive and compatibility with base oils*	*Tribological Performance: under boundary layer conditions*	*Additional Remarks: environmentally acceptable, cost of manufacturing, and scale up*
Gas phase reactions, (*15*); Solid gas reaction between WO_3 & H_2S (*20*)	Less than 100 nm	Passive thus insoluble in other hydrocarbons	Tendency of fullerenes to clump & compress into high shear strength layers increasing COF at high load, with increasing plastic deformation of hollow particles into oval shape.	Uses harmful chemicals and high temperature, expensive, low yield
Ion modification (*21*); Microemulsion based (*22*)	DDP coated MoS_2 nanoparticles	Active but needs temperature activation, compatible	Provides good tribological properties	Uses harmful chemicals and high temperature, expensive, low yield
Inverse micellar reactions, (*23*); Micro-emulsion method (*24*)	4-8 nm, Bare MoS_2 Nanoparticles	Passive	Agglomerated	Uses harmful chemicals, expensive, low yield
Reactive ball milling (*25*)	hBN 20-100 nm	Passive	Not tested for tribological applications	Environment friendly, low cost, can be scaled up
Arc evaporation, laser ablation (*26*)	10-100 nm, WS_2, MoS_2, niobium sulfide (NbS_2), tantalum sulfide (TaS_2), zirconium sulfide (ZrS_2), nanotubes, polyhedra, agglomerated	Passive	Not tested for tribological applications	Uses harmful chemicals, expensive, low yield

Synthesis Technique	*Average Size (nm), shape, and de-agglomeration of synthesized nanoparticles*	*Their chemical character: active or passive and compatibility with base oils*	*Tribological Performance: under boundary layer conditions*	*Additional Remarks: environmentally acceptable, cost of manufacturing, and scale up*
Chemical surface modification method (*25*, *27*)	4nm, 25nm (di-n-hexadecyl dithiophosphate) DDP coated ZnS,CeF_3	Active but needs temperature activation	Improvement observed,	Difficult synthetic route, expensive and toxic reagents, F- ions are causing corrosion.
Treated nanoparticles (*28*)	40-80nm Acryl amide copolymer	Passive	Improvement observed, high	Unknown
Arc discharge in water (*29*)	MoS_2-graphite hybrid structures, 5-30 nm, polyhedra, curled nanoparticles	Passive	Not tested for tribological performance	Environmentally friendly process, low cost, poor yield
High energy ball milling in air followed by ball milling in oil and post-milled (*13*)	15-70 nm, open ended – oval shaped, deagglomerated due to polar capping layer	Active, compatible with mineral oil and other base stocks	Withstand severe friction and wear conditions, as the additives are not hollow and compacted using dynamic high-energy milling. Organic molecules surface capping provides active molecules for extreme pressure conditions and durable transfer films formation	Environmentally friendly and cost effective process. Capable of large scale production

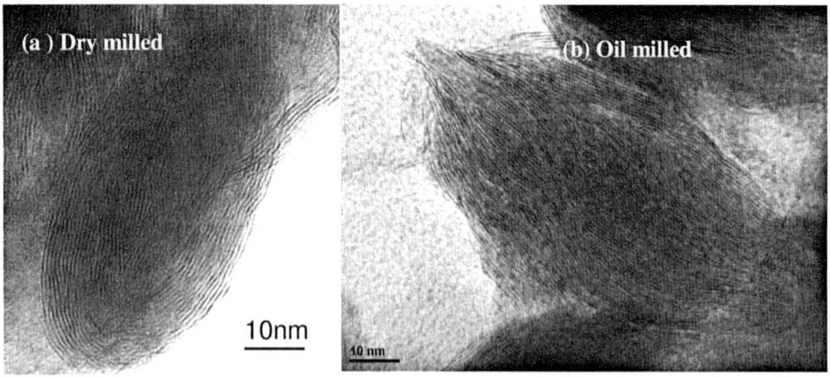

Figure 1. A typical, (a) close caged dense oval shaped nanoparticle of MoS_2 [like a football; referred by others as fullerene-like] architecture, when milled MoS_2 in dry air ambient, and (b) open-ended oval shaped nanoparticles of MoS_2 [like a coconut] architecture, when milled MoS_2 in air followed by in oil medium

In a recent research development (*17*) demonstrating the synthesis of active lubricating nanoparticles, it is strongly indicated that the current limitations of boundary area lubricants can be overcome. A processing window in a simple but unique high energy mechanical ball milling process (top-down nanomanufacturing) has been developed where one of the world's softest and most slippery materials: MoS_2 (Mohs scale 1-1.5) has been ball milled and impacted with very high impact force (Stress- 621 MPa; load- 14.9 N (*18*)) in air and oil media using hardened steel balls.

Crystalline MoS_2 nanoparticles (15-70 nm) have been synthesized with two distinct architectures, namely (1) close caged dense oval shaped [*like a football*] when milled in air medium, shown in Figure 1a, and (2) open ended oval shaped [*like a coconut*] (*19*) when milled in air followed by oil medium (Figure 1b).

The open-ended architecture in Figure 1b is of particular interest to derive pressure sensitivity due to its inherent tendency for inter-planar slippage and exfoliation, as well as its ability to supply reactive transfer films and also act as a carrier for delivering metal-organic agents derived from oil and phospholipids which are integrated using mechanical ball milling. It is believed that ball milling depletes sulfur (S) from the surface of MoS_2 particles (*30*, *31*) creating Mo terminated "active" sites for bonding of organic molecules through polar groups.

3. Advanced Lubrication

Historically, all lubricants were liquid oil-based materials. Industries have effectively implemented liquid organic materials such as mineral and synthetic oils, fatty acids and others in liquid phases as effective lubricants. However, lubrication can be also offered through solid phases of layered materials such as molybdenum sulfide (MoS_2), tungsten sulfide (WS_2), and sp-bonded carbon such as graphite, fullerene, nanotubes, etc. (*15*, *32*). Further progress has resulted in

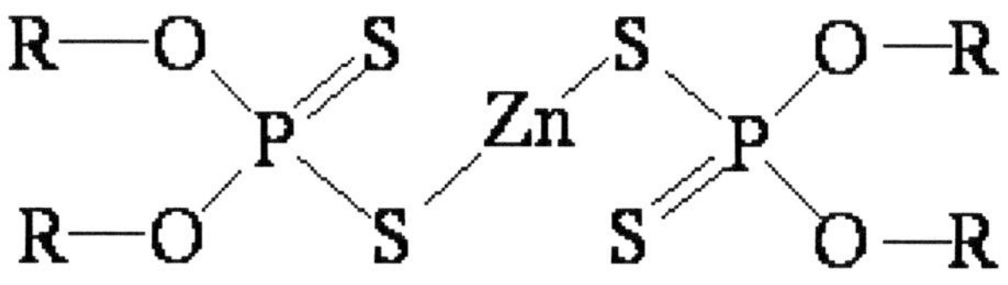

Figure 2. Zinc dialkyldithiophosphate formula (RO – alkyl groups)

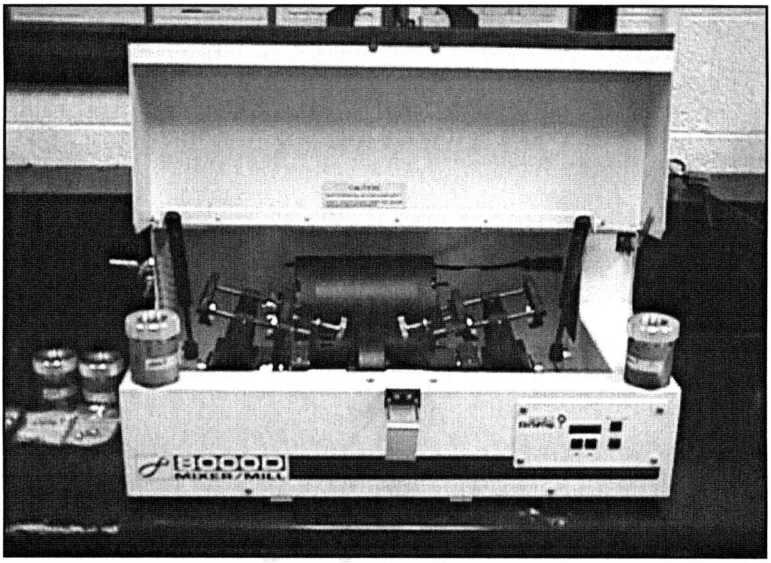

Figure 3. High energy ball mill SPEX8000D

combining inorganic and organic phases and their usage in hybrid forms such as pastes, greases, suspensions, etc.

Different lubrication additives (friction modifiers, antioxidants, viscosity modifiers, extreme pressure, dispersants, etc.) can also be added to base oils to improve lubrication performance, reliability, productivity and energy savings in engineering systems. One of the most commonly and extensively used extreme pressure and antioxidant additives is zinc salt of dialkyl dithiophosphate groups (ZDDP) schematically shown in Figure 2. This additive provides unique lubrication in boundary regime through decomposition at friction sites and formation of polyphosphate glassy film isles that decrease wear and friction. However, its application in lubrication oils is strictly controlled by EPA due to harmful effect on environment and poisoning of catalytic sites in engine exhaust converters. New environmental regulations require significant reduction of sulfur and phosphorus concentrations in lubrication oils (*33*). While this additive needs to be removed from the market, there is no significant competitor which can replace it currently. The environmental challenges could be resolved by application of molybdenum sulfide (MoS_2) nanoparticles integrated with DDP groups in oil that will have less environmental impact by reduction of toxic DDP additive concentration in formulated oils.

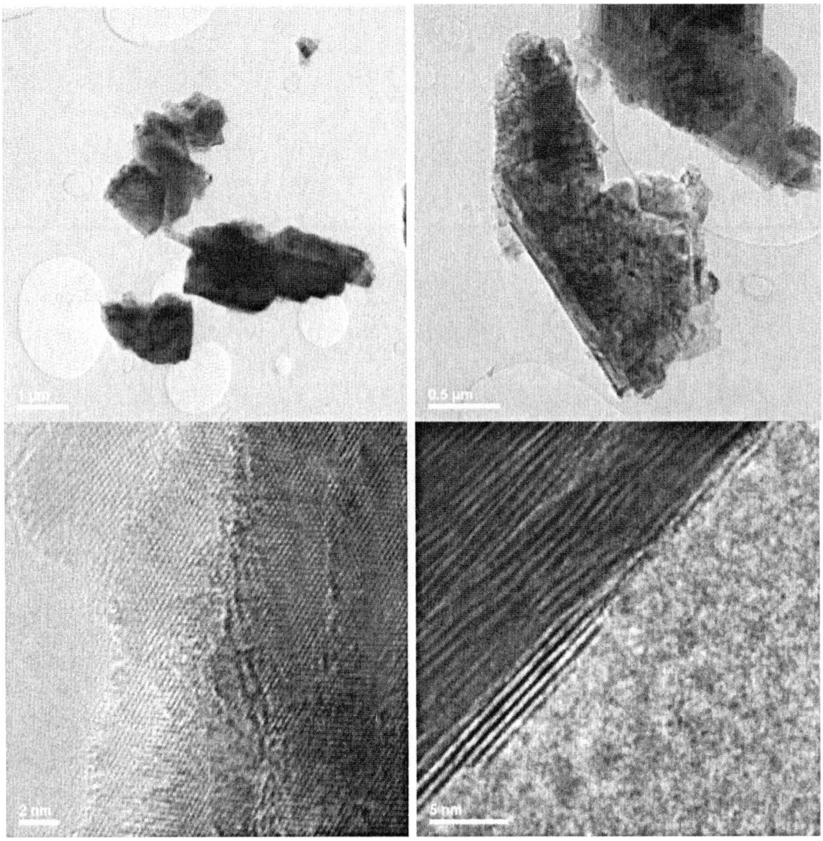

Figure 4. TEM/HRTEM graphs of commercially available MoS_2 (Alfa Aesar) and lamellar structure of MoS_2 planes

In the past few years, a breakthrough has delivered the ability to synthesize layered inorganic solid phase particles of nanometer size capped with organic molecular structures, particularly as an additive to the base oil (*34*, *35*). The tribological design of these new class materials has allowed this incorporation of lubrication performance and other functions causing important scientific challenges and opportunities in synthesizing and the understanding of these new classes of inorganic-organic nanoengineered lubricants.

Thus, the development of experimental procedures for the synthesis of nanoparticles of different chemical compositions, sizes/shapes and controlled dispersion is essential for its advancement. As far as the synthesis of nanoparticles is concerned, a number of chemical methods exist in the literature that use toxic chemicals in the synthesis protocol, which raises great concern for environmental and sustainability reasons.

The described below is a hybrid chemo-mechanical milling process for molybdenum sulfide integrated with second element component to realize multi-component nanoparticle additives, for example, dialkyldithiophosphate (DDP)-functionalized molybdenum sulfide nanoparticles.

Figure 5. TEM graphs of MoS_2 sample prepared by dry milling process and agglomeration of primary crystallites into secondary particles

4. Hybrid Chemo-Mechanical Milling

The preparation method for nanoparticulate additive is based on high energy milling of MoS_2 involving fracture and extrusion mechanism. The milling process involves the shearing of the bulk particles to form smaller particles. The shape of these particles strongly depends on the milling process and the timelength of processing.

Ball milling (SPEX 8000D milling equipment, Figure 3) was performed in stainless steel vials in selected media, namely air for dry milling, liquid chemical for wet milling, and a hybrid of air and liquid, i.e., milling in air followed by addition of chemical at room temperature.

The commercially available MoS_2 was purchased from Alfa Aesar and used for milling as received (Figure 4). This material was characterized as slightly agglomerated particles of irregular shapes with average sizes of 4 μm, and

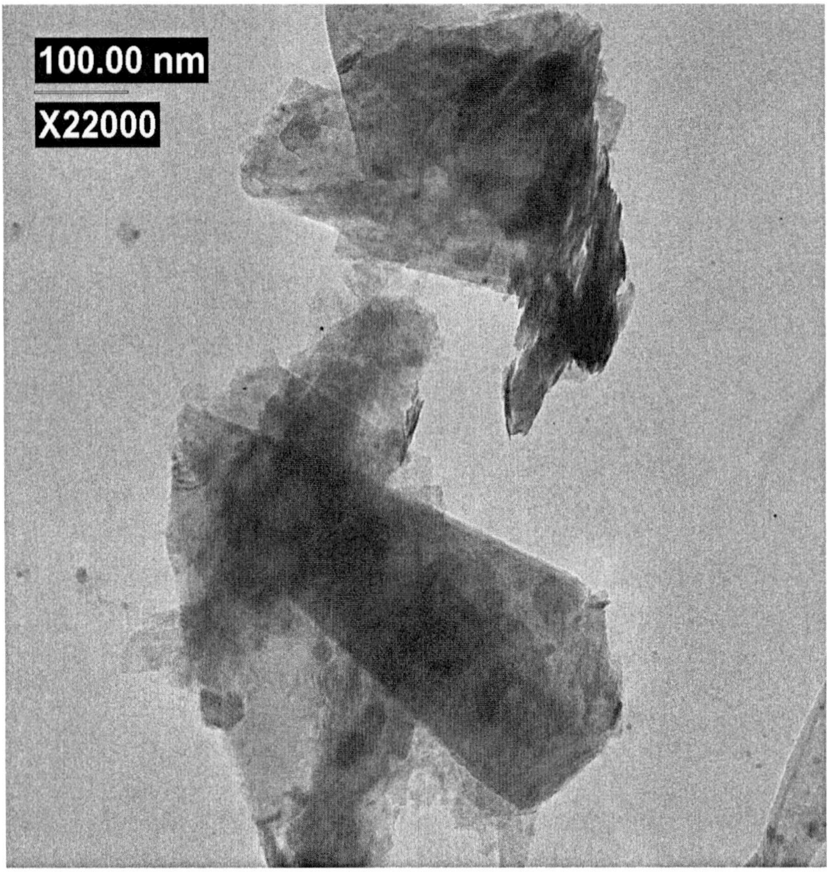

Figure 6. TEM graphs of MoS_2 sample of thin platelets exfoliated during wet milling process

confirmed by data provided by Alfa Aesar (sieves method, 3.5 μm) and data from the particle size analysis from Horiba particle size analyzer (4.5 μm mean).

Zinc dialkyl dithiophosphates from Rhein Chemie and Infinium as a source of DDP groups were used as the starting materials for all the synthesis processes. A constant ball-to-powder mass ratio, commonly called charge ratio, of 2:1 was selected. During the ball milling, the vials (stainless steel) containing MoS_2 powder and balls (stainless steel) were moved in three orthogonal directions in a figure "8" pattern to assure homogeneous processing. To assure cleanliness and minimize cross contamination, before and after milling, the vials and the balls were cleaned using standard laboratory cleaning agents starting with soft soap followed by clean water, and then, acetone and isopropanol sonication steps. After milling, the samples were transferred in a glove box and stored in closed desiccators under room temperature and pressure. Powder samples, before and after processing, were analyzed for morphology and structure using high resolution transmission electron microscopy (TEM; JEOL JEM-2100F, operating voltage: 200kV).

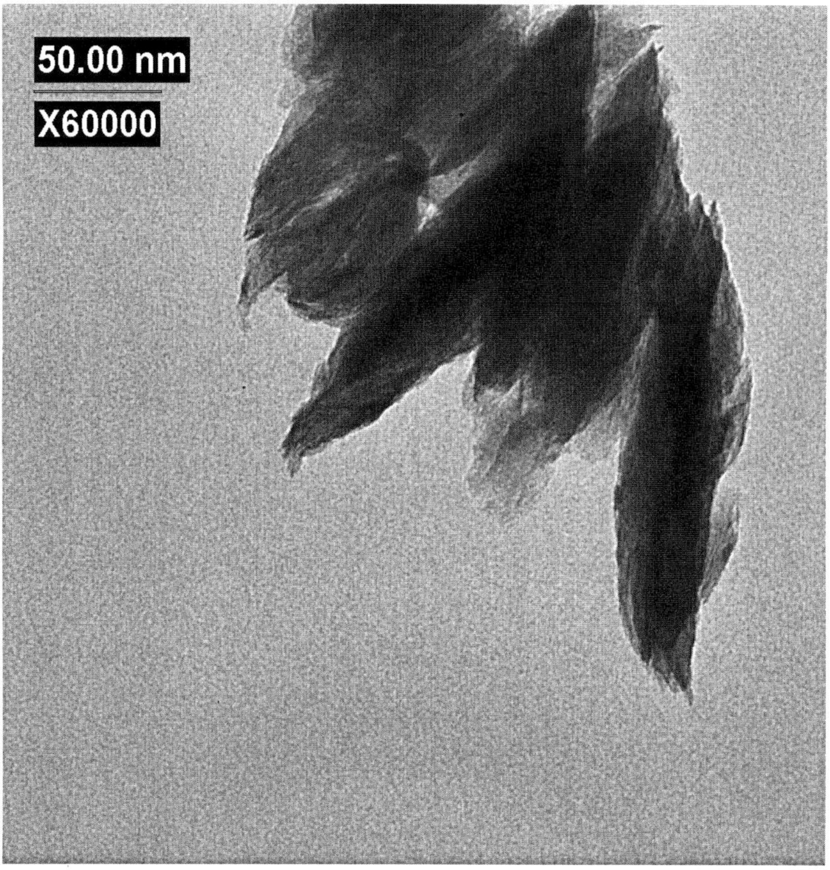

Figure 7. TEM graphs of MoS_2 sample prepared by hybrid milling process with dry milling step followed by wet milling step

4.1. Dry Milling Process

MoS_2 samples were dry milled for different time lengths of processing (12 hour intervals). The milled samples were studied and compared with commercially available MoS_2. TEM of dry milled samples include big agglomerates of small primary particles (crystallites) (Figure 5). From TEM analysis it was confirmed a significant decrease of crystallite sizes during dry milling. However, these particles were fused caused by dry milling. The fractured small primary particles (crystallites) were aggregated into larger agglomerates (secondary particles) during the milling process due to an increase in the particle surface energy and the tendency of nanoparticles to decrease surface energy through agglomeration. The milled samples showed big agglomerates of crystallites of non-uniformed shapes, and the sizes of these agglomerates were significantly larger than the initial particles of commercial MoS_2.

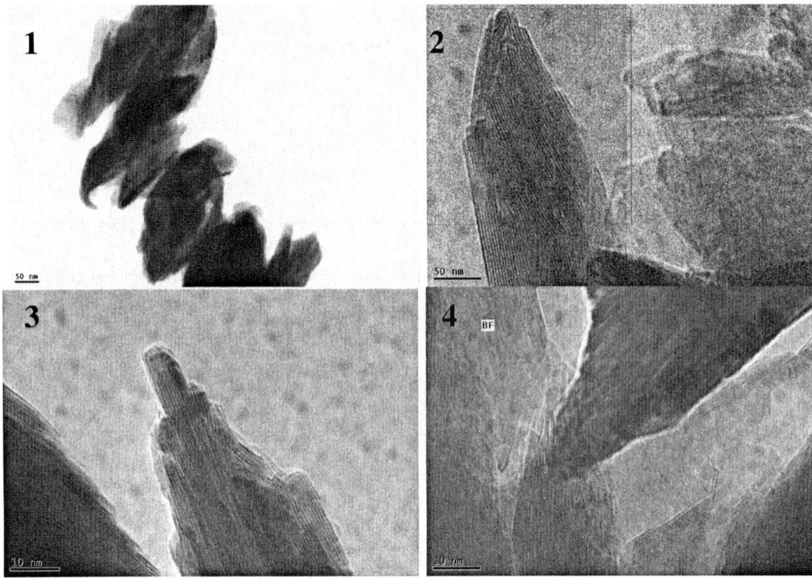

Figure 8. TEM graphs of milling mechanism for MoS_2 sample prepared by hybrid milling process (1 through 4 mechanism scenarios)

During dry milling, two process mechanisms were active, simultaneously: 1) the fracture of the initial micro particles and 2) fusing of the particles in big agglomerates. The latter process dominated over the former and leads to the formation of big agglomerates of small particles (*36*). The study of the shapes for these small particles showed that dry milling for 12 and 24 hours have caused high dispersion of small as well as big crystallites, probably due to insufficient time for complete milling of big particles. While the 36 hours of milling showed crystallites of uniform size and shape across the sampled milled powder.

4.2. Wet Milling Process

MoS_2 samples were wet milled with zinc dialkyldithiophosphate for different time length of processing (12 hour intervals). A liquid-to-powder weight ratio of 1:1 was selected. The milled samples were studied and compared with unmilled commercially available as well as dry milled molybdenum sulfide particle samples. The wet milled samples contained particles of different shapes in comparison with dry milled samples and were not strongly agglomerated into big secondary particles (Figure 6).

Wet milling process exfoliated planar layers of MoS_2 particles and formed separated film-like platelets. The shapes of crystallite are similar to platelets and are very thin in one dimension, so when they stand on the edge they show distinct needle shapes in TEM graphs. During wet milling, there were no significant decreases in size, but oil media helped to keep them from aggregation.

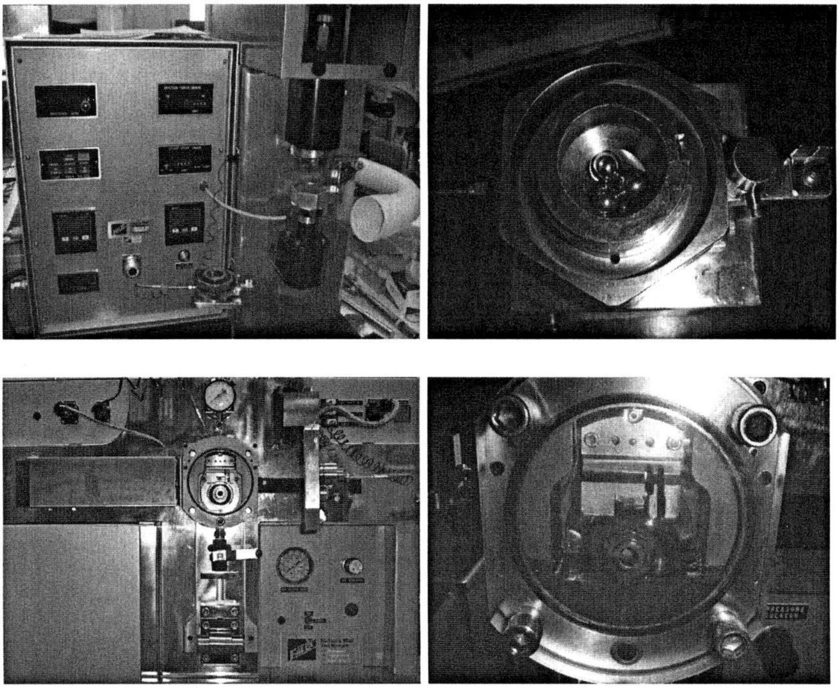

Figure 9. Four –ball tribometer and testing cell (top images) and Block-on-ring tribometer and testing cell (bottom images)

4.3. Hybrid Milling Process

The hybrid milling process was a sequential combination of dry milling followed by a wet milling process (milling in liquid media). The combination of high energy milling to decrease the particle size in dry air ambient followed by milling in liquid ambient to prevent the particles from fusing, delivered samples containing ellipsoidal-shaped nanoparticles dispersed uniformly (Figure 7).

The hybrid milling showed formation of uniform ellipsoidal particles with morphology resembling, "extruded" ends of the particle. The shape of particles may be in favor of their movement in the liquid media (similar to a fish). With longer milling time, particles have an increased extrusion like morphology, and in some cases are elongated significantly.

4.4. Milling Mechanism of Hybrid Milling Process

Detailed information on the milling mechanism for dry and wet steps, and hybrid process in full was described elsewhere in a separate manuscript prepared for publication. In general, the 8-shaped rotation of vials, along with the path and interaction of balls in the vials causes the formation of spherical, elongated, open structure particles due to the squeezing and extrusion mechanisms. Initial particles underwent four mechanism scenarios during the milling process (Figure 8): (1)

mechanical deformation of planar layers in lamellar structure on the MoS_2 on the tip of the particle, (2) breakage of the surface layers which were sticking out on the end of particle and extrude them to form squeezed endings, (3) conical shape of the particle tip with a long base interlayer and shorter layers closer to the surface caused by grinding and extrusion by milling media (vial walls and balls), and (4) formation of interlayered defects and layer dislocations at the particle endings.

Three zinc dialkyldithiophospate commercial additives with different primary and secondary alkyl groups were tested as a liquid phase in wet milling that is expected to form a continuous film on the surface of MoS_2 nanoparticles acting as a capping agent, protecting nanoparticles from aggregating, but also giving additional lubrication properties to MoS_2 nanoparticles. Despite different viscosity and chemical composition of ZDDPs, particles have similar sizes and shapes after hybrid milling.

The same hybrid approach can be used for the synthesis of the multi-component system when one of the components is liquid material. Hybrid milling with typical commercial additives favors: 1) formation of nanoparticles of uniformed ellipsoidal shapes and well-dispersed in the liquid media, 2) improvement in lubrication efficiency as anti-ware and extreme pressure additives, 3) solving the dispersion problems related to aggregation and settling of MoS_2 nanoparticles in base oil.

5. Tribological Testing

To evaluate tribological performance of prepared nanoparticles and compare it with performance of commercially available MoS_2 (starting material for milling) and just pure ZDDP additive in base oil, samples were tested using Four-ball and Block-on-ring tribometers (Figure 9). The samples were dispersed in base oil (non-formulated poly-α-olefin PAO, 10 cSt at 100°C) at 1.0 % concentration by weight of MoS_2. In the case of just pure ZDDP in oil, concentration of ZDDP was selected the same as was selected for ZDDP in hybrid milled MoS_2 - ZDDP sample to compare their performances.

Tribological properties of following samples were studied and compared on these bases: 1) pure base oil; 2) base oil with commercially available MoS_2 microparticles; 3) base oil with just pure ZDDP additive; 4) base oil with MoS_2 nanoparticles prepared by dry milling; 5) base oil with MoS_2 nanoparticles prepared by hybrid milling with PAO oil; and 6) base oil with MoS_2 nanoparticles prepared by hybrid milling with ZDDP.

The tribological behaviors (wear scar and coefficient of friction) of all samples were studied using a Four-ball tribometer. The wear scar diameter (WSD) and coefficient of friction (COF) were examined by following the ASTM standard D4172 with the test conditions: 75°C, 1200 rpm, AISI 52100 steel balls with a 12.7 mm diameter and a hardness of HRC 64–66. The WSD and the COF were measured within the test duration of 60 minutes and under a load of 40 kg, a maximum contact pressure of 4.6 GPa. The wear scar diameters of specimen balls were measured by optical microscope.

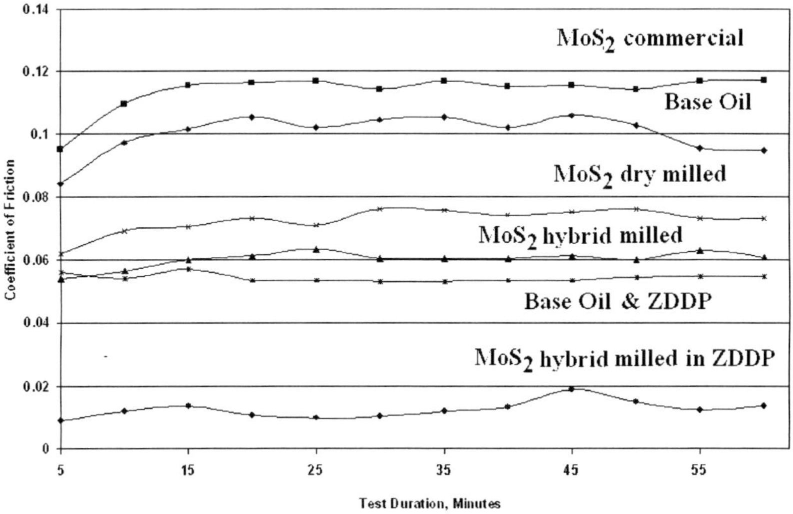

Figure 10. Four-ball test coefficient of friction

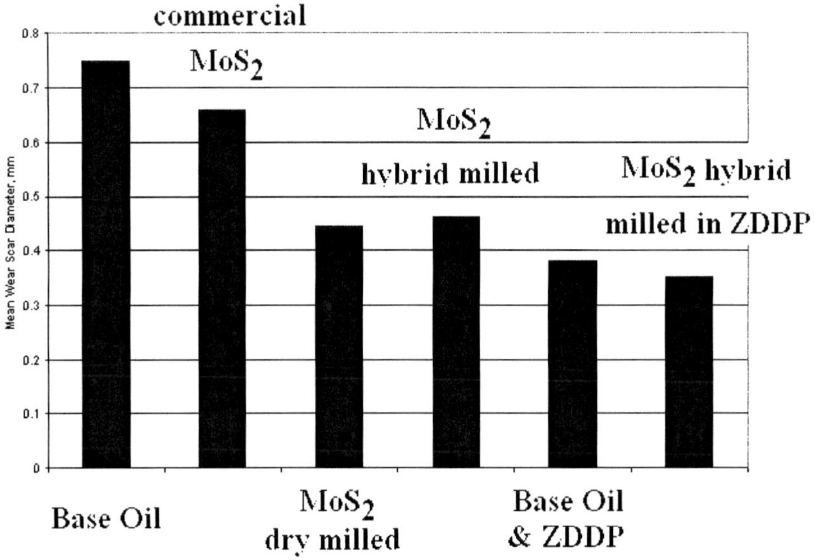

Figure 11. Four-ball test wear scar diameter (mm)

The wear scar volume (WSV) and coefficient of friction (COF) were examined by Block-on-ring test following the Caterpillar standard. Selection of the Block-on-ring setup for this study was specifically made with consideration for the potential applications, where extreme contact pressure is a major concern. The test conditions were 25°C, ring speed was 500 rpm, ring and block material was AISI-E52100 (64–66 HRc). The WSV and the COF were measured within the test duration of 30 minutes and under a load of 34 kg (75 lb). The wear scar volumes of specimen blocks were measured by optical profilometer.

5.1. Four-Ball Test Tribological Testing

This test was performed to compare mean scar diameter (Figure 10) and the coefficient of friction (Figure 11) for different samples. The background testing of oil and commercial MoS_2 were also performed. Samples with the lowest coefficient of friction and smallest wear scar were potential candidates to provide better lubrication toward anti-wear and extreme pressure applications. MoS_2 samples prepared by dry milling and hybrid milling gave smaller mean wear scars in comparison with the commercial MoS_2 sample. The analysis of coefficient of friction for sample prepared with zinc dialkyl dithiophosphate (ZDDP) by hybrid milling revealed that this sample showed extraordinarily low friction. The value of COF for this sample is significantly lower than COF values for pure ZDDP in oil and hybrid milled MoS_2 in oil. The DDP-functionalized MoS_2 hybrid milled sample had a synergetic lubrication effect between nanoparticles and dialkyldithiophosphate groups to lower the coefficient of friction value.

The addition of second component liquid phase (ZDDP) to MoS_2 during hybrid milling gave the smallest wear scars in comparison with MoS_2 commercials and dry milled samples. The presence of dialkyldithiophosphate groups in the sample favored lower friction and smaller wear scars.

5.2. Block-on-Ring Tribological Testing

Testing of samples was done at heavy loading on a block (34 kg) and a low ring rotation (500 rpm) to achieve a high wear force. The addition of MoS_2 based additives to the oil was supposed to lubricate better and decrease the wear between block and ring. The samples with more efficient lubrication properties at high loading protected the surface of the block effectively from aggressive wear, resulting in smaller wear volume of surface scar. By measuring this wear volume for different samples, we have assessed lubrication efficiency of different formulations.

Several prepared samples revealed advanced lubrication activity and a significant decrease in wear volume on the blocks (Figure 12). The samples prepared by hybrid milling with ZDDP showed extraordinary results again. The hybrid milled MoS_2 showed a significant improvement in lubrication and decreasing values not only for wear volume but also for COF. The chemical analysis of formed tribofilms on the specimen during tribotesting could explain the interaction between molybdenum sulfide and DDP groups and will be discussed in the next section.

All tribotesting samples were freshly prepared by the same operation procedure (sample of 1% weight by MoS_2 in oil was stirred for 15 min and sonicated for 10 min), and these samples were directly loaded into tribometers. The issue of the long-term settling of MoS_2 nanoparticles was not studied. However, the settling is an important parameter and needs to be addressed.

For these studies only 1:1 weight ratio of ZDDP to MoS_2 was used and the amount of the additive to oil was kept constant for 1% weigtt by MoS_2. It meant that in hybrid ZDDP/MoS_2 sample had not only 1% of MoS_2 but also 1% of ZDPP. The different ratios and concentrations of ZDDP and MoS_2 components

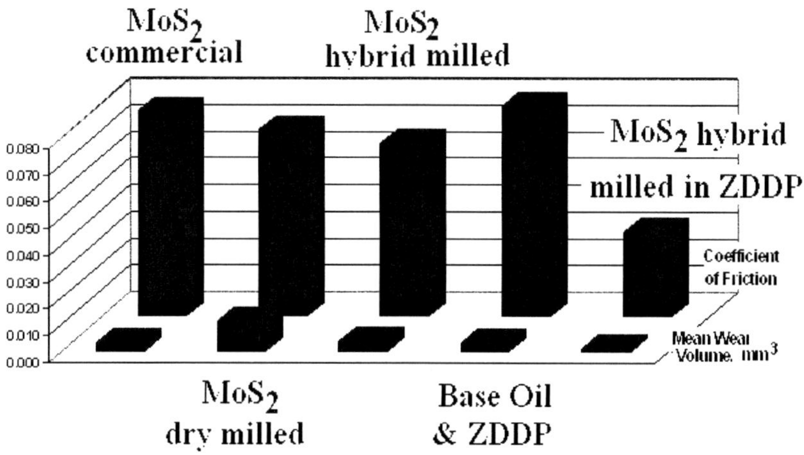

Figure 12. Block-on-ring coefficient of friction and wear scar volume (mm^3)

were studied along with their performance in base oils, formulated industrial oils, and formulated oils with low concentration of free ZDDP and will be presented elsewhere.

The general research goals would be to reduce the concentration of DDP in formulated industrial oils while keeping the hybrid milled DDP-functionalized MoS_2 nanoparticles as a source of thiophosphate groups to form polyphosphate film at extreme pressure friction points and to exfoliate molybdenum sulfide layers to form tribofilms for high lubrication performance.

6. Tribofilm Analysis

The MoS_2 nanoparticles formed low friction tribofilms through the shearing of nanoparticle layers at the surface friction points during tribotesting which decreased the coefficient of friction and wear scar of tested specimen. The DDP-functionalized MoS_2 nanoparticles possessed much higher friction reduction activity and better results with respect to wear reduction due to combined performance of molybdenum sulfide and dialkyldithiophosphate groups during tribological testing. The interaction between DDP and MoS_2 during the friction process was studied through the analysis of formed tribofilms.

The morphology and chemical composition of tribofilms (Figure 13) generated during tribotesting for different lubricant samples were analyzed using the following techniques: a) scanning electron microscopy (SEM), b) Auger electron spectroscopy (AES), c) X-ray photoelectron spectroscopy (XPS), and d) time-of-flight secondary ion mass spectrometry (TOF-SIMS).

6.1. Auger Analysis of Tribofilms

Tribofilms formed during the Four-ball test on specimen balls and during the Block-on-ring test on specimen blocks were analyzed using Auger spectroscopy.

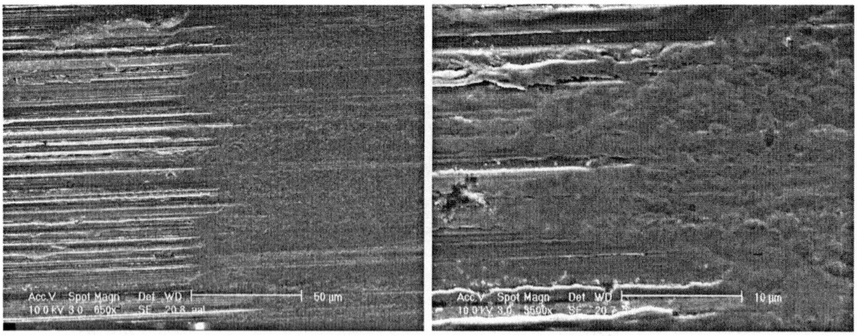

Figure 13. SEM graphs of tribofilms formed on block specimen (scale bar: left – 50 µm and right - 10 µm)

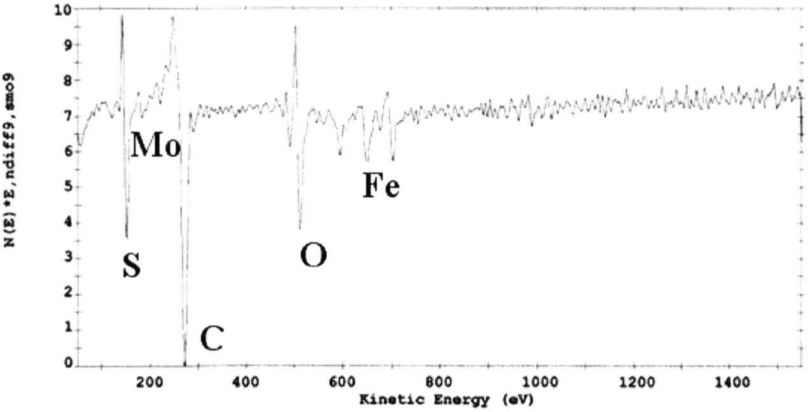

Figure 14. Auger analysis of hybrid milled MoS_2 tribofilm in wear area of ball specimen before Ar spattering

The Auger analysis was specifically selected to study the elemental analysis of tribofilm and the mapping each element to assign their distribution on the wear scar. The Physical Electronics model PHI 660 Scanning Auger Microprobe (SAM) was used at Center for Microanalysis of Materials (UIUC). It had elemental mapping with high spatial resolution and is capable of performing scanning electron microscopy (SEM).

The Auger analysis of hybrid milled MoS_2 (milled without ZDDP) tribofilm (Figure 14) was used to scan for elements in the wear surface and mapping them on the ball specimen after the Four-ball test (Figure 15).

The initial Auger scan showed a presence of residual oil (C and O elements) on the wear surface, however almost all oil was removed from the wear surface after several steps of argon sputtering. It is clear from mapping scans that the MoS_2 nanoparticles (Mo and S elements) that got between the mating surfaces were trapped in the asperities and grooves of the wear scar during friction testing. They formed tribofilms by sheering and exfoliation of MoS_2 layers and caused a decrease in the coefficient of friction and in wear. It was also noticed that some molybdenum sulfide has oxidized into molybdenum oxide during tribotesting.

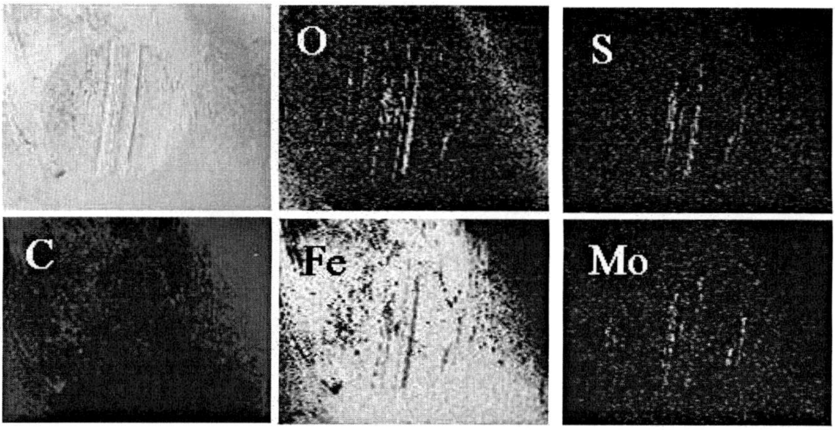

Figure 15. Auger analysis elemental mapping for hybrid milled MoS_2 tribofilm on the ball specimen

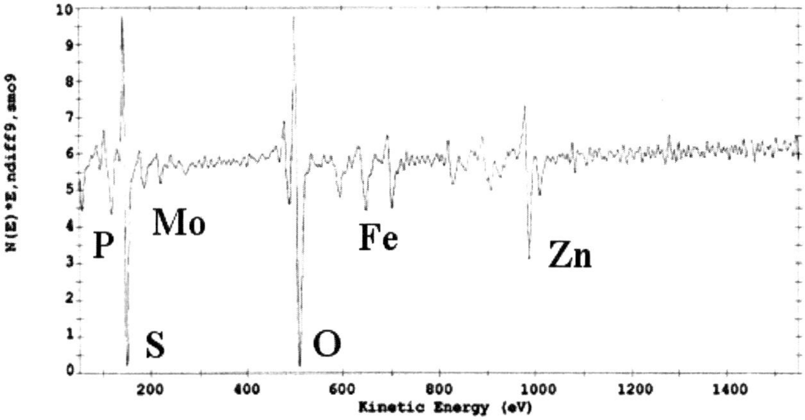

Figure 16. Auger analysis of hybrid milled MoS_2 –ZDDP tribofilm in wear area of block specimen after argon sputtering

The Auger analysis of hybrid milled DDP-functionalized MoS_2 (milled with ZDDP) tribofilm (Figure 16) was used to scan for elements in the wear surface and mapping them on the block specimen after the Block-on-ring test (Figure 17). The initial Auger scan also showed a presence of residual oil (C and O elements) on the wear surface and this oil was removed from the wear surface by Ar sputtering. In this case, MoS_2 nanoparticles (Mo and S elements) were not only trapped in the asperities and grooves of the wear scar, but were also well distributed in the tribofilm on the surface of the wear scar. The tribofilm in the wear area also showed a large presence of polyphosphate (P and O elements), and zinc sulfide/oxide from decomposition of zinc dialkyl dithiophosphate. The lubrication role of zinc compounds formed during tribotesting was not fully understood, as they were embedded into the polyphosphate tribofilm and had insignificant influence on the friction. The best scenario for future studies would be to use molybdenum-based dialkyl dithiophosphates to prevent the presence and formation of zinc compounds.

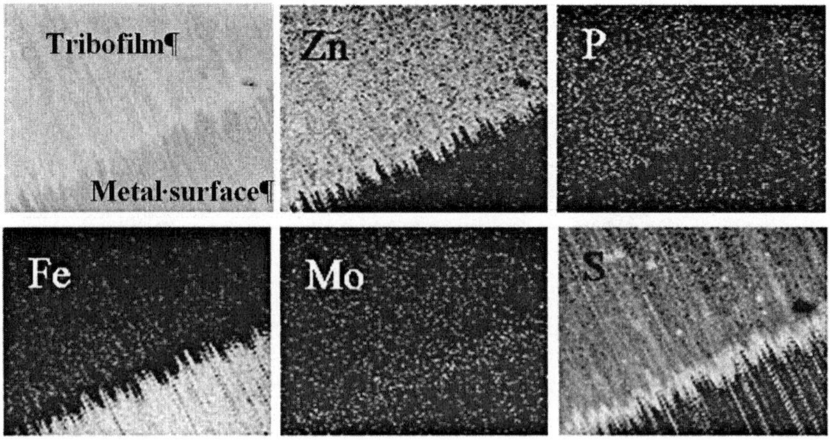

Figure 17. Auger analysis elemental mapping for hybrid milled MoS_2 –ZDDP tribofilm on the block specimen

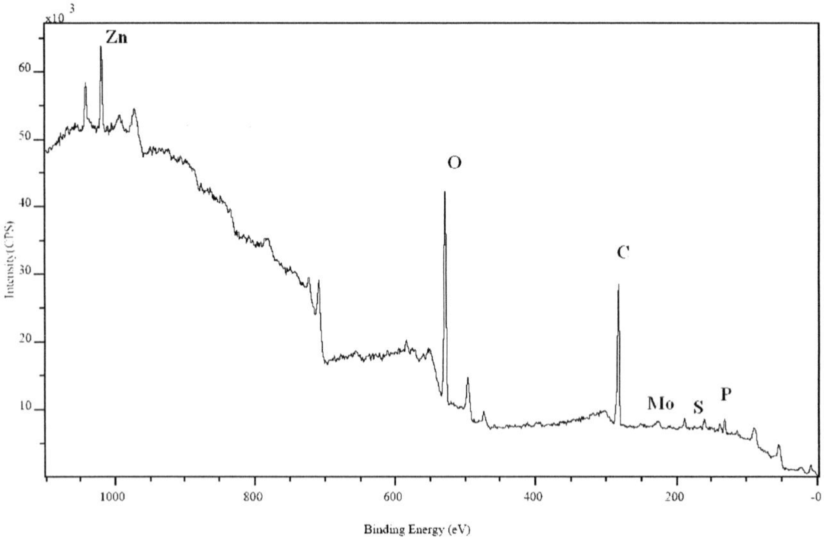

Figure 18. XPS analysis of hybrid milled MoS_2 –ZDDP tribofilm in wear area of block specimen

The hybrid milled DDP-functionalized MoS_2 tribofilm showed an extraordinary decrease in friction and wear scar, and it could be attributed to the formation of a very smooth glassy film that filled all asperities and grooves of the specimen surface (Figure 17).

6.2. XPS Analysis of Tribofilms

The XPS analysis was specifically chosen to study oxidation states of elements and chemical composition of formed tribofilms for the hybrid milled

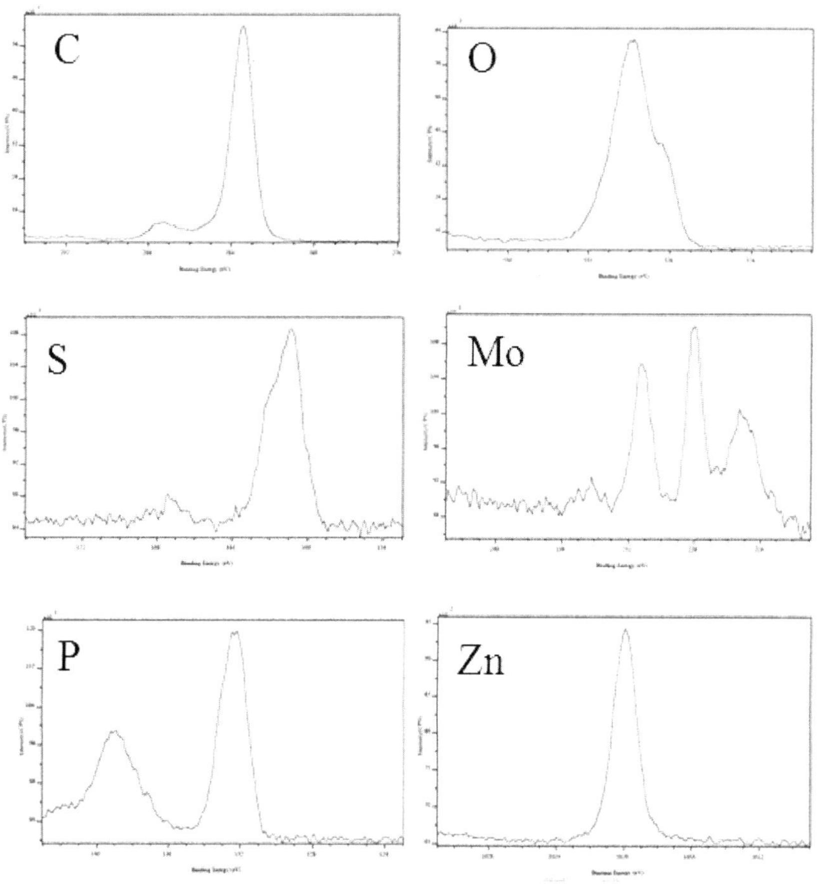

Figure 19. XPS elemental analysis (C, O, S, Mo, P and Zn elements) for hybrid milled MoS_2 –ZDDP tribofilm on the block specimen

DDP-functionalized MoS_2 sample. The Kratos Axis ULTRA high performance X-ray photoelectron spectrometer for photoelectron spectroscopy (XPS) and parallel imaging (imaging XPS) was used in this analysis. The system had a differentially pumped argon ion gun for sputter cleaning and depth profiling.

The full XPS scan showed a presence of molybdenum, sulfur, phosphorus, zinc, carbon, and oxygen elements on the wear surface (Figure 18). The presence of carbon and some of oxygen (C and O elements) was assigned to residual oil on the wear surface, as these peaks were decreasing after argon sputtering. More detailed analysis of binding energies values for element peaks were performed to analyze the oxidation states and chemical compositions of the tribofilm (Figure 19). Molybdenum binding energy was matching the oxidation state (IV) and chemical presence in the form of disulfide MoS_2 and/or dioxide MoO_2. This confirms a partial oxidation of molybdenum sulfide to oxide which was also noticed for Auger analysis. The sulfur analysis showed a partial presence in form of sulfate and sulfide. The sulfate was formed by the partial oxidation of sulfide into sulfate during tribotesting. It was difficult to assign sulfate species

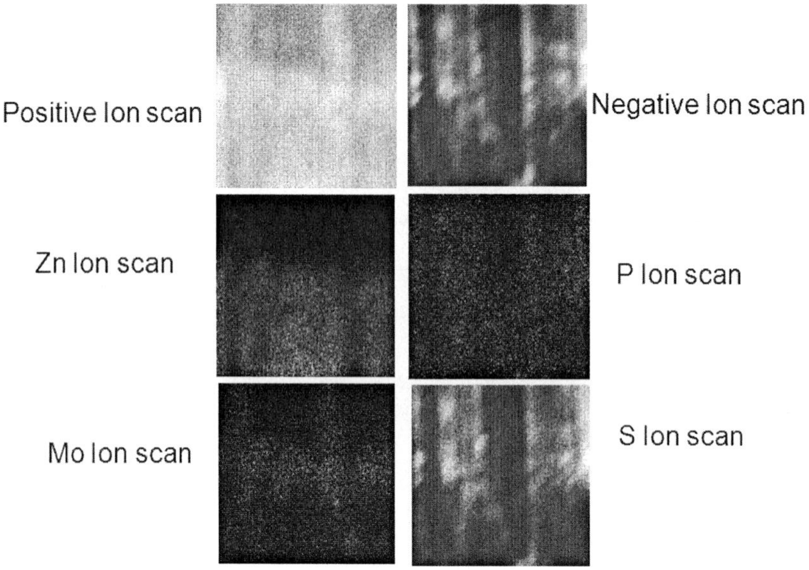

Figure 20. TOF-SIMS analysis ion mapping for hybrid milled MoS_2 –ZDDP tribofilm on the block specimen

to molybdenum or zinc, however both molybdenum sulfate and zinc sulfate could be present in the tribofilm. The most important result was obtained for phosphorous analysis, where phosphorous binding energies were assigned to the presence of polyphosphate groups (139 eV) together with the presence of thiophosphate groups. There is a strong evidence that DDP-functionalized MoS_2 nanoparticles delivered dialkyl dithiophosphate groups to wear points between mating surfaces during friction where these groups decomposed under high pressure and temperature to form protective polyphosphate tribofilm integrated with MoS_2.

6.3. Time-of-Flight Secondary Ion Mass Spectrometry (TOF-SIMS) Analysis of Tribofilms

The TOF-SIMS analysis was selected to complement the XPS and Auger findings for the tribofilm morphology and composition by using Ion Mass Spectroscopy. The Physical Electronics PHI Trift III was used for the analysis of positive and negative elemental ions at the tribofilm surface (Figure 20).

From TOF-SIMS analysis, it was found that polyphosphate tribofilm was formed at the wear points to protect from wear scarring and molybdenum nanoparticles were integrated with tribofilm but also trapped in the asperities and grooves of the metal specimen block. This combination of phosphate tribofilm formation at wear points and the presence of molybdenum sulfide nanoparticles at the friction points gave advanced lubrication properties to DDP-functionalized MoS_2 nanoparticles and extraordinary lubrication performance during tribotesting.

7. Conclusions

Ellipsoidal MoS_2 nanoparticles with squeezed endings were synthesized by hybrid chemo-mechanical milling by mechanical deformation, exfoliation, and extrusion mechanisms. The hybrid milling process is a combination of dry milling followed by a wet milling process (milling in liquid media). The combination of high energy dry milling to decrease the particle sizes and oil milling process to de-agglomerate them delivered the best samples uniform in the size of nanoparticles.

Tribological properties of MoS_2 samples were studied using Four-ball and Block-on-ring tests. The tribotesting results confirmed that hybrid milled MoS_2 nanoparticles significantly improve the anti-wear properties of base oil, and there is a friction reduction in comparison with that of base oil, commercial MoS_2, or non-modified MoS_2 nanoparticles. Addition of 1% by weight of hybrid milled nanoparticles to base oil caused a significant reduction in the coefficient of friction (COF) and significant decrease in the size of wear scar.

The DDP-functionalized MoS_2 nanoparticles significantly improved anti-wear properties and delivered a friction reduction with twice less of DDP groups in the base oil. The tribofilm formation was studied using XPS, Auger, and TOF-SIMS techniques and demonstrated synergetic interaction of molybdenum sulfide and dialkyldithiophosphate groups during friction and showed the formation of tribofilms of polyphosphates with inclusions of MoS_2 nanoparticles that decreased the coefficient of friction and resultant wear.

Acknowledgments

The authors are grateful to the Arkansas Analytical Laboratory (AAL) and Materials and Manufacturing Research Laboratories (MMRL) at the University of Arkansas in Fayetteville, AR for SEM and particle size analysis facilities, Technical Center of Caterpillar Inc. in Peoria IL for tribological facilities, Center of Materials Microanalysis at the University of Illinois in Urbana-Champaign, IL for TEM, XPS, Auger, and TOF-SIMS facilities.

This study was financially supported by Caterpillar Inc. and U.S. Department of Energy (National Energy Technology Laboratory, Office of Freedom CAR & Vehicle Technologies) under grant DE-PS26-07NT43103-02. Tribofilm analysis was carried out in part in the Frederick Seitz Materials Research Laboratory Central Facilities, University of Illinois, which are partially supported by the U.S. Department of Energy under grants DE-FG02-07ER46453 and DE-FG02-07ER46471.

References

1. Bartz, W. J. *Wear* **1978**, *49* (1), 1.
2. Kovalev, E. P.; Ignatiev, M. B.; Semenov, A. P.; Smirnov, N. I.; Nevolin, V. N.; Fominskii, V. Yu. *J. Friction Wear* **2004**, *25* (3), 78.
3. Jost, P. *Tribol. Lubr. Technol.* **2005**, *61* (10), 18.

4. Fenske, G.; Robert, E.; Layo, A.; Erdemir, A.; Eryilmaz, O. *Parasitic Energy Loss Mechanisms Impacton Vehicle System Efficiency*; Argonne National Laboratory, 2006 (http://www1.eere.energy.gov/vehiclesandfuels/pdfs/hvso_2006/07_fenske.pdf).
5. Koelsch, J. R. *Manuf. Eng.* **1997**, *118* (5), 6.
6. Hsu, S. M.; Gates, R. S. *Tribol. Int.* **2005**, *38* (3), 305.
7. Matveevskii, R. M. *Tribol. Int.* **1995**, *28* (1), 51.
8. Bakunin, V. N.; Suslov, A. Yu.; Kuzmina, G. N.; Parenago, O. P. *J. Nanopart. Res.* **2004**, *6*, 273.
9. Ishibashi, A.; Ezoe, S.; Tanaka, S. *IMechE Conference Publications (Institution of Mechanical Engineers)* **1987**, 845.
10. Spikes, H. *Tribol. Lett.* **2004**, *17* (3), 469.
11. Choudhary, R. B.; Jha, M. K. *Lubr. Sci.* **2004**, *17* (1), 75.
12. Li, B.; Wang, X.; Liu, W.; Xue, Q. *Tribol. Lett.* **2006**, *22* (1), 79.
13. U.S. Patent Applications Publ. US 2008/031211 A1 and US 2008/0234149 A1.
14. Chen, Y.; Chadderton, L. T.; Gerald, J. F.; Williams, J. S. *Appl. Phys. Lett.* **1999**, *74* (20), 2960.
15. Tenne, R.; Margulis, L.; Genut, M.; Hodes, G. *Nature (London)* **1992**, *360* (6403), 444.
16. Apnano News Release. http://www.apnano.com.
17. Verma, A.; Jiang, W.; Abu Safe, H.; Brown, W. D.; Malshe, A. P. *Tribol. Trans.* **1992**, *51*, 673.
18. Schaffer, G. B.; Forrester, J. S. *J. Mater. Sci.* **1997**, *32*, 3157.
19. Pedraza, F.; Cruz-Reyes, J.; Acosta, D.; Yanez, M. J.; Avalos-Borja, M.; Fuentes, S. *J. Phys.: Condens. Matter* **1993**, *5* (33a), A219.
20. Rapoport, L.; Feldman, Y.; Homyonfer, M.; Cohen, H.; Cohen, S.; Tenne, R. *Tribol. Series* **1999**, *36* (Lubrication at the Frontier), 567–573.
21. Zhang, Z. J.; Zhang, J.; Xue, Q. Ji. *J. Phys. Chem.* **1994**, *98* (49), 12973.
22. Bakunin, V. N. *Neftekhimiya* **1994**, *41* (4), 317.
23. Wilcoxon, J. P.; Samara, G. A. *Phys. Rev. B* **1995**, *51* (11), 7299.
24. Boakye, E.; Radovic, L. R.; Osseo-Asare, K. *J. Colloid Interface Sci.* **1999**, *163* (1), 120.
25. Chen, Y.; Chadderton, L. T.; Gerald, J. F.; Williams, J. S. *Appl. Phys. Lett.* **1999**, *74* (20), 2960.
26. Rao, C. N. R.; Nath, M. *Dalton Trans.* **2003** (Inorganic Nanotubes), 1.
27. Qiu, S. Q.; Dong, J. X.; Chen, G. X. *Wear* **1999**, *2301*, 35.
28. Jiang, G.; Guan, W.; Zheng, Z. *Wear* **2005**, *258* (11–12), 1625.
29. Hu, J. J.; Sanders, J. H.; Zabinski, J. S. *J. Mater. Res.* **2006**, *21* (4), 1033.
30. Faye, P.; Payen, E.; Bougeard, D. *J. Mol. Model.* **1999**, *5*, 63.
31. Helveg, S.; Lauritsen, J. V.; Lægsgaard, E.; Stensgaard, I.; Nørskov, J. K.; Clausen, B. S.; Topsøe, H. T.; Besenbacher, H. T. *Phys. Rev. Lett.* **2000**, *84*, 951.
32. Feldman, Y.; Wasserman, E.; Srolovitz, D. J.; Tenne, R. *Science* **1995**, *267*, 222.
33. Collins, N. R.; Twigg, M. V. *Top. Catal.* **2007**, *42–43*, 323.

34. Bakunin, V. N.; Kuzmina, G. N.; Kasrai, M.; Parenago, O. P.; Bancroft, G. M. *Tribol. Lett.* **2006**, *22* (3), 289.
35. Zhang, Z. J.; Xue, Q. J.; Zhang, J. *Wear* **1997**, *209* (1–2), 8.
36. Suryanarayana, C. *Prog. Mater. Sci.* **2001**, *46*, 1.

Chapter 9

Environmental Applications of Zerovalent Metals: Iron vs. Zinc

Paul G. Tratnyek,* Alexandra J. Salter, James T. Nurmi, and Vaishnavi Sarathy

Division of Environmental and Biomolecular Systems, Oregon Health & Science University, 20000 NW Walker Road, Beaverton, OR 97006
***tratnyek@ebs.ogi.edu**

The reactivity of particulate zero-valent metals in solution is affected by the metal type (e.g., Fe vs. Zn), particle size (nano vs. micro), surface conditions (passivation by coatings of oxides), and solution conditions (including the type and concentration of oxidants). Comparing the reactivity of various types of Fe^0 and Zn^0 with carbon tetrachloride (CCl_4) shows that the intended effect of properties engineered to give enhanced reactivity can be obscured by effects of environmental factors. In this case, rates of CCl_4 reduction by Zn^0 are more strongly affected by solution chemistry than particle size or surface morphology. Under favorable conditions, however, Zn^0 reduces CCl_4 more rapidly—and more completely—than Fe^0, regardless of particle size. The suitability of nano-sized Zn^0 for environmental remediation applications is uncertain.

Introduction

The use of zero-valent metals for removal of contaminants from environmental media (groundwater, sediments, soils, waste water, drinking water, and air) arose—apparently independently—in several forms. Some of these have been in use for a long time, and have been well studied, such as the processes known as "cementation" (*1*, *2*) and "electrocoagulation" (*3*, *4*). Currently, however, the most significant use of zero-valent metals involves emplacement of granular zero-valent iron (ZVI for the bulk material) into the subsurface to form reactive treatment zones known as permeable reactive barriers (PRBs) (*5*). Just as not all remediation applications of ZVI involve PRBs, not all PRBs are based on ZVI (Figure 1); however, it was the timely confluence of developments in these two areas that made the emergence of ZVI PRBs into one of the landmark developments in environmental engineering and a major focus of recent research on the fate and remediation of environmental contaminants.

Recently, there has been convergence between the applications of ZVI, PRBs, and what is now categorized as "nanotechnology". While not all remediation applications of nanotechnology involve ZVI or PRBs (Figure 1), the potential advantages of using nano-sized ZVI (nZVI) to create PRBs has made it another major development in environmental engineering science (*6–8*). However, the basis for this application of nanotechnology owes a great deal to earlier (and continuing) work on the environmentally-relevant reactions of other types of nanoparticles, especially (alkali) metal oxides such as MgO, CaO, TiO_2, ZnO, Fe_2O_3, etc. (*9, 10*). For metals that undergo oxidative dissolution, like Fe, Zn, and Sn, the tie is particularly strong because the surface of the metal is covered with a film of the corresponding oxide, under all environmentally-relevant conditions. These oxides mediate reactions with the medium (*11*), making their structure and composition among the factors that determine whether particles of any particular type of metal are effective for degradation of any particular contaminant.

Although most research on contaminant degradation by ZVMs has been focused on iron (*12*), other metals have been studied, especially zinc (ZVZ for the bulk granular material). The main rationales that have been given for studying contaminant reduction by ZVZ are: (*i*) since Zn^0 is a stronger reductant than Fe^0, ZVZ might result in more rapid and complete degradation, thereby expanding the scope of treatable contaminants and/or lessening the accumulation of undesirable intermediates, and (*ii*) the absence of a valence state analogous to Fe^{III} leaves only one oxidation step ($Zn^0 \rightarrow Zn^{II}$) to consider in the reaction mechanism, thereby making ZVZ complementary to ZVI as a model system for studying mechanistic aspects of contaminant degradation by ZVMs. With respect to these criteria, studies of contaminant reduction by ZVZ have been quite successful, but they have not yet led to any pilot- or full-scale applications of ZVZ at contaminated field sites.

Most research using ZVZ to degrade contaminants has focused on dehalogenation, with successful results having been reported for the whole range of environmentally-significant organic structures, including halogenated methanes (*13–18*), ethanes (*19–21*), ethenes (*22–25*), propanes (*26*), phenols (*27*), biphenyls (*28*), and dioxins (*29*). Among the halogenated methanes, most of the data is for carbon tetrachloride (CCl_4), which is commonly used as a model contaminant and probe compound in studies of environmental reduction reactions. The pathway and kinetic models for CT degradation that are used in these studies vary somewhat, but the model we used in our recent work on contaminant degradation by nZVI (*30, 31*) is among the more complete and quantitative (Scheme 1). In this scheme, "unk" represents unknown products, consisting of the combination of less easily quantified products (HCOOH, CO, CH_4) that are believed to be responsible for the sometimes substantial quantity of CCl_4 that is not accounted for by the products from sequential hydrolysis ($CHCl_3$, CH_2Cl_2, etc.).

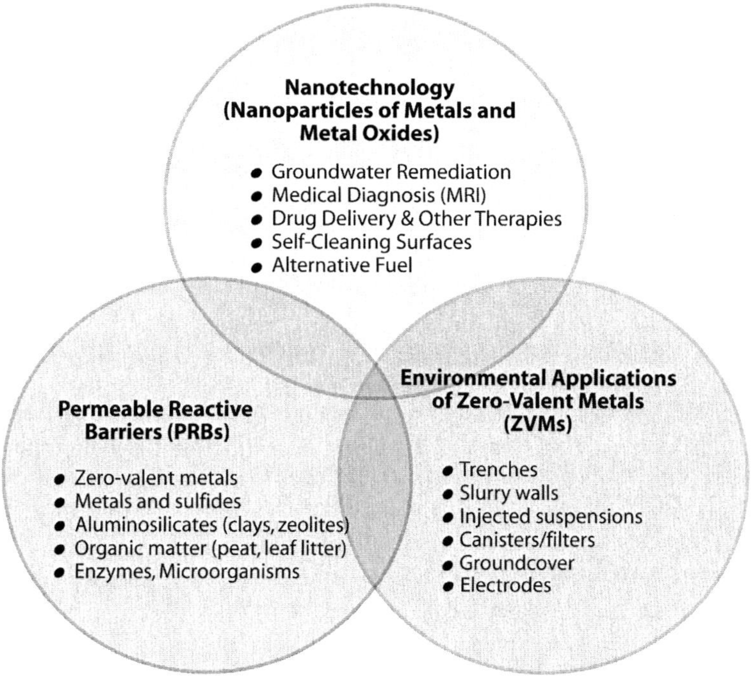

Figure 1. Overlap between emerging areas of environmental technology. Adapted from figures in (5, 12).

In this study, we apply our pathway/kinetic model (Scheme 1) to reinterpret previously published data on reduction of CCl_4 with ZVZ (*14–16*), and put these results into a larger context provided by comparison with new data for this system and literature data on CCl_4 reduction by other ZVMs, including nZVI.

Methods

Batch experiments were performed using two types of ZVZ: "Fisher powder" (Certified grade, 99.3%, Fisher Scientific, Fairlawn, NJ; specific surface area by BET gas adsorption (a_s) = 0.23±0.07 m^2/g), a reagent-grade product similar to that used in previous studies; and "Zinc Dust 64" (mean particle size = 4.5 um, 99.3% zinc, 96.5% Zn^0, a_s = 0.62±0.01 m^2/g) from an industrial-scale supplier of recycled zinc (Horsehead Corp., Monaca, PA).

The procedure was selected to combine characteristics of the method used by Boronina (*13*, *15*, *16*)—especially the high initial concentration of CCl_4—with features that are now generally consided desirable for batch experiments with ZVMs, such as a period of exposure of the particles to the solution before addition of the contaminant. Batch reactors were prepared in 60 mL, Arflushed serum bottles by adding 2-3 g of ZVZ and 34-38 mL of deoxygenated deionized water, crimp-sealing with a Hycar septa and rotating at 32 rpm for 20-28 hr. After this "preexposure" period, 4-8 mL of saturated CCl_4 was injected directly into the solution—using a second needle to vent the excess headspace gas—resulting in

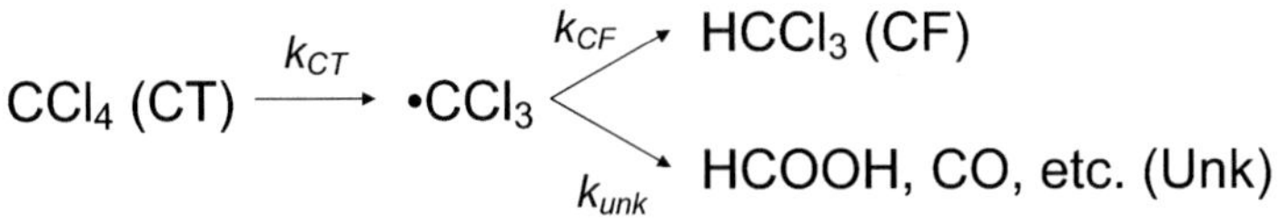

$$\left.\begin{aligned}[CT]_t &= [CT]_0 e^{-k_{CT}t} \\ [CF]_t &= [CT]_0 Y_{CF}\left(1 - e^{-k_{CT}t}\right)\end{aligned}\right\} \quad \begin{aligned} k_{CT} &= k_m \rho_m = k_{sa} a_s \rho_m \\ Y_{CF} &= k_{CF} / (k_{CF} + k_{unk}) \end{aligned}$$

Scheme 1. Pathway and kinetic model for reduction of carbon tetrachloride. Adapted from (30, 31).

a total liquid volume of 42 mL in each case and initial concentrations of CCl_4 ranging from 0.5 to 1.0 mM. Bottles were hand-shaken for 1 min, and then rotated at 32 rpm for the duration of the experiment. Starting about 1.5 min after the CCl_4 injection, 10 μL samples were withdrawn from the reactors and extracted in 1 mL of hexane. The hexane extracts were analyzed by gas chromatography, using a DB-624 column (J&W/Agilent) with electron capture detection.

Results and Discussion

Rates of CCl_4 Reduction by (n)ZVMs

Rates of any reaction with particles in a fluid medium can be summarized in a plot of surface area normalized rate constants (k_{SA}, $Lm^{-2}hr^{-1}$) versus the corresponding mass normalized rate constants (k_M, $Lg^{-1}hr^{-1}$), where the two types of rate constants are related by the specific surface area of the particles (a_s, m^2g^{-1}) (*32*). A generic version of such a plot is shown in Figure 2A to clarify the nature of the putative "nano-size effect" on reactivity of Fe^0 with contaminants. Relative to the point labeled *1* (chosen to be typical for CCl_4 reduction by high-purity ZVI), decreasing particle size with no change in the intrinsic reactivity of the particle surface (at this level of approximation represented by k_{SA}) increases k_M along the line labeled *2*. Increasing the intrinsic reactivity of the particle surface without changing the specific surface area (a_s) increases both k_{SA} and k_M proportionately along the line labeled *3*. Only the area enclosed by these lines (and shaded in gray) represents an increase in k_{SA} that exceeds the expected effect of increased a_s due to decreased particle size. Thus, only data that fall in the area shaded gray would support the idea that nanoparticles of Fe^0 have a greater intrinsic reactivity than larger particles.

We first used this approach to characterize the effect of ZVI particles size on CCl_4 in Nurmi et al. (*31*), using data we measured—and the kinetic model summarized in Scheme 1—for two types of nano Fe^0 with one micro Fe^0. Subsequently, we expanded this analysis to include more of our own data and a compilation of literature data for CCl_4 vs. various types of ZVI (*33*), a version of which is shown in Figure 2B. The results confirm that, for CCl_4 under the conditions of conventional batch experiments, k_M is greater for nano Fe^0

(asterisks) than micro Fe^0 (circles) but that there is no nano-size effect on k_{SA}. Another conclusion that can be drawn from Figure 2B is that k_M and k_{SA} are both smaller for low-purity iron (solid circles) than high purity iron (nano or micro).

The perspective on CCl_4 reduction kinetics provided by Figure 2B for ZVI, can be expanded to include other ZVMs, such as ZVZ. Data from Boronina et al. (*13*, *15*) for batch experiments with CCl_4 and several types of ZVZ (Table 1) are shown in Figure 3 as black symbols. The results are surprising in two respects: (*i*) they do not support the expectation that disappearance of CCl_4 will be more rapid for ZVZ than ZVI, and (*ii*) they do not show the hypothesized increase in k_{SA} from the cryo synthesis method, which was supposed to generate more reactive sites on the particle surfaces. Nevertheless, the overall accuracy of Boronina's results is supported by their approximate agreement with the kinetic data from another study (*17*), which used a different reagent-grade ZVZ but otherwise similar experimental conditions (deionized water without buffer or pH adjustment, no acid-washing or other cleaning of the ZVZ, without preexposure of the ZVZ to solution before introduction of the CCl_4).

Among the group of data from (*13*, *15*, *17*) shown in Figure 3, one point (marked ⊡) suggests significantly smaller values of k_M and (especially) k_{SA}. This result is also contrary to the intended effect of the treatment: in this case, because the ZVZ was prepared by a high pressure pelletization process that gave the material higher surface area (*13*). The observed result is most likely because the added surface area of the primary particles (measured by BET gas adsorption) was mostly in gas-filled pores within the pellets during batch experiments and therefore not available for reaction with the CCl_4 in solution.

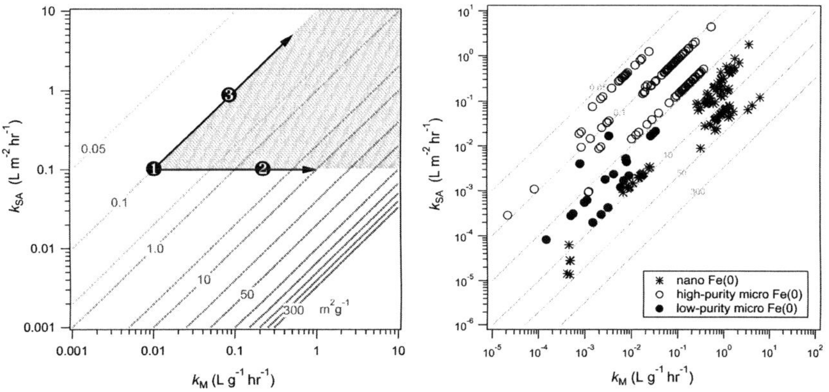

Figure 2. (A) Generic k_{SA} vs. k_M plot showing general features and the primary effects of particle size on rate constants. (B) Comparison of rate constants for reduction of CCl_4 by nano Fe^0 and two types of micro-sized Fe^0 (high-purity laboratory-grade and low-purity construction-grade). (Adapted from (7, 32))

Table 1. Experimental Data for CCl_4 Reduction by Zero-Valent Zinc

Ref	*Zn Type*	*Zn* (g)	ρ_m ($g\ L^{-1}$)	a_s ($m^2\ g^{-1}$)	ρ_a ($m^2\ L^{-1}$)	CCl_4 (mM)	Γ_0 ($mmol\ m^{-2}$)	k_{CT} (hr^{-1})	k_M ($L\ g^{-1}\ hr^{-1}$)	k_{SA} ($L\ m^{-2}\ hr^{-1}$)
(*13*)	Cryo	0.5	17.86	5.23	93.3	1.15	0.012	0.47±0.02	0.026	0.0051
(*13*)	Cryo	0.1	3.571	5.23	18.7	1.15	0.061	0.084±0.006	0.023	0.0045
(*15*)	Cryo	1.0	9.99	5.23	52.25	10	0.19	1.07±0.06[a]	0.1071	0.020
(*13*)	FD	2.2	78.57	0.243	19.1	0.925	0.048	0.54±0.08	0.0068	0.028
(*13*)	FD	0.5	17.86	0.243	4.33	1.15	0.27	0.208±0.009	0.0116	0.048
(*13*)	FD	0.1	3.571	0.243	0.87	1.15	1.32	0.026±0.003	0.0073	0.030
(*15*)	FD	1.0	9.99	0.243	2.43	10	4.11	lag only	--	--
(*15*)	FD	5.0	49.95	0.243	12.14	10	0.82	2.14±0.05[a]	0.0428	0.176
(*13*)	FDP	0.11	3.93	3.34	13.1	1.15	0.089	0.0144±0.0004	0.0037	0.0011
(*18*)	FD	1.8[b]	10[b]	0.47	4.7[b]	0.65	0.14[b]	27[b]	2.7	5.74
(*17*)	WD	5.0	125.0	0.401	50.12	0.13	0.0026	2.29	0.018	0.0457
T.S.[d]	FD	3.00	71.43	0.231	16.50	1.0	0.061	1.9±0.2	0.027	0.11
T.S.[d]	H64	2.06	49.05	0.62	30.4	1.0	0.033	65±3[c]	1.34	2.16
T.S.[d]	H64	2.09	49.76	0.62	30.85	1.1	0.035	78±7[c]	1.57	2.53

Ref	*Zn Type*	*Zn* (g)	ρ_m ($g\ L^{-1}$)	a_s ($m^2\ g^{-1}$)	ρ_a ($m^2\ L^{-1}$)	CCl_4 (mM)	Γ_0 ($mmol\ m^{-2}$)	k_{CT} (hr^{-1})	k_M ($L\ g^{-1}\ hr^{-1}$)	k_{SA} ($L\ m^{-2}\ hr^{-1}$)
T.S.[d]	H64	3.04	72.38	0.62	44.87	1.0	0.022	98±6[c]	1.35	2.18
T.S.[d]	H64	1.98	47.14	0.62	29.23	0.5	0.017	168±11[c]	3.55	5.73

[a] Zinc Type: Cryo = Cryogenically prepared, FD = Fisher dust, FDP = Pelletized Fisher dust, WB = Waso dust, H64 = Horsehead zinc dust 64. From the pseudo-first-order portion of data extracted from Fig. 9 in (*15*). [b] Median value for range of 7 concentrations of ZVZ. [c] Estimated as $V_{max}/K_{1/2}$ from fit to progress curve data, as described in (*34*). [d] T.S. = this study.

In contrast to the above results with CCl_4, most of the available data for reduction of polychlorinated ethenes, ethanes, and propanes by ZVZ (*19*, *23*, *25*, *26*) suggest considerably faster reaction rates than expected with ZVI. To help resolve this apparent inconsistency, we measured CCl_4 disappearance kinetics in batch reactors under conditions designed to straddle some of the potentially-important differences between the prior studies on dechlorination with ZVZ. Preliminary results (not shown) obtained without preexposing the ZVZ to water included a time lag of 0.5 to >6 hr before initiation of CCl_4 reduction. In other studies (with ZVI) where lag phases have been observed (e.g., (*35*, *36*)), it usually has been possible to avoid them by preexposing the particles to the medium before adding the contaminant.

Adopting the preexposure step here usually eliminated the lag phase, but the resulting kinetics of CCl_4 disappearance still were not always simple pseudo first-order. For example, one set of experiments, shown in Figure 4, consistently showed what is most likely the transition from zero- to first-order kinetics due to relaxation of reactive site saturated conditions. We observed such "progress curves" in a previous study with ZVI (*34*) and described the fitting and interpretation of such results in detail. Applying the same methods here gave the smooth curves shown in Figure 4, and the fitting constants were used to calculate the values of k_{CT} given in Table 1.

Progress curve kinetics should arise only where the concentration of reactant is high relative to the amount of reactive surface area. However, the only practical way to define this is relative to the total surface area determined by BET gas adsorption on dry samples (i.e., initial mmoles CCl_4 per m^2 of ZVZ, or Γ_0). Values of Γ_0 calculated in this way (given in Table 1) do not predict which data show progress-curve-like kinetics, which is probably because other factors such as changes in the surface composition during preexposure can be more significant. However, within the data obtained under the conditions used by Boronina et al. (*13*, *15*), the three with the highest Γ_0 are least adequately described by simple pseudo-first-order kinetics.

Using values of k_{obs} obtained by fitting the progress curves in Figure 4, we calculated k_M and k_{SA} and have presented the results in Table 1 and Figure 3 (marked ⌑). The rate constants we measured are much greater (2-3 orders of magnitude) than the group of data from (*13*, *15*, *17*), but are in good agreement with the result (marked ✧) from an independent, previously published study (*18*). For the latter, we calculated k_{SA} from the value of k_M determined by the slope of the straight line fit to their k_{obs} vs. ρ_m data ((*18*), *Fig. 6*). Reducing the data in this way gives only one point on Figure 3, but it has greater statistical significance than the other points, which represent individual experiments.

Another conclusion that can be drawn from Figure 3, is that the group of data showing rapid reduction of CCl_4 with ZVZ lies about an order of magnitude higher than the median values of k_M and k_{SA} for CCl_4 reduction by high purity and/or nano-size ZVI. This is consistent with the expected trend in reactivity (ZVZ > ZVI) based on analogy to prior work with other polychlorinated ethanes, ethenes, and propanes. The largest values of k_{SA} for CCl_4 vs. ZVZ are roughly 5 L m^{-2} hr^{-1}, which is similar to the values reported previously for reduction of polychlorinated ethanes with ZVZ (*19*). We expect k_{SA}'s for the polychlorinated

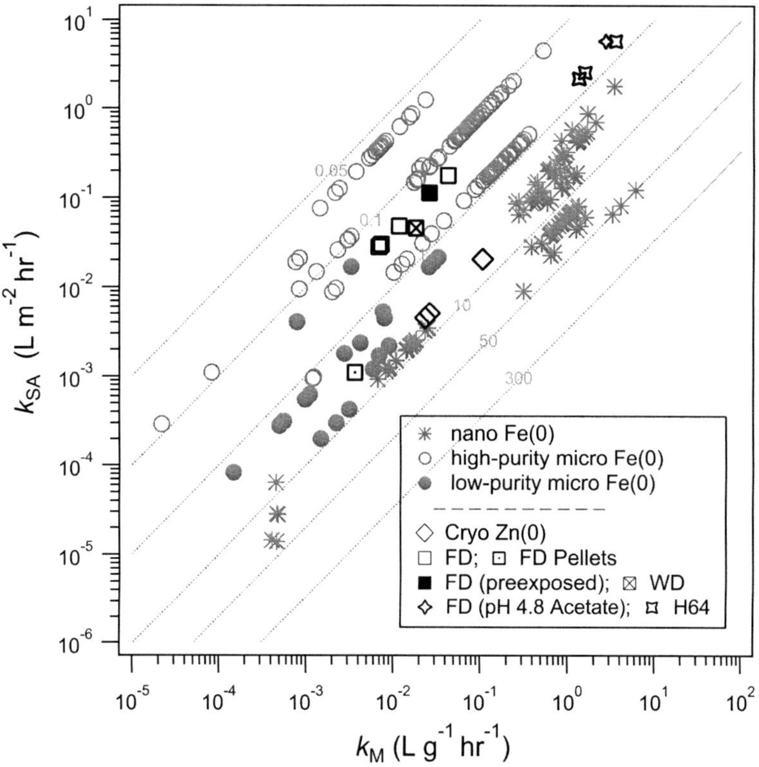

Figure 3. k_{SA} vs. k_M plot comparing CCl_4 reduction by ZVZ (black) with ZVI (gray). Data for ZVZ are also given in Table 1.

methanes and ethanes to be comparable because as k_{SA} approaches 10 L m^{-2} hr^{-1}, the contaminant reduction rates that are observed under the batch experimental conditions that apply here become increasingly limited by mass transport and therefore less sensitive to differences in molecular structure (*19*).

The difference between the fast and slow groups of data for CCl_4 reduction by ZVZ does not appear to be due to any one experimental variable, but does follow a pattern where the faster rates are associated with ZVZ that was acid-washed or pre-exposed to solution and reacted in the solution resulting from preexposure or buffer/electrolyte solutions at circum-neutral pH, whereas the slow rates are associated with ZVZ that was used without precleaning or preexposure and the contaminant degradation was initiated in solutions of DI water. We think these two groups reflect a difference in surface conditions on the ZVZ, where the former develops a surface film—probably dominated by Zn hydroxides (*37*)—that is activated with respect to reduction of solutes like CCl_4, and the latter surface type reflects the early stages of passive film (ZnO) breakdown but not the formation of the phases that result in rapid reduction of contaminants. Note that both cases appear to be "depassivated" with respect to corrosion of Zn, because chrono-potentiograms obtained using powder disk electrodes prepared with all of the ZVZ types used in our study all show a rapid decay to open-circuit potentials

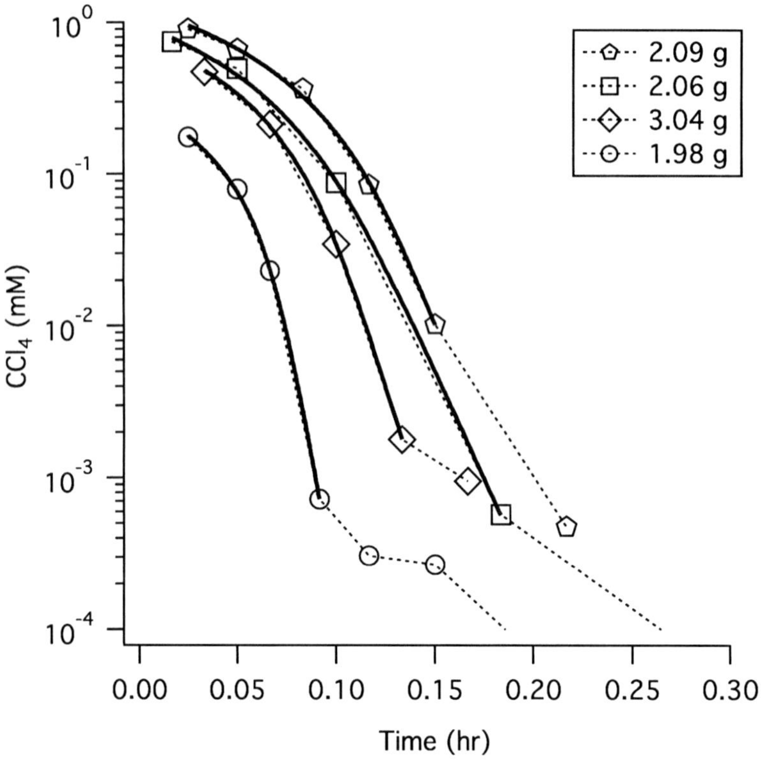

Figure 4. Kinetics of CCl_4 reduction by ZVZ (H64) that had been preexposed to DO/DI water for ~24 hr.

that are consistent with active corrosion. These results are not shown here, but the methods, results, and interpretation are analogous to what we have reported previously for ZVI (*38*, *39*).

To test our characterization of the difference between fast and slow rates of CCl_4 reduction in Figure 3, we performed batch experiments with 0 and 4.8 days of preexposure to DI water using ZVZ dust from Fisher (although we can not be sure this is exactly the same product used by Boronina et al. (*13*, *15*), the BET a_s we measured is very close to the value they reported). Without preexposure, CCl_4 degradation was negligible, but with 4.8 days preexposure, CCl_4 showed first-order disappearance. Calculating k_M and k_{SA} from the data for 4.8 days preexposure gave the result shown in Table 1 and Figure 3 (marked ■). This point plots higher than most of the previously reported data for un-preexposed ZVZ, and therefore is consistent with the hypothesis that changes in ZVZ that occur during preexposure to solution contribute to the conditions that result in rapid degradation of CCl_4. Preliminary results (not shown) suggest a similar sensitivity of contaminant reduction by ZVZ to preexposure time for two other organic contaminants: 1,2,3-trichloropropane and nitrobenzene. Preexposure to solution also has been found to affect contaminant reduction by ZVI (e.g, (*30*)), presumably also due to changes in ("aging" of) the structure and/or composition of the film of oxides and other phases that mediate reaction between the metal and solutes.

Products of CCl_4 Reduction

The original work on CCl_4 reduction by ZVZ (*13*, *15*) recognized that the distribution of products is as important as the rate of contaminant degradation in determining the potential utility of any remediation process, and devoted considerable effort to product distributions. Under the conditions of their experiments, CCl_4 reduction by ZVZ gave less $CHCl_3$ and more CH_2Cl_2 and CH_4 compared to what is typically observed with ZVI (*30*). This was attributed mainly to "deep" sequential hydrogenolysis, with a lesser portion of the reaction occurring by a concerted multiple-dechlorination mechanism (such as reductive elimination). Other work on CCl_4 vs. ZVZ (including the results first reported here) has not included as much detail on products, but the results are qualitatively consistant with those of Boronina et al. (*13*, *15*). Establishing whether this apparent advantage of ZVZ—that it tends to produce less accumulation of $CHCl_3$ than ZVI—can be sustained under realistic environmental conditions will require further investigation of the numerous differences in reaction conditions. Of particular interest, in light of the effects on k_{SA} discussed above, is whether aging of ZVZ in solution will create time-dependent changes in the distribution of products, like we have described in previous work with ZVI (*30*).

Conclusions

Many factors that influence the reactivity of zero-valent metals with contaminants have been investigated with the aim of improving on traditional uses of ZVI for remediation. These factors include the type of metal (e.g., ZVZ vs. ZVI), the quantity of surface area (e.g., as increased by porous or nano-sized particles), and the reactivity of the accessible surface area (e.g., which may be greater for nano-sized or nano–structured particles). Within any series of controlled experiments designed to isolate a particular effect, differences will be observed, as has been the case in many recent studies that have found faster contaminant degradation with nano-sized ZVMs compared with larger particles of the same type of metal. Such comparisons, however, have limited scope and therefore do not generally provide much perspective on the relative significance of the various factors.

In this study, we put unusual emphasis on comparing data from multiple sources, which has the advantage and disadvantage of introducing variability (in this case, into rate constants for dechlorination of CCl_4). This approach may obscure small effects, but it is necessary to evaluate which effects are dominant over the ranges of relevant operating conditions. Our analysis demonstrates a case where most of the factors that were varied by design (including particle size and mechanical activation of their surface area) have effects that are relatively small compared to the unanticipated side-effect of the overall solution chemistry and experimental design (e.g., preexposure to solution). It is likely that this situation applies to environmental remediation applications of nanoparticles, so that the improved reactivity with contaminants that can be demonstrated with controlled experiments in the laboratory is overwhelmed by other effects that are determined by environmental conditions in situ.

Acknowledgments

This work was supported by grants from the Strategic Environmental Research and Development Program (SERDP Project No. ER-1457), and the U.S. Department of Energy, Office of Science (DE-AC05-76RLO 1830) and Environmental Management Sciences Program (DE-FG07-02ER63485). This report has not been reviewed by any of these sponsors and therefore does not necessarily reflect their views and no official endorsement should be inferred.

References

1. Khudenko, B. M.; Gould, J. P. Specifics of cementation processes for metals removal. *Water Sci. Technol.* **1991**, *24*, 235–246.
2. Khudenko, B. M. Feasibility evaluation of a novel method for destruction of organics. *Water Sci. Technol.* **1991**, *23*, 1873–1881.
3. Lakshmanan, D.; Clifford, D. A.; Samanta, G. Ferrous and ferric ion generation during iron electrocoagulation. *Environ. Sci. Technol.* **2009**, *43*, 3853–3859.
4. Moreno, C. H. A.; Cocke, D. L.; Gomes, J. A. G.; Morkovsky, P.; Parga, J. R.; Peterson, E.; Garcia, C. Electrochemical reactions for electrocoagulation using iron electrodes. *Ind. Eng. Chem. Res.* **2009**, *48*, 2275–2282.
5. Tratnyek, P. G.; Scherer, M. M.; Johnson, T. J.; Matheson, L. J. Permeable reactive barriers of iron and other zero-valent metals. In *Chemical Degradation Methods for Wastes and Pollutants: Environmental and Industrial Applications*; Tarr, M. A., Ed.; Marcel Dekker: New York, 2003; pp 371–421.
6. Zhang, W.-X. Nanoscale iron particles for environmental remediation. *J. Nanoparticle Res.* **2003**, *5*, 323–332.
7. Tratnyek, P. G.; Johnson, R. L. Nanotechnologies for environmental cleanup. *NanoToday* **2006**, *1*, 44–48.
8. Lowry, G. V. Nanomaterials for groundwater remediation. In *Environmental Nanotechnology*; Wiesner, M. R.; Bottero, J.-Y., Eds.; McGraw Hill: New York, 2007; pp 297–336.
9. Klabunde, K. J. *Nanoscale Materials in Chemistry*; Wiley: New York, 2001.
10. Ranjit, K. T.; Medine, G.; Jeevanandam, P.; Martyanov, I. N.; Klabunde, K. J. Nanoparticles in environmental remediation. *Environ. Catal.* **2005**, 391–420.
11. Scherer, M. M.; Balko, B. A.; Tratnyek, P. G. The role of oxides in reduction reactions at the metal-water interface. In *Mineral-Water Interfacial Reactions: Kinetics and Mechanisms*; ACS Symposium Series 715; American Chemical Society: Washington, DC, 1998; pp 301–322.

12. Tratnyek, P. G. Keeping up with all that literature: the IronRefs database turns 500. *Ground Water Monit. Rem.* **2002**, *22*, 92–94.
13. Boronina, T. N.; Lagadic, I.; Sergeev, G. B.; Klabunde, K. J. Activated and nonactivated forms of zinc powder: Reactivity toward chlorocarbons in water and AFM studies of surface morphologies. *Environ. Sci. Technol.* **1998**, *32*, 2614–2622.
14. Boronina, T. N.; Klabunde, K. J.; Sergeev, G. B. Dechlorination of carbon tetrachloride in water on an activated zinc surface. *Mendeleev Commun.* **1998**, 154–155.
15. Boronina, T.; Klabunde, K. J.; Sergeev, G. Destruction of organohalides in water using metal particles: Carbon tetrachloride/water reactions with magnesium, tin, and zinc. *Environ. Sci. Technol.* **1995**, *29*, 1511–1517.
16. Boronina, T. N.; Dieken, L.; Lagadic, I.; Klabunde, K. J. Zinc-silver, zinc-palladium, and zinc-gold as bimetallic systems for carbon tetrachloride dechlorination in water. *J. Hazard. Subst. Res.* **1998**, *1*, 6–1, 6–23.
17. Feng, J.; Lim, T.-T. Pathways and kinetics of carbon tetrachloride and chloroform reductions by nano-scale Fe and Fe/Ni particles: comparison with commercial micro-scale Fe and Zn. *Chemosphere* **2005**, *59*, 1267–1277.
18. Warren, K. D.; Arnold, R. G.; Bishop, T. L.; Lindholm, L. C.; Betterton, E. A. Kinetics and mechanism of reductive dehalogenation of carbon tetrachloride using zero-valence metals. *J. Hazard. Mater.* **1995**, *41*, 217–227.
19. Arnold, W. A.; Ball, W. P.; Roberts, A. L. Polychlorinated ethane reaction with zero-valent zinc: Pathways and rate control. *J. Contam. Hydrol.* **1999**, *40*, 183–200.
20. Fennelly, J. P.; Roberts, A. L. Reaction of 1,1,1-trichloroethane with zero-valent metals and bimetallic reductants. *Environ. Sci. Technol.* **1998**, *32*, 1980–1988.
21. Totten, L. A.; Jans, U.; Roberts, A. L. Alkyl bromides as mechanistic probes of reductive dehalogenation: reactions of vicinal dibromide stereoisomers with zerovalent metals. *Environ. Sci. Technol.* **2001**, *35*, 2804–2811.
22. Li, W. F.; Klabunde, K. J. Ultrafine zinc and nickel, palladium, silver coated zinc particles used for reductive dehalogenation of chlorinated ethylenes in aqueous solution. *Croat. Chem. Acta* **1998**, *71*, 853–872.
23. Arnold, W. A.; Roberts, A. L. Pathways of chlorinated ethylene and chlorinated acetylene reaction with Zn(0). *Environ. Sci. Technol.* **1998**, *32*, 3017–3025.
24. Roberts, A. L.; Totten, L. A.; Arnold, W. A.; Burris, D. R.; Campbell, T. J. Reductive elimination of chlorinated ethylenes by zero-valent metals. *Environ. Sci. Technol.* **1996**, *30*, 2654–2659.
25. Cheng, S.-F.; Wu, S.-C. The enhancement methods for the degradation of TCE by zero-valent metals. *Chemosphere* **2000**, *41*, 1263–1270.
26. Sarathy, V.; Tratnyek, P. G.; Nurmi, J. T.; Johnson, R. L.; Johnson, G. O. B. Degradation of 1,2,3-trichloropropane (TCP): Hydrolysis, elimination, and reduction by iron and zinc. *Environ. Sci. Technol.* **2010**, *44*, 787–793.
27. Kim, Y. H.; Carraway, E. R. Dechlorination of chlorinated phenols by zero valent zinc. *Environ. Technol.* **2003**, *24*, 1455–1463.

28. Kim, Y. H.; Shin, W. S.; Ko, S. O. Reductive dechlorination of chlorinated biphenyls by palladized zero-valent metals. *J. Environ. Sci. Health, Part A* **2004**, *39*, 1177–1188.
29. Wang, Z. Y.; Huang, W. L.; Fennell, D. E.; Peng, P. A. Kinetics of reductive dechlorination of 1,2,3,4-TCDD in the presence of zero-valent zinc. *Chemosphere* **2008**, *71*, 360–368.
30. Sarathy, V.; Tratnyek, P. G.; Nurmi, J. T.; Baer, D. R.; Amonette, J. E.; Chun, C.; Penn, R. L.; Reardon, E. J. Aging of iron nanoparticles in aqueous solution: effects on structure and reactivity. *J. Phys. Chem. C* **2008**, *112*, 2286–2293.
31. Nurmi, J. T.; Tratnyek, P. G.; Sarathy, V.; Baer, D. R.; Amonette, J. E.; Pecher, K.; Wang, C.; Linehan, J. C.; Matson, D. W.; Penn, R. L.; Driessen, M. D. Characterization and properties of metallic iron nanoparticles: Spectroscopy, electrochemistry, and kinetics. *Environ. Sci. Technol.* **2005**, *39*, 1221–1230.
32. Tratnyek, P. G.; Sarathy, V.; Kim, J.-H.; Chang, Y.-S.; Bae, B. Effects of particle size on the kinetics of degradation of contaminants. In *International Environmental Nanotechnology Conference—Applications and Implications*; EPA 905-R09-032; U.S. Environmental Protection Agency: Chicago, IL, 2009.
33. Tratnyek, P. G.; Sarathy, V.; Bae, B. Nanosize effects on the kinetics of contaminant reduction by iron oxides, Preprints of Extended Abstracts. In *230th ACS National Meeting*; American Chemical Society, Division of Environmental Chemistry: Washington, DC, 2005; Vol. 45, pp 673−677.
34. Agrawal, A.; Ferguson, W. J.; Gardner, B. O.; Christ, J. A.; Bandstra, J. Z.; Tratnyek, P. G. Effects of carbonate species on the kinetics of 1,1,1-trichloroethane by zero-valent iron. *Environ. Sci. Technol.* **2002**, *36*, 4326–4333.
35. Miehr, R.; Tratnyek, P. G.; Bandstra, J. Z.; Scherer, M. M.; Alowitz, M. J.; Bylaska, E. J. Diversity of contaminant reduction reactions by zerovalent iron: Role of the reductate. *Environ. Sci. Technol.* **2004**, *38*, 139–147.
36. Alowitz, M. J.; Scherer, M. M. Kinetics of nitrate, nitrite, and Cr(VI) reduction by iron metal. *Environ. Sci. Technol.* **2002**, *36*, 299–306.
37. Zhang, X. G. *Corrosion and Electrochemistry of Zinc*; Plenum: New York, 1996.
38. Nurmi, J. T.; Tratnyek, P. G. Electrochemical studies of packed iron powder electrodes: Effects of common constituents of natural waters on corrosion potential. *Corros. Sci.* **2008**, *50*, 144–154.
39. Nurmi, J. T.; Bandstra, J. Z.; Tratnyek, P. G. Packed powder electrodes for characterizing the reactivity of granular iron in borate solutions. *J. Electrochem. Soc.* **2004**, *151*, B347–B353.

Chapter 10

Visible and UV Light Photocatalysts in Environmental Remediation

Kenneth J. Klabunde*

Kansas State University, Manhattan, KS 66506
***kenjk@ksu.edu**

The use of nanoscale particles as photocatalysts for remediation of air/water pollutants is briefly reviewed in this chapter. Emphasis is given to new materials that possess the ability to absorb visible light, not only UV light. These visible light photocatalysts should prove more useful in indoor air cleaning and remediation.

Introduction

The use of nanotechnology as a tool to help destroy toxic chemicals, chemical warfare agents, and biological threats has mainly involved destructive adsorption using reactive, high surface area nanostructured powders (*1*), and the use of devices for detection of such dangers (*2*).

In order to enhance the abilities of nanoscale destructive adsorbents, we considered employing light harvesting nanoscale materials, such as semiconducting titanium dioxide (TiO_2) (*3–6*). The idea was to maintain high surface areas and reactivities, but also absorb light for the benefit of enhancing reactivity, cleaning the surface, and converting the whole destructive adsorption process into a catalytic process. Recently, we have called this "photo boosting" (*7*).

We began with the study of the closest simulant to mustard gas [bis-(2-chloroethyl)sulfide], or HD. The simulant chosen was 2-chloroethyl ethyl sulfide [half-mustard, 2-CEES, $ClCH_2CH_2SCH_2CH_3$] (*8*).

There have been earlier studies of deep photooxidation of CWA simulants in air under UV light (*9–12*). We hoped to add to these investigations by the use of dry nanopowders (rather than slurries or solutions) in contact with gaseous 2-CEES in air and under UV irradiation. This approach would be most relevant regarding

TiO_2 in coatings, paints, or as dry powders. Also, special emphasis was given to identification and evaluation of the lifetime of toxic gaseous intermediates as compounds posing a major threat to humans.

Gaseous products given off at 25°C or 80°C were monitored over time. Indeed, many products were formed, and Figure 1 shows the apparent quantum yields for several of the volatile products, especially at the higher temperature 80°C.

Less volatile and nonvolatile products were extracted from the surface and analyzed by FT-IR and other means. Overall, photocatalytic oxidation of 2-CEES led to numerous intermediates and final products. Some of the intermediates, such as bis-(2-chloroethyl) disulfide are believed to be toxic. However, this disulfide is of low volatility and quickly suffers further oxidation to less toxic products at the TiO_2 surface. Deactivation of the TiO_2 surface was caused by non-volatile products, especially sulfate ions. Extended photocatalysis resulted in mineralized products CO_2, H_2O, SO_2, HCl, SO_4^{2-} and Cl^- salts. However, at least 15 intermediate products were detected (*8*).

Some details of the mechanism of this chemistry were reported by Thomson, et. al. (*13*) Using surface sensitive IR, UV, and GC-MS of desorbed products, it was shown that the TiO_2 lattice provides oxygen for 2-CEES oxidation (photooxidation in the absence of oxygen gas). It was postulated that free radical species, produced from the 2-CEES molecules by electron or hole attack, would lead to a sequence of reactions to extract lattice oxygen.

The 2-CEES work was difficult to carry out, and the major part of the work was in product analysis. However, since we were primarily interested in testing the activities of new nanomaterials, we devised several more convenient photochemical test reactions:

(1) acetaldehyde photooxidation

$$CH_3CHO + O_2 \xrightarrow[\text{photocat.}]{h\nu} CO_2 + H_2O$$

(2) trichloroethylene photooxidation

$$CHCl = CCl_2 + O_2 \xrightarrow[\text{photocat.}]{h\nu} CO_2 + H_2O + HCl$$

(3) carbon monoxide photooxidation

$$CO + O_2 \xrightarrow[\text{photocat.}]{h\nu} CO_2$$

Using these convenient test procedures, we began an in-depth study of different nano-formulations. We believed that mixed oxides of high surface area, containing metal or non-metal dopants, might lead to some very active photocatalysts under both UV and visible light (*14–19*).

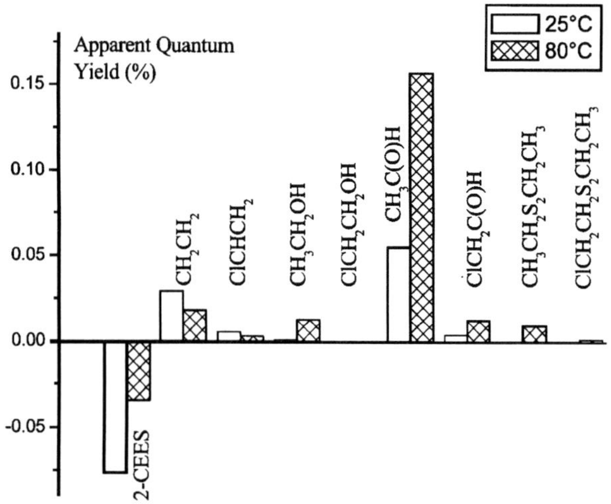

Figure 1. Apparent quantum yields of gaseous organic compounds in the beginning of the reaction of 2-CEES with UV light, oxygen, and TiO_2. (Reproduced with permission from reference (8).)

At this point, it is worthwhile to consider the steps that are believed to be necessary for an effective photocatalyst. First, consider a metal ion such as Cr^{6+} (as CrO_3) doped onto a semiconductor such as TiO_2.

Photon absorption

$$Cr^{6+} - O^{2-} \xrightarrow{h\nu} [\dot{C}r^{5+} - O^{1-}]^{*}$$

Electron-hole pair separation

$$[\dot{C}r^{5+} - O^{1-}] + \text{conduction band of } (TiO_2)_n \longrightarrow Cr^{6+} - O^{2-} + {}^{+}(TiO_2)_n^{-}$$

Reduction of oxygen gas

$${}^{+}(TiO_2)_n^{-} + O_2 \longrightarrow O_2^{-} + {}^{+}(TiO_2)_n$$

Oxidation of hydroxide

$${}^{+}(TiO_2)_n + {}^{-}OH \longrightarrow (TiO_2)_n + \cdot OH$$

The presence of O_2^- and $\cdot OH$ allows a cascade of oxidation-reduction reactions to occur, and thereby, organic species can be oxidized/reduced (*20*).

However, if the main component is not a semiconductor, for example SiO_2 instead of TiO_2, the electron hole-pair separation would not so readily occur, and so the localized $[\dot{C}r^{+5} - O^{-}]^{*}$ would have to carry out the desired chemistry, such as in a dispersed Cr^{6+} (as CrO_3) in SiO_2. It would be expected that electron-hole

Table 1. Surface areas and pore structures of Co-Al-MCM-41 system. (Reproduced with permission from reference (*21*))

Material	A_{BET} (m^2/g)	V_{BJH} (cm^3/g)	D_{BJH} (nm)
$Co_{0.025}$-$Al_{0.04}$-MCM-41	1192	0.86	2.9
$Co_{0.050}$-$Al_{0.04}$-MCM-41	1077	0.73	2.7
$Co_{0.075}$-$Al_{0.04}$-MCM-41	1068	0.65	2.7
$Co_{0.1}$-$Al_{0.04}$-MCM-41	1074	0.68	2.7
$Co_{0.125}$-$Al_{0.04}$-MCM-41	980	0.67	2.7
$Co_{0.25}$-$Al_{0.04}$-MCM-41	910	0.64	2.9

recombination would occur more readily, and therefore, the Cr/SiO_2 photocatalyst might be expected to be less effective than a Cr/TiO_2 system.

With these concepts in mind, we set out to prepare a wide series of such mixed oxide nanomaterials by advanced sol-gel/aerogel techniques, as well as surfactant assisted syntheses (such as to obtain MCM-41 and MCM-48 materials). We set as a goal to obtain very high surface materials with homogeneous, or near homogeneous mixing of the dopant metal ions with the host TiO_2 or SiO_2 or mixed TiO_2-SiO_2 systems.

We began by preparing a series of cobalt-silica systems with small amounts of aluminum ions for stabilization. Table 1 shows that very light loadings of cobalt were used, and huge surface areas were achieved (*21*).

To our surprise, the photoactivity of these catalysts under visible light irradiation was independent of the loading level of cobalt ions in the Co-Al-MCM-41 framework. However, under UV light there was a rough correlation of activity with the amount of cobalt.

An initial comparison of different metal ion dopants (Cr, Mn, and Co) (*22*) was made (*23*). Under UV light the activity for acetaldehyde conversion to CO_2 was in the order Cr > Mn > Co.

Diffuse-reflectance UV-vis spectra showed the presence of Co^{3+}, Co^{2+}, Cr^{6+}, Cr^{5+}, and Cr^{3+} in these catalysts. The higher oxidations states were implicated as important, especially with visible light (*23*).

These results led to another study that utilized a series of transition metal ion dopants Cr, V, Mn, Fe, Co, and Ni dispersed throughout high surface area TiO_2 as well as TiO_2-SiO_2 (*23*). These were prepared by a modified aerogel procedure designed to ensure rapid gelation and good mixing (*24*), which was especially important for the TiO_2-SiO_2 mix (*25–27*).

Again, surprising results were observed. The best visible light catalysts were for loadings of Cr > Co > Ni > V > Fe > Mn for the conditions of 5% loading transition metal oxide, 20% TiO_2 and 75% SiO_2. Thus, our early work showed that silica based catalysts with transition metal ion doping (well mixed) were active visible light catalysts. The next step was to look at an aerogel type SiO_2 system (*26*).

Table 2. Comparisons of apparent first order rate constants for decrease of acetaldehyde under visible light (*23, 26, 28*). k (min^{-1}), while utilizing different photocatalyst supports. Pure TiO_2 did not show any activity

Doping Ion	*Al-MCM-41*	*SiO_2 Aerogel*	*TiO_2-SiO_2 Aerogel*	*ETS-10 (templated SiO_2-TiO_2)*
Cr	0.016 (0.068 UV)	0.013	0.018	0.002
Co	0.025 (0.098 UV)	0.011	0.015	0.004
Ni		Low	0.006	
Fe		0.0015	0.004	
V		0.0057	0.002	
Mn			0.004	
Cu		Low		
Ag				0.003

Once again, a range of transition metal ions were employed (*27*), but only 1% loadings were used. Surface areas of over 800 m^2/g were achieved. Table 2 summarizes rate constants observed for these aerogel metal ion/SiO_2 and /TiO_2-SiO_2 systems.

Once again, loadings of metal ion over the range of 1-5% did not make much difference. What does seem surprising though, is that SiO_2 based systems seemed to be about equivalent to TiO_2-SiO_2 systems. Another conclusion was that high oxidation states (Cr^{6+} and Co^{3+}) were important in visible light induced photocatalysis.

Up to this point in 2004 we had enjoyed some success producing visble light photocatalysts with transition metal ion doped aerogel SiO_2, aerogel TiO_2-SiO_2, and MCM-41 systems. It was decided to pursue another template system made from TiO_2-SiO_2 (ETS-10) (*28–32*). In our hands, after final heat treatment, surface areas of about 200 m^2/g were achieved (*31*), and pore volumes were rather small (0.12-0.15 cm^3/g).

In this work it was determined that visible light photoactivities were in the same range as with transition metal ion doped SiO_2 and TiO_2-SiO_2 aerogels, and MCM-41 materials (see Table 2). In all cases, when UV light irradiation was compared with visible light irradiation, the photoactivities were much higher with UV, generally 4 to 10 fold higher. As expected, even pure TiO_2 gave good photoactive response under UV light.

One unusual result from the ETS work was that silver ion doping was also investigated, and, although visible light photoactivity was not very high, upon UV irradiation, the rate constant jumped more than 10 fold. This finding did inspire us to look more closely at silver doped TiO_2 systems, which will be covered in a later chapter of this book.

A short summary is now perhaps in order. We investigated a series of high surface area mesoporous oxides with transition metal doping from chromium through silver. Several interesting conclusions were derived:

1. Regarding visible light photoactivities, not much difference was observed comparing aerogels and MCM-41 materials, when all were doped with transition metal ions.
2. Either SiO_2, TiO_2, or TiO_2-SiO_2 can serve as catalyst supports.
3. The best transition metal ions to obtain good visible light photoactivities are Cr and Co.
4. Under UV light, photoactivities jump 4-10 fold over visible light.
5. High oxidation states, such as Cr^{6+} and Co^{3+}, are important in the visible light photochemistry.
6. Silver ion doping appears to be very interesting and worthy of further work.

Of course, we and others have shown that pure TiO_2 is an active UV photocatalyst, but pure SiO_2 is not. And yet SiO_2 can serve as a support for transition metal ions, and photochemistry can take place. Therefore, the photoactive site must be a local site, since we have shown that photoactivities for doped TiO_2, SiO_2, and TiO_2-SiO_2 are very similar.

We continued to seek a mixed oxide, high surface area system that could take advantage of (a) the high surface area of a mesoporous template silica system, (b) the visible light absorbing ability of a selected transition metal ion, and (c) the semiconducting nature of titania. Therefore, our next attempt was to employ an MCM-48 silica structure that was mesoporous and had a continuous 3-dimensional pore system favorable for mass transfer (*33*). We chose ion hexavalent chromium ion as the transition metal, and titania as a semiconducting component.

A synthesis procedure was devised by a combination of Schumacker, et. al. (*34*) and Stöber, et. al. (*35*) methods. Loadings of Cr ion were 1-2% by weight and the SiO_2/TiO_2 ratio was about 3.4 (best results), but ranged as high as 200. The starting reagents $Cr(NO_3)_3$, $Ti(OET)_4$, and $Si[OCH(CH_3)_2]_4$ were mixed, hydrolyzed, and calcined to obtain an MCM-48 (ordered 2 nm pores), surface area of $1400 m^2/g$ and $0.9 cc^3/g$ pore volume. Examination by XRD, ESR, Raman, and diffuse reflectance UV-vis indicated that the structure consisted of (*33*) mostly Cr^{6+} isolated ions with some Cr^{3+}. A UV band at 350 nm is believed to be due to the $O^{2-} \rightarrow Cr^{6+}$ charge transfer band, and substantial absorption was observed in the 400 – 600 nm range as well. No evidence of Cr_2O_3 was found. A UV-band at 250 nm is probably due to $O^{2-} \rightarrow Ti^{4+}$. Raman spectra suggested the presence of very small TiO_2 nanocrystals that were probably on the inner walls of the silica pores [Ti-Cr-MCM-48].

A similar photocatalyst composition was prepared, except the TiO_2 was added by an impregnation method after the formation of the Cr-MCM-48 structure to give [TiO_2-Cr-MCM-48]. According to the sum of characteristic data, this material has the unusual structure of a Cr ion doped MCM-48 support with titania nanoparticles on the exterior of the pores.

These catalysts were examined for the photocatalytic degradation of acetaldehyde. As expected, a pure MCM-48 sample showed no activity under visible light. Two samples Cr-MCM-48 and TiO_2-Cr-MCM-48 showed moderate activity. However, the Ti-Cr-MCM-48 sample, where the Cr ions and TiO_2 were in the same vicinity within the pores, showed greatly enhanced activity (Figure 2) (*33*).

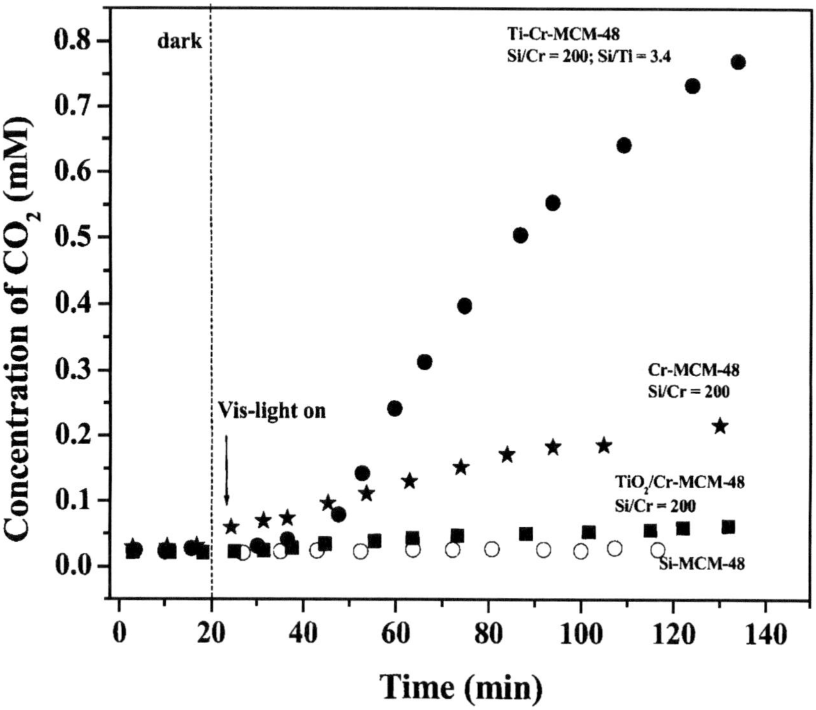

Figure 2. Photocatalytic activity of MCM-48 mesoporous materials for the degradation of acetaldehyde under visible light. Production of CO_2 is plotted vs. time. (Reproduced with permission from reference (33).)

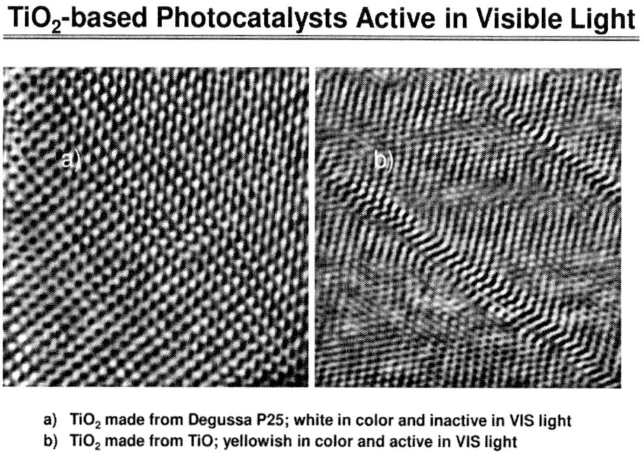

TiO2-based Photocatalysts Active in Visible Light
a)
b)
a) TiO2 made from Degussa P25; white in color and inactive in VIS light
b) TiO2 made from TiO; yellowish in color and active in VIS light

Figure 3. HRTEM of (a) ordered surface of TiO_2, and (b) visible light active TiO_2 sample.

Thus, we gained evidence that chromium ions working in conjunction with nearby titania nanoparticles was an advantageous structure for visible light photocatalysts.

So, through judicious choices of metal oxide combinations, intimately mixed, high surface, and mesoporous, prepared by aerogel or surfactant templating, we learned what was most important for visible light activity.

For another example, a Cr-Al-MCM-41 composition was active for degradation of recalcitrant organohalides under visible and UV light (*36*). The products of photooxidation under visible light were similar to those under UV light, and isolated Cr^{6+} ions supported on a predominantly SiO_2 framework were believed to be the localized photocatalytic sites. The highly dispersed Cr^{6+} ions can be excited to form the corresponding charge-transfer excited state involving an electron transfer from O^{2-} to Cr^{6+}.

These charge transfer excited states have high reactivities due to the presence of electron-hole pairs localized next to each other, compared with such excited states in semiconductors (TiO_2, ZnO, CdS) (*37*), and may be capable of initiating unique photocatalytic activity, especially if surface – OH groups are nearby (*38*).

$$OH^- + h^+ \rightarrow \cdot OH$$

In this work (*36*), we detected hydroxyl radicals using spin traps and ESR, and proposed the following initial reactions:

$$Cl_2C = CClH + \cdot OH \rightarrow Cl_2\dot{C} - CHClOH$$

$$Cl_2\dot{C} - CHClOH \rightarrow Cl_2C = CHOH + Cl\cdot$$

Eventually ·OH, O_2, and Cl· completely degrade trichloroethylene in the presence of hydroxyl groups and water vapor:

$$Cl_2C = CClH + 3/2\ O_2 + H_2O \rightarrow 2CO_2 + 3HCl$$

As time went on, more and more active catalysts were prepared. Indeed, two important discoveries came about recently: (a) non-metal doping such as with carbon, nitrogen, or sulfur could be important, especially in combination with transition metal ion doping (*39*, *40*), and (b) silver salts, such as halides, sulfates, oxides, and even nanocrystalline silver metal (*41*, *42*) led to very active photocatalysts, where visible and UV light were almost equivalent in effectiveness. These interesting more recent discoveries are dealt with in later chapters of this book.

Summary

This brief review has helped put things in perspective especially regarding the nature of the photocatalytic site. Localized electron-hole pairs are highly reactive and efficient if they are on mesoporous high surface area supports. With this type of geometric structure, the short-lived electron-hole pairs are accessible by nearby reagents, and material transfer is probably not rate determining. This is especially important for visible light photocatalysts.

Finally, it needs to be stressed that there is still much to be learned. More and more active catalysts are being discovered, and the true nature of the active sites is still to be elucidated.

Indeed, to show that there is still much to learn, pure TiO_2 has always been described as an active photocatalyst under UV light (3.2 ev band gap), not visible light. This has been confirmed by many research groups.

However, if the TiO_2 is prepared by rapid heating and oxidation of other titanium compounds (such as Ti_2O_3, TiN, TiO, or TiC), in order to maximize crystalline defects, visible light photoactivity is the result (*43*, *44*). What is the structure of these active sites? It is possible that Ti^{3+} and O^- defect sites are involved, but certainly, much more work to understand this phenomenon is necessary. What is fascinating to observe is the high resolution TEM images of the defective (visible light active) TiO_2 and normal (on UV active) TiO_2 (Figure 3). Note the distinct surface defect ridges. How these affect or produce surface photoactive sites is still unclear.

Acknowledgments

The financial support of the Army Research Office for many years is acknowledged with gratitude. Also, for the numerous research students and postdocs in my laboratories over these years, I am deeply grateful and thankful.

References

1. Klabunde, K. J. In *Nanoscale Materials in Chemistry*; Klabunde, K. J., Ed.; Wiley Interscience: New York, NY, 2001; pp 1−13, 223−261.
2. Tok, J. B. H., Ed.; *Nano and Microsensors for Chemical and Biological Surveillance*; Royal Society of Chemistry: Cambridge, U.K., 2008.
3. Fox, M. A.; Dulay, M. T. *Chem. Rev.* **1993**, *93*, 341–57.
4. Linsebigler, A.L.; Lu, G.U.; Yates, J.T., Jr. *Chem. Rev.* **1995**, *95*, 735–58.
5. Hoffmann, M.R.; Martin, S.T.; Choi, W.Y.; Bahnemann, D.W. *Chem. Rev.* **1995**, *95*, 69–96.
6. Anpo, M. *Pure Appl. Chem.* **2000**, *72*, 1265–70.
7. Army Research Office Contractors Meeting. Organic and Inorganic Chemistry, Dynamics and Chemistry of Surface Chemistry and Interface Basic Research Workshop, Savannah, GA, June 23−25, 2009.
8. Martyanov, I. N.; Klabunde, K. J. *Environ. Sci. Technol.* **2003**, *37*, 3448–53.
9. Davidson, R. S.; Pratt, J. E. *Tetrahedron Lett.* **1983**, *24*, 5903–6.
10. Vorontsov, A. V.; Savinov, E. V.; Davydov, L.; Smirniotis, P. G. *Appl. Catal., B* **2001**, *32*, 11–24.
11. Fox, M. A.; Kim, Y. S.; Abel-Wahab, A. A.; Dulay, M. *Catal. Lett.* **1990**, *5*, 369–76.
12. Vorontsov, A. V.; Davydov, L.; Reddy, E. P.; Lion, C.; Savinov, E. N.; Smirniotis, P. G. *New J. Chem.* **2002**, *26*, 732–44.
13. Thompson, T. L.; Panayotov, D. A.; Yates, J. T., Jr.; Martyanov, I.; Klabunde, K. J. *J. Phys. Chem. B* **2004**, *108*, 17857–65.

14. Anpo, M.; Dohshi, S.; Kitano, M.; Hu, Y.; Takeuchi, M.; Matsuoka, M. *Ann. Rev. Mater. Res.* **2005**, *35*, 1–27.
15. Yin, S.; Yamaki, H.; Komatsu, M.; Zhang, Q.; Wang, J.; Tang, Q.; Saito, F.; Sato, T. *J. Mater. Chem.* **2003**, *13*, 2996–3001.
16. Irie, H.; Watanabe, Y.; Hashimoto, K. *J. Phys. Chem. B* **2003**, *107*, 5483–6.
17. Klosek, S.; Raftery, D. *J. Phys. Chem. B* **2001**, *105*, 2815–9.
18. Wu, J. C. S.; Chen, C. H. *J. Photochem. Photobiol., A* **2004**, *163*, 509–15.
19. Joung, S.; Amemiya, T.; Murabayashi, M.; Itoh, K. *Chem. Eur. J.* **2006**, *12*, 5526–34.
20. Rodrigues, S.; Ranjit, K. T.; Uma, S.; Martyanov, I. N.; Klabunde, K. J. *Adv. Mater.* **2005**, *17*, 2467–71.
21. Rodrigues, S.; Uma, S.; Martyanov, I. N.; Klabunde, K. J. *J. Photochem. Photobiol., A* **2004**, *165*, 51–8.
22. Anpo, M. *Stud. Surf. Sci. Catal.* **2000**, *130*, 157–66.
23. Wang, J.; Uma, S.; Klabunde, K. J. *Appl. Catal., B* **2004**, *48*, 151–4.
24. Diao, Y.; Walawender, W.; Sorensen, C.; Klabunde, K. J.; Rieker, T. *Chem. Mater.* **2002**, *14*, 362–8.
25. Dutoit, D. C. M.; Schneider, M.; Hutter, R.; Baiker, A. *J. Catal.* **1996**, *161*, 651–8.
26. Wang, J.; Uma, S.; Klabunde, K. J. *Microporous Mesoporous Mater.* **2004**, *75*, 143–7.
27. Yamashita, H.; Yoshizawa, K.; Ariyuki, S.; Higshimoto, M.; Che, M.; Anpo, M. *Chem. Commun.* **2001**, 435–6.
28. Uma, S.; Rodrigues, S.; Martyanov, I. N.; Klabunde, K. J. *Microporous Mesoporous Mater.* **2004**, *67*, 181–7.
29. Kuznicki, S. M. U.S. Patent 4,853,202, 1989.
30. Anderson, M. W.; Terasaki, O.; Ohsuna, T.; Phiippou, A.; Mackay, S. P.; Ferreira, A.; Rocha, J.; Lidin, S. *Nature* **1994**, *367*, 347–51.
31. Rocha, J.; Ferriera, A.; Lin, Z.; Anderson, M. W. *Microporous Mesoporous Mater.* **1998**, *23*, 253–63.
32. Weckhuysen, B. M.; Wachs, I. E.; Schoonmheydt, R. A. *Chem. Rev.* **1996**, *96*, 3327–49.
33. Rodrigues, S.; Ranjit, K. T.; Sitheraman, U.; Martyanov, I. N.; Klabunde, K. J. *Adv. Mater.* **2005**, *17*, 2467–71.
34. Schumacher, K.; Du Fresne von Hohenesche, C.; Unger, K. K.; Ulrich, R.; Du Chesne, A. D.; Wiesner, U.; Spiess, H. W. *Adv. Mater.* **1999**, *11*, 1194–8.
35. Stöber, W.; Fink, A.; Bohn, E. *J. Colloid Interface Sci.* **1968**, *26*, 62–9.
36. Rodrigues, S.; Ranjit, K. T.; Uma, S.; Martyanov, I. N.; Klabunde, K. J. *J. Catal.* **2005**, *230*, 158–65.
37. Matsuoka, M.; Anpo, M. *J. Photochem. Photobiol., C* **2003**, *3*, 225–52.
38. Zhao, X. S.; Lu, G. Q.; Whittaker, A. K.; Millar, G. J.; Zhu, H. Y. *J. Phys. Chem. B* **1997**, *101*, 6525–31.
39. Yang, X. X; Cao, C.; Hohn, K.; Erickson, L.; Maghirang, R.; Hamal, D.; Klabunde, K. J. *J. Catal.* **2007**, *252*, 296–302.
40. Yang, X. X.; Cao, C.; Erickson, L.; Hohn, K.; Maghirang, R.; Klabunde, K. *J. Catal.* **2008**, *260*, 128–33.
41. Hamal, D.; Klabunde, K. J. *J. Colloid Interface Sci.* **2007**, *311*, 514–22.

42. Rodrigues, S.; Uma, S.; Martyanov, I. N.; Klabunde, K. J. *J. Catal.* **2005**, *233*, 405–10.
43. Martyanov, I. N.; Uma, S.; Rodrigues, S.; Klabunde, K. J. *Chem. Commun.* **2004**, 2476–7.
44. Martyanov, I. N.; Savinov, E. N.; Klabunde, K. J. *J. Colloid Interface Sci.* **2003**, *267*, 111–6.

Chapter 11

Heterogeneous Photocatalysis over High-Surface-Area Silica-Supported Silver Halide Photocatalysts for Environmental Remediation

Dambar B. Hamal* **and Kenneth J. Klabunde**

Department of Chemistry, Kansas State University, Manhattan, KS 66506, USA
***dhamal@ksu.edu**

High-surface area (971 -1232 m^2/g) $AgX-SiO_2$ (X = Cl, Br, I) photocatalysts were synthesized by using hexadecyltrimethylammonium bromide (CTAB) as structure-directing template. Photocatalyst samples were characterized by nitrogen adsorption-desorption isotherm, X-ray diffraction analysis (XRD), and UV-Vis measurements. To demonstrate the environmental remediation processes, photodegradation reactions of acetaldehyde (gas-phase) and rhodamine blue (liquid-phase) were tested as model reactions under UV (320 – 400 nm) and visible light (420 nm) irradiation. It is concluded that silver iodide-silica system is the most active photocatalyst due to its stability, high-surface area, and low band gap as compared with silver bromide and chloride-silica systems for solar remediation of polluted air and water.

1. Introduction

It is well-known that silver and its compounds exhibit both antibacterial and photocatalytic properties (*1–7*). Besides its use as antibacterial agents, research interest in silver based photocatalysts for environmental remediation has been well documented. In this context, Liu et al. (*8*) reported silver ion doped TiO_2 nanoparticles (10 nm) for the degradation of methyl orange (MO) dye in aqueous solution under UV light. It was found that the degradation of MO dye was favored over all silver doped TiO_2 samples and optimum photocatalytic activity was observed when silver doping is 0.05%. Based on the X-ray diffraction investigation, the authors confirmed that silver doping changed the lattice parameters of TiO_2 producing oxygen vacancies, which act as sites for photocatalysis. Gupta et al. (*9*) studied the photocatalytic degradation of a mixture of water soluble dyes (Crystal Violet and Methyl Red) under UV light irradiation over 1.0 wt % silver doped titania photocatalyst. They assumed that the doped silver ions scavenge conduction band electrons, increasing the life time of the photoinduced charge-carriers and enhancing the degradation of dyes oxidatively. Chen et al. (*10*) found the degradation rate of o-cresol with 0.50 wt % Ag/TiO_2 increased two-fold under visible light irradiation. The authors concluded that the formation of a Schottky barrier between the silver and titania prevented the recombination of photoinduced holes and electrons. The other two factors that contributed to improved photoactivity of Ag/TiO_2 are the presence of the defect sites (Ti^{3+}) and reduced band gap due to the presence of Ag and Ag_2O. Likewise, Seery et al. (*11*) reported 5 mol% Ag-TiO_2 nanomaterials for enhanced visible light photocatalysis of Rhodamine 6G dye. This enhancement effect in the dye degradation was attributed to the increased visible absorption capacity due to the presence of silver nanoparticles. Silver doping not only improves the oxidative degradation of water soluble dye but also improves the reductive degradation of N_2O into N_2 over highly dispersed 0.16 wt% Ag^+ ions on TiO_2 (*12*). Lu et al. (*13*) prepared hierarchical 3D Ag-ZnO hollow microspheres for the photocatalytic degradation of Orange G (an anionic azo dye) under UV light. The photocatalytic performance of Ag-ZnO photocatalysts has been attributed to Ag deposits, which not only act as electron sinks to enhance the separation of photoinduced electrons from holes, but also elevates the concentration of the surface hydroxyl radicals (OH) for the higher photodegradation efficiency of the dye.

In addition to Ag-TiO_2 and Ag-ZnO systems, another class of silver based photocatalysts has also been reported. Ouyang et al. (*14*) prepared $AgAlO_2$ (2.8 eV) as visible light active photocatalyst for the photodegradation of the dye alizarin red (AR) and acetaldehyde. Likewise, Maruyama et al. (*15*) prepared delafossite structured alpha-$AgGaO_2$ powder through a cation exchange reaction. The first-principle calculation of alpha-$AgGaO_2$ indicated that the delocalized and dispersed valence band facilitates efficient hole conduction, leading to higher oxidation activity for the decomposition of 2-propanol into acetone and carbon dioxide under both UV (300 – 400 nm) and visible light (420 – 530 nm) irradiation. Hu et al. (*16*) synthesized monoclinic structure silver vandate (Ag_3VO_4) by hydrothermal process. High-photoactivity of silver vandate for the decolorization of azo dye acid red B (ARB) under visible light irradiation was observed when excess vanadium was used. Their results showed that excess vanadium in the preparation increased the crystallinity and suppressed the formation of grain boundaries, while Ag^0 species on the surface of the photocatalyst promoted the electron-hole separation and interfacial charge transfer thereby resulting in an enhanced dye degradation activity. Moreover, NiO loading increased the activity of silver vanadate for the decomposition of phenol and aniline by 11 times. This has been ascribed to the formation of a short-circuited microphotoelectrochemical cell on the surface of NiO-Ag_3VO_4, which enhanced the separation of photogenerated electron-hole pairs. From these findings, it turns out that silver species (Ag^0 or Ag^+) present in single or multimetal oxides plays a central role in the photodegradation efficiency of volatile organics as well as water soluble dyes.

Besides multimetal-oxide silver photocatalysts, silver halides have also been reported as photocatalytic materials due to their abilities to absorb photons. Reddy et al. (*17*) prepared AgCl (3.3 eV)-zeolite A and zeolite L as photocatalysts for the production of oxygen from water. Wang et al. (*18*) reported Ag@AgCl as plasmonic photocatalysts for the photodecomposition of methyl orange (MO), rhodamine blue (RB), and methylene blue (MB) under visible light irradiation. The authors claimed that the enhanced activity of the plasmonic photocatalyst for the degradation of dyes was attributed to the fact that Ag nanoparticles (NPs) absorb visible light photons (the localized surface plasmon state of a Ag nanoparticle lies in the visible region) and hole transfer to the AgCl surface oxidize chloride anions to the reactive chlorine atoms, which are responsible for the dye degradation. The stability of the photocatalyst was maintained by the reduction of reactive chlorine atoms to chloride ions to form AgCl on the photocatalyst surface. Kakuta et al. (*19*) reported silver bromide supported on silica as photocatalyst for hydrogen production from methanol-water solution under UV light. They observed that hydrogen was continuously evolved for 200 h without destruction of AgBr even though Ag^0 species was detected after the reaction. They believed that Ag^0 species formed at the early stage of the reaction acts as sites for hydrogen generation from alcohol radicals formed by photoinduced holes and the silica support enhanced the adsorption of methanol. Furthermore, Zang et al. (*20*) investigated the photocatalytic activity of AgBr-TiO_2 system for degradation of methyl orange (MO) in water under simulated sunlight illumination. The results of X-ray diffraction analysis indicated that Ag^0 species are formed at the early stage of the photocatalytic reaction. It seems that both Ag^0 and AgBr on TiO_2 support contribute to the photodegradation of the MO dye. Hu et al. (*21*) reported AgI-TiO_2 photocatalyst prepared by the deposition-precipitation method. They found that AgI-TiO_2 system demonstrated high efficiency for the degradation of the nonbiodegradable azodyes (reactive red KD-3G, reactive brilliant red X-3B, reactive red KD-2BP, and reactive yellow KD-3G) under visible light irradiation (wavelength > 420 nm). In this system, the authors did not observe the formation of Ag^0 species after the photoreaction of the dyes, indicating the photostability of the AgI-TiO_2 system. From the electron spin resonance studies, they concluded that the hydroxyl radicals formed by the reaction of the valence band holes created on the photoexcited AgI-TiO_2 with the surface hydroxyl group or adsorbed water are the main reactive oxygen species for the photodegradation of the azodyes under visible light. This implies that the formation of Ag^0 species is not necessary to achieve photocatalysis over AgI-TiO_2 system. Most of these early studies did not report the surface areas of silver photocatalysts, except Ag-TiO_2 (< 50 m^2/g) and AgBr-TiO_2 (< 60 m^2/g).

In heterogeneous photocatalysis, however, surface areas are also important for the adsorption of a large amount of reactant molecules and for the desorption of the product molecules. One good strategy to increase the surface area is to employ high-surface area silica support for active photocatalysts. Wang et al.

(*22*) prepared transition metal incorporated titania-silica aero-gels as visible light active photocatalysts that have surface area > 800 m^2/g. These titania-silica mixed systems are very active for photooxidation of gas-phase acetaldehyde under both UV and visible light irradiation. Rodrigues et al. (*23*, *24*) synthesized visible light active photocatalysts of transition metal incorporated MCM-41 (> 900 m^2/g) and MCM-48 (> 1400 m^2/g). These systems were found to be active for gas-phase acetaldehyde decomposition under visible light. The same group of researchers also reported AgBr/Al-MCM-41 (1150 m^2/g) visible light photocatalyst for gas-phase decomposition of acetaldehyde (*25*).

To the best of our knowledge, there has been no report in the field of heterogeneous photocatalysis that involves a comparative study of high-surface area silica supported silver halide photocatalysts from the environmental remediation perspective. In this chapter, synthesis of high-surface area AgX-SiO_2 (X = Cl, Br, I) photocatalysts by using a cationic surfactant was carried out. The photoactivities of silver halide-silica photocatalyst system in both gas-phase (acetaldehyde degradation) and liquid-phase (Rhodamine B degradation) under light irradiation were also investigated.

2. Synthesis of Silica-Supported Silver Halide Photocatalysts

Yang et al. (*26*) reported a TiO_2-SiO_2 mixed system for the photodegradation of methylene blue (MB) dye under UV light irradiation. The main purpose of addition of silica is to achieve high-surface area for photocatalysis. Herein, pure SiO_2 and AgX-SiO_2 samples were prepared by a modified procedure published elsewhere (*27*). Wang et al. (*28*) found that F^- ion has the abilities to lead the formation and growth of CTAB (n-hexadecyltrimethylammonium bromide) micelles and to accelerate the hydrolytic condensation of silicates due to the high electronegativity of fluorine. They proposed that fluoride ions located on the CTA^+ (hexadcyltrimethylammonium cation) is advantageous to the polymerization of silicates around the micellar shell, which facilitates the interaction between CTA^+ and silicate anions. As a result, they found the formation of MCM-48 without destructing the pore ordering with fluoride-induced reduction of CTAB template. Therefore, it is interesting to make use of fluoride ions in the synthesis of AgX-SiO_2 mesoporous samples. In a typical experimental procedure, 4.8972 g of CTAB was dissolved into a solution containing 200 mL water and 100 mL ethanol, followed by addition of 20 mL diethylamine and 5 mL tetrabutyl ammonium fluoride. After stirring vigorously for 10 min, 7.2 mL tetraethyl orthosilicate (98 %) was quickly added to the surfactant solution and the contents of the reaction were stirred for 4h at room temperature. The sample was filtered, washed with an excess amount of deionized water, and placed in a drying cabinet over night. The dried sample was then annealed at 550° C for 12h in air (heating rate 2° C/min) to remove the template. The 5% AgI-SiO_2 photocatalyst was synthesized by the same method, except for the addition of KI solution before the addition of silver nitrate solution and silica precursor. For 5% AgBr-SiO_2 system, the amount of CTAB is enough to

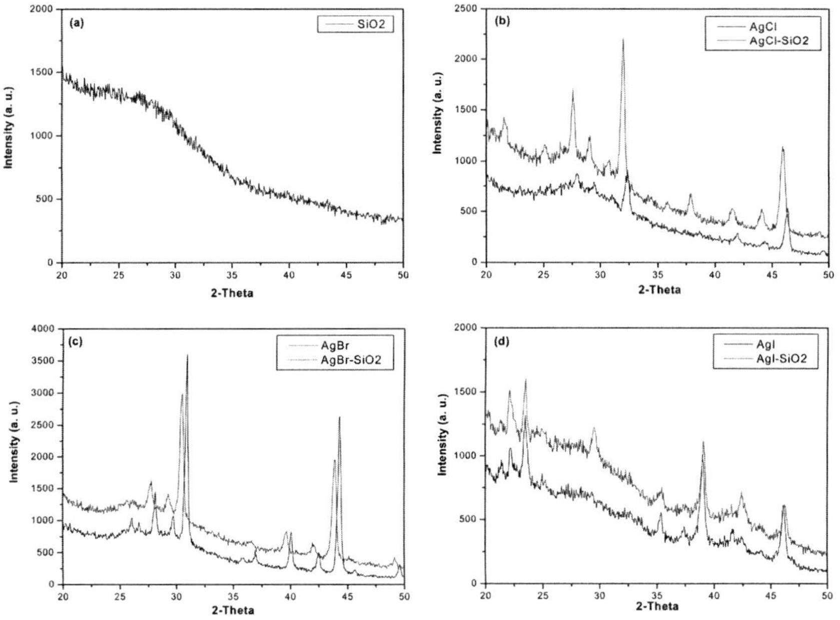

Figure 1. Powder X-ray diffraction analysis of naked and 5% silver halide loaded silica

precipitate Ag ions, while for 5% $AgCl-SiO_2$ system, n-Hexadecyl Trimethyl Ammonium Chloride (CTAC) was used as surfactant and chloride ion source. The rest of synthesis steps and annealing conditions remains the same.

3. Characterization of Silica-Supported Silver Halide Photocatalysts

The annealed samples were characterized with powder X-ray diffraction (XRD), UV-Vis spectroscopy, and N_2 adsorption-desorption measurements. A Scintag XDS 2000 d8 diffractometer with CuKα radiation of wavelength of 0.15406 nm was used to record powder X-ray diffraction patterns of the samples and analyzed from $2\theta = 20°$ to $50°$ with a step size of 0.05° increment. A Carry 500 Scan UV-Vis-NIR Spectrometer was used to measure UV-Vis optical absorption spectra of samples from 200 nm to 800 nm using PTFE as a reference. The band gap energy was determined from the equation, Eg (eV) = $1240/\lambda$(nm), obtained by substituting the values of the velocity of light ($c = 2.99792458 \times 10^8\ ms^{-1}$), Planck constant ($h = 6.62607733 \times 10^{-34}$ Js), and two conversion factor ($1\ eV = 1.60217733 \times 10^{-19}$ C and 1 nm threshold wavelength = 10^{-9} m) in $E = hc/\lambda$. N_2 adsorption-desorption isotherms of the samples (degassed at 150° C in vacuum for 1h) were measured at liquid nitrogen temperature (77 K) on a Quantachrome Instrument (NOVA 1200 series) and specific surface area were calculated by the Branauer-Emmett-Teller (BET) method. The pore-size distribution was derived from desorption isotherms by using the Barrett-Joyner-Halenda (BJH) method.

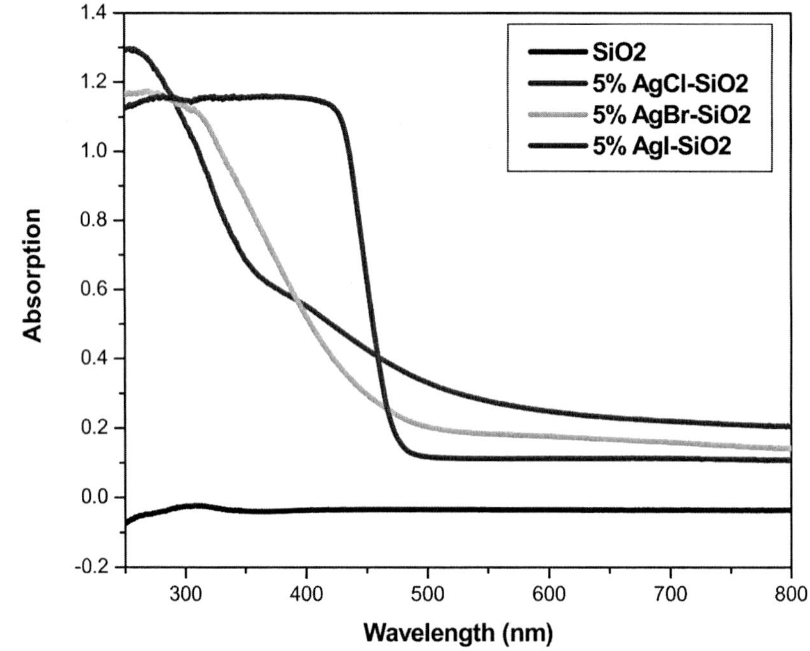

Figure 2. UV-Vis optical absorption spectra of 5% silver halide supported on silica

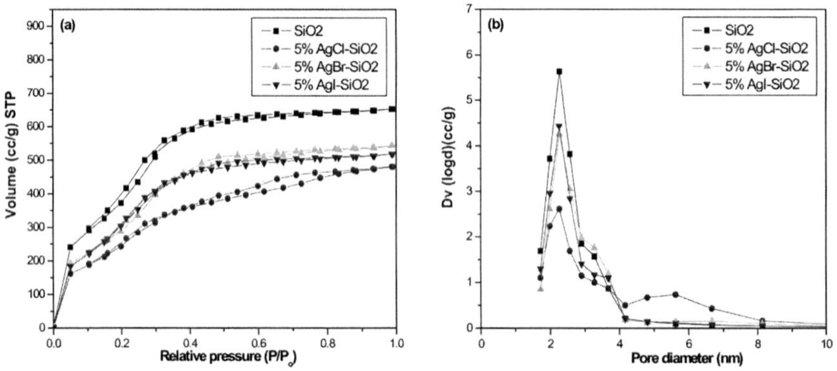

Figure 3. (a) Nitrogen adsorption-desorption isotherm (P_0 = 760 mm Hg) and (b) BJH pore-size distribution of 5% silver halide supported on silica.

Figure 1 shows the X-ray diffraction (XRD) patterns of SiO_2 and AgX-SiO_2 samples annealed at 550° C in air for 12h. No definite XRD patterns were observed for pure silica (Figure 1a), whereas for AgX-SiO_2 samples well defined XRD patterns were observed for AgCl, AgBr, and AgI samples (Figure 1b, 1c, and 1d). This confirmed the formation of silver halides on silica matrix. Furthermore, in an attempt to elucidate the pore ordering, samples were also analyzed at the low angle ($2\theta = 1°$ to $10°$). Unfortunately, no XRD patterns that were characteristic to MCM-41 and MCM-48 (*25, 27*)were observed for the samples at low diffraction

Table 1. Textural properties of silver halide photocatalysts supported on silica

Sample	*BET surface area (m^2/g)*	*Pore volume (cc/g)*	*Mean pore diameter (nm)*
SiO_2	1532	1.13	2.26
5% AgCl-SiO_2	971	0.82	2.27
5% AgBr-SiO_2	1192	0.96	2.26
5% AgI-SiO_2	1232	0.92	2.26

angle. This indicates the absence of the pore-ordering in SiO_2 and AgX-SiO_2 samples.

Figure 2 displays the UV-Vis absorption spectra of 5% silver halides supported on silica. It is clear that pure silica does not show any optical absorption in the range of 200 nm – 800 nm, indicating that pure silica cannot generate electron-hole pairs, when excited with photons of wavelength greater than 200 nm. However, 5% AgCl-SiO_2 and 5% AgBr-SiO_2 samples supported on silica displayed the onset of the optical absorption edge shifted to the visible region, without any pronounced surface plasmon resonance peak for Ag^0 species (*18*). But, it is likely that some Ag_2O species could be formed due to the thermal decomposition of AgCl (m. p. 455° C) and AgBr (m. p. 423° C) during annealing step at 550° C for 12h in air. Nevertheless, the results indicate that these samples have some abilities to absorb visible light photons ($\lambda \geq 420$ nm) to some extent for the generation of excitons. Compared with 5% AgCl-SiO_2 and 5% AgBr-SiO_2 samples, 5% AgI-SiO_2 sample displays a pronounced direct band-gap absorption profile in visible region, indicating that this system has great promise for visible-light-induced photocatalytic pollution remediation as reported in the literature (*21*). Moreover, the nature of the optical absorption profiles of silver halides supported on silica indicates that these are photosensitive and have the abilities to absorb photons to create electron-hole pairs, which can be used in photocatalysis.

Figure 3a and 3b show the nitrogen adsorption-desorption isotherms measured at liquid nitrogen temperature (77 K) and the BJH (Barret-Joyner-Halenda) pore-size distribution derived from the desorption isotherm, respectively. The adsorption-desorption isotherms are reversible for all samples and are of type IV (*24, 27*). A sharp inflection at relative pressures (P/P_o) between 0.1 and 0.4 was observed due to capillary condensation within the mesopores. The nature of the isotherms and BJH pore-size distribution indicate that pure SiO_2, 5% AgBr-SiO_2, and 5% AgI-SiO_2 samples exhibit the presence of nearly uniform cylindrical mesopores as compared with 5% AgCl-SiO_2 sample that shows a broad pore-size distribution.

Table 1 shows the BET (Brunauer-Emmett-Teller) specific surface area, pore volume, and pore diameter of silver halides supported on silica. It was found that pure silica has the highest surface area > 1500 m^2/g and pore volume > 1 cc/g. The high-surface area and pore volume of pure silica sample results from the structure directing nature of the surfactant (CTAB). Even though silver halide loaded silica

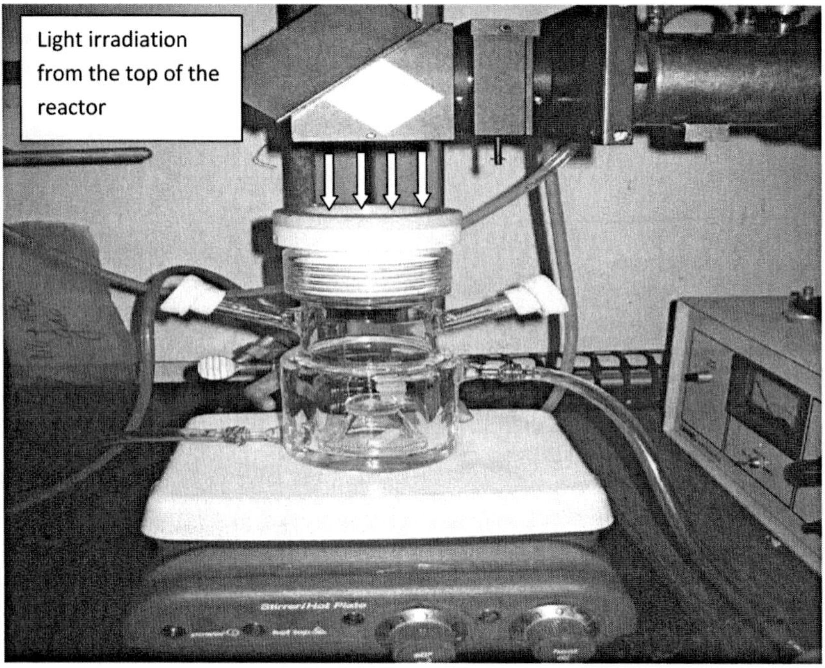

Figure 4. Reactor used for both gas-phase and liquid-phase photocatalytic reactions

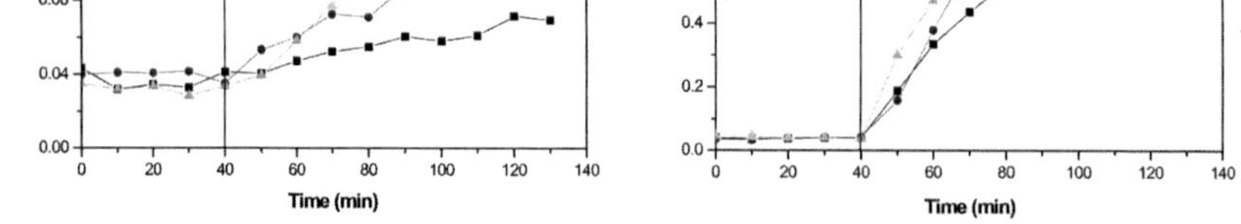

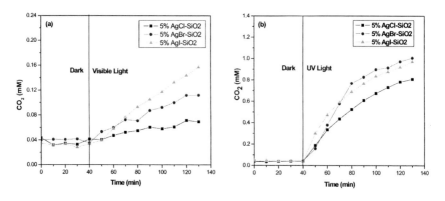

Figure 5. Evolution of carbon dioxide from the degradation of gas-phase acetaldehyde over 5% silver halide photocatalysts supported on high-surface area silica host (a) under visible and (b) UV light irradiation (λ = 320 – 400 nm).

samples have surface areas and pore volumes lower than that of pure silica, the formation of silver halides still retains the structural integrity of the host.

Table 2. Threshold wavelength and band gap energies of silver halides

Sample	*Onset of excitation (nm)*	*Band gap (eV)*
5% $AgCl-SiO_2$	395	3.14
5% $AgBr-SiO_2$	447	2.77
5% $AgI-SiO_2$	474	2.61

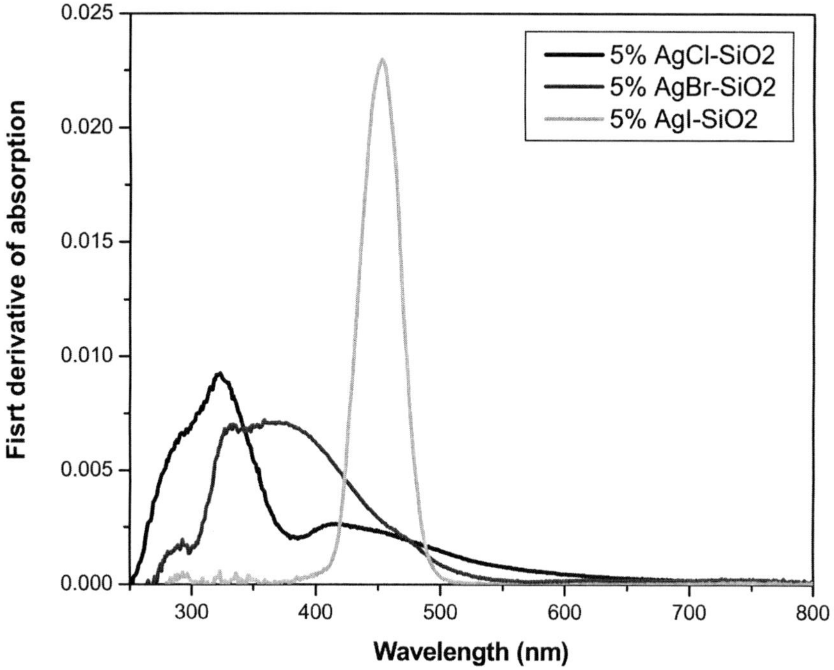

Figure 6. First derivative curves of UV-Vis absorption spectra of silver halide supported on silica photocatalysts

4. Gas-Phase Heterogeneous Photocatalysis over Silica-Supported Silver Halide Photocatalysts

It is well-known that heterogeneous photocatalysis has been widely studied for solar environmental remediation. In the past years, the photocatalytic degradation of acetaldehyde over the doped titanium dioxide semiconductor photocatalysts has been reported because acetaldehyde is a volatile organic compound and is considered as an indoor air pollutant (*29–31*). Now, it seems interesting to investigate the photocatalytic properties of silver halides supported on silica matrix for solar environmental remediation. In this chapter, the use of $AgX-SiO_2$ as photocatalysts has been explored for the photodegradation of a volatile organic compound, acetaldehyde. Figure 4 shows the photocatalytic reactor used in the current study. 100 mg sample in a circular glass disc holder was placed in 305 mL cylindrical reactor with quartz window at the top and 100 μL liquid CH_3CHO injected at the bottom of the reactor. Before light irradiation, acetaldehyde-air mixture was constantly stirred in the dark for a certain time period at 298 K so that adsorption-desorption equilibrium was achieved. Then a 35 μL sample containing gaseous acetaldehyde was injected into the injection port of Shimadzu GC-MS QP5000. To follow the photocatalytic degradation of acetaldehyde in gas-phase, the amount of evolved CO_2 *vs.* irradiation time was demonstrated for simplicity.

Figure 5a shows the evolution of carbon dioxide from the photodegradation of gas-phase acetaldehyde under visible light ($\lambda \geq 420$ nm) over silver chloride, bromide, and iodide supported silica photocatalysts.

The visible light activity of these systems follows the order: AgI-SiO_2 > AgBr-SiO_2 > AgCl-SiO_2. The high-photoactivity of silver iodide-silica system under visible light could be attributed to high-surface area (Table 1), low band gap (Table 2), and increased absorption of visible light photons (Figure 6). However, under UV light irradiation, 5% AgBr-SiO_2 and 5% AgI-SiO_2 displayed comparable but higher photoactivities for acetaldehyde degradation than 5% AgCl-SiO_2 (Figure 5b).

5. Liquid-Phase Heterogeneous Photocatalysis over Silver Halide-Silica Photocatalysts

Colored dyes released into the environment from textile and dyestuff industries have polluted effluent water streams. In this context, heterogeneous photocatalysis over mixed titania and silica photocatalyst systems has also been employed for the purification of dye polluted water under UV light irradiation (*32*, *33*). Therefore, it is also interesting to explore the potential use of silver halide photocatalysts for the purification of polluted water. For this purpose, we consider Rhodamine B (RB) dye as model organic dye, a water pollutant, and study its photodegradation over AgCl, AgBr, and AgI photocatalyst supported on SiO_2 host matrix by using a photocatalytic reactor shown in Figure 4. Rhodamine B gives the absorption maxima at 554 nm in its UV-vis spectrum profile. If the dye undergoes photodecolorization, the absorption peak at 554 nm declines over irradiation time (*33*). From the UV-vis measurement, it was found that silver halide photocatalysts supported on high-surface area silica matrix have the abilities to degrade the Rhodamine B dye under broad band UV light illumination ($\lambda = 320 - 400$ nm). However, the photocatalytic degradation results showed that the dye degradation is least favored in presence of silver chloride as compared to silver bromide and iodide-silica systems (Figure 7). Conceivably, this difference in photodecolorization of RB dye could be ascribed to the surface-area-effect because specific surface area of AgCl-SiO_2 is lower than those of AgBr-SiO_2 and AgI-SiO_2 systems (Table 1), leading to a less amount of RB dye adsorbed and lower activity under UV light excitation. On the basis of UV-light-mediated photodegradation of the RB dye aqueous solution, silver halide-silica systems follows the photoactivity order: AgI-SiO_2 > AgBr-SiO_2 > AgCl-SiO_2, which is similar to the gas-phase photodegradation of acetaldehyde under visible light

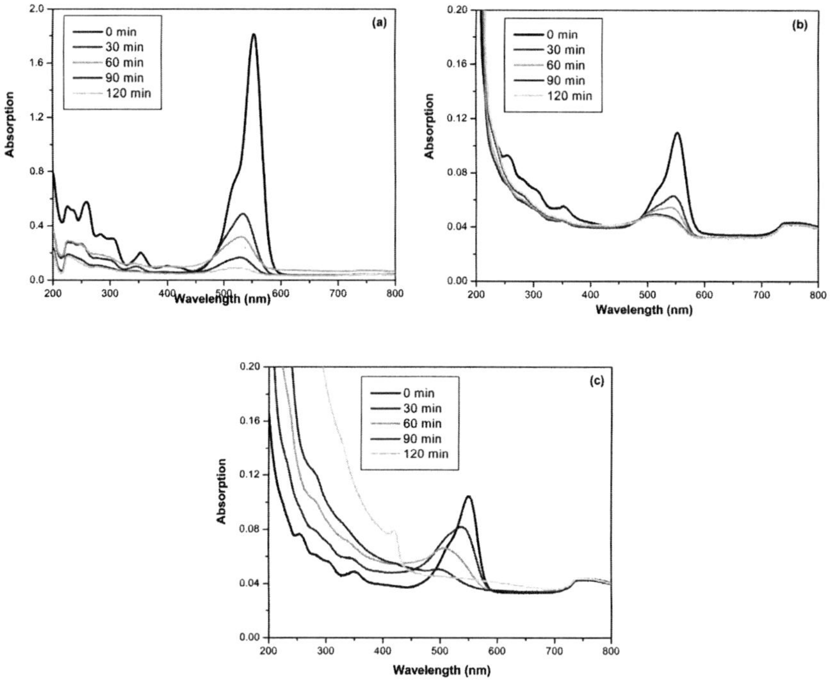

Figure 7. UV-Vis absorption spectra of Rhodamine B aqueous solution in presence of (a) 5% AgCl-SiO$_2$, (b) 5% AgBr-SiO$_2$, and (c) 5% AgI-SiO$_2$ under UV light illumination. Experimental conditions: RB dye concentration = 2 x 10^{-5} M, photocatalyst = 0.1 g, total volume = 150 mL

(Figure 5a). Interestingly, only AgI-SiO$_2$ system has the ability to degrade RB dye under visible light (Figure 8), and other two systems (AgBr-SiO$_2$ and AgCl-SiO$_2$) are inactive for RB dye degradation under visible light. One plausible reason could be due to the refractory nature of the Rhodamine B dye, a non-azo xanthene dye that contains four aromatic nuclei as compared with acetaldehyde. Another reason could be due to the ability of AgI-SiO$_2$ system to absorb a large amount of visible light photons (Figure 6).

In the present study, it was observed that AgI-SiO$_2$ photocatalyst is the most active and the most stable system in both gas-phase (acetaldehyde) and liquid-phase (Rhodamine B) heterogeneous photocatalytic degradation reaction conditions. This was confirmed with the fact that the surface of both silver bromide and chloride-silica photocatalysts instantly turns black in color during the initial stages of light excitation. It is believed that the appearance of a black color on the surface of silver bromide and chloride photocatalysts is due to the presence of metallic silver particles formed by the photocatalytic reduction of silver ions from silver bromide and chloride. The physical and electrochemical properties of silver halide compounds (Table 3) further confirm that the reduction of silver ions from AgI is the least favorable. Consequently, AgI-SiO$_2$ system retains its commendable photocatalytic activity over time.

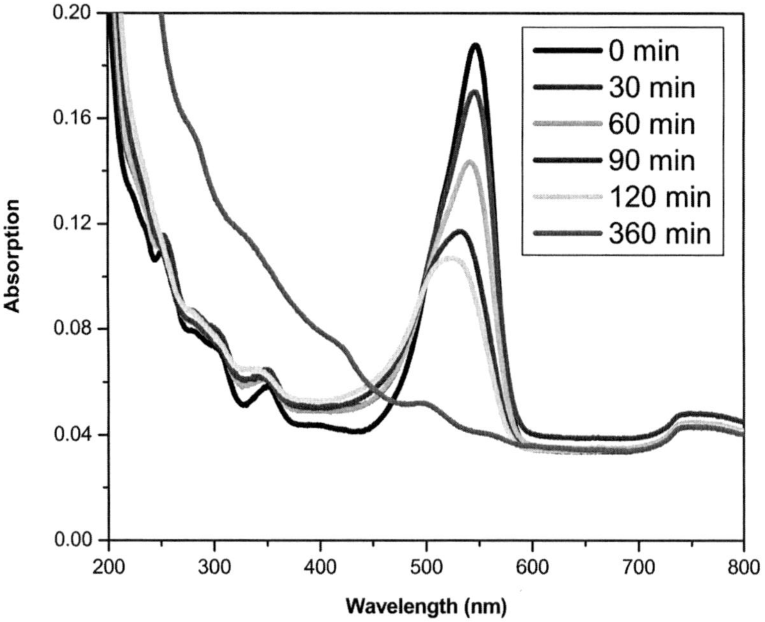

Figure 8. UV-Vis absorption spectra of Rhodamine B aqueous solution in presence of 5% AgI-SiO_2 photocatalyst under visible light illumination ($\lambda \geq 420$ nm). Experimental conditions: RB concentration = 5.4×10^{-5}M, photocatalyst = 0.15 g, total volume = 150 mL

Table 3. Physical and electrochemical properties of pure silver halides

Silver halide	*Melting point*	*Solubility product*	*Reduction potential*
AgCl	455° C	1.8×10^{-10}	0.22 V
AgBr	432° C	5.4×10^{-13}	0.07 V
AgI	558° C	8.9×10^{-17}	-0.15 V

6. Conclusions

High-surface-area silver halide photocatalysts supported on mesoporous silica host prepared by a surfactant templated method were introduced and their textural and photocatalytic characteristics were demonstrated. The characterization and photodegradation results (gas-phase and liquid-phase) revealed that the modification of high-surface area silica with silver halides leads to photocatalytically active mesoporous materials for solar environmental remediation. Among the chosen silver halides, silver iodide has beneficial effects on maintaining the structural integrity of the host and exhibits high photoactivities for pollution remediation of polluted air and water environments under both UV and visible light illumination.

References

1. Nair, L. S.; Laurencin, C. T. Silver nanoparticles: synthesis and therapeutic applications. *J. Biomed. Nanotechnol.* **2007**, *3*, 301–316.
2. Smetana, A. B.; Klabunde, K. J.; Marchin, G. R.; Sorensen, C. M. Biocidal Activity of Nanocrystalline Silver Powders and Particles. *Langmuir* **2008**, *24*, 7457–7464.
3. Buckley, J. J.; Gai, P. L.; Lee, A. F.; Olivi, L.; Wilson, K. Silver carbonate nanoparticles stabilized over alumina nanoneedles exhibiting potent antibacterial properties. *Chem. Commun.* **2008**, *34*, 4013–4015.
4. Feng, Q. L.; Wu, J.; Chen, G. Q.; Kim, T. N.; Kim, J. O. A mechanistic study of the antibacterial effect of silver ions on Escherichia coli and Staphylococcus aureus. *J. Biomed. Mater. Res.* **2000**, *52*, 662–668.
5. Lan, Y.; Hu, C.; Hu, X.; Qu, J. Efficient destruction of pathogenic bacteria with $AgBr/TiO_2$ under visible light irradiation. *Appl. Catal., B* **2007**, *73*, 354–360.
6. Hu, C.; Guo, J.; Qu, J.; Hu, X. Photocatalytic Degradation of Pathogenic Bacteria with AgI/TiO_2 under Visible Light Irradiation. *Langmuir* **2007**, *23*, 4982–4987.
7. Elahifard, M. R.; Rahimnejad, S.; Haghighi, S.; Gholami, M. R. Apatite – coated $Ag/AgBr/TiO_2$ visible-light photocatalyst for destruction of bacteria. *J. Am. Chem. Soc.* **2007**, *129*, 9552–9553.
8. Liu, Y.; Liu, C.-y.; Rong, Q.-h.; Zhang, Z. Characteristics of the silver-doped TiO_2 nanoparticles. *Appl. Surf. Sci.* **2003**, *220*, 7–11.
9. Gupta, A. K.; Pal, A.; Sahoo, C. Photocatalytic degradation of a mixture of Crystal Violet (Basic Violet 3) and Methyl Red dye in aqueous suspensions using Ag^+ doped TiO_2. *Dyes Pigm.* **2006**, *69*, 224–232.
10. Chen, H. W.; Ku, Y.; Kuo, Y. L. Photodegradation of o-cresol with Ag deposited on TiO_2 under visible and UV light irradiation. *Chem. Eng. Technol.* **2007**, *30*, 1242–1247.
11. Seery, M. K.; George, R.; Floris, P.; Pillai, S. C. Silver doped titanium dioxide nanomaterials for enhanced visible light photocatalysis. *J. Photochem. Photobiol., A* **2007**, *189*, 258–263.
12. Sano, T.; Negishi, N.; Mas, D.; Takeuchi, K. Photocatalytic Decomposition of N_2O on Highly Dispersed Ag^+ Ions on TiO_2 Prepared by Photodeposition. *J. Catal.* **2000**, *19*, 71–79.
13. Lu, W.; Gao, S.; Wang, J. One-Pot Synthesis of Ag/ZnO Self-Assembled 3D Hollow Microspheres with Enhanced Photocatalytic Performance. *J. Phys. Chem. C* **2008**, *112*, 16792–16800.
14. Ouyang, S.; Zhang, H.; Li, D.; Yu, T.; Ye, J.; Zou, Z. Electronic Structure and Photocatalytic Characterization of a Novel Photocatalyst $AgAlO_2$. *J. Phys. Chem. B* **2006**, *110*, 11677–11682.
15. Maruyama, Y.; Irie, H.; Hashimoto, K. Visible Light Sensitive Photocatalyst, Delafossite Structured alpha-$AgGaO_2$. *J. Phys. Chem. B* **2006**, *110*, 23274–23278.
16. Hu, X.; Hu, C.; Qu, J. Preparation and visible-light activity of silver vanadate for the degradation of pollutants. *Mater. Res. Bull.* **2008**, *43*, 2986–2997.

17. Reddy, V. R.; Currao, A.; Calzaferri, G. Zeolite A and zeolite L monolayers modified with AgCl as photocatalyst for water oxidation to O_2. *J. Mater. Chem.* **2007**, *17*, 3603–3609.
18. Wang, P.; Huang, B.; Qin, X.; Zhang, X.; Dai, Y.; Wei, J.; Whangbo, M.-H. Ag@AgCl: a highly efficient and stable photocatalyst active under visible light. *Angew. Chem., Int. Ed.* **2008**, *47*, 7931–7933.
19. Kakuta, N.; Goto, N.; Ohkita, H.; Mizushima, T. Silver Bromide as a Photocatalyst for Hydrogen Generation from CH_3OH/H_2O Solution. *J. Phys. Chem. B* **1999**, *103*, 5917–5919.
20. Zang, Y.; Farnood, R. Photocatalytic activity of $AgBr/TiO_2$ in water under simulated sunlight irradiation. *Appl. Catal., B* **2008**, *79*, 334–340.
21. Hu, C.; Hu, X.; Wang, L.; Qu, J.; Wang, A. Visible-light-induced photocatalytic degradation of azodyes in aqueous AgI/TiO_2 dispersion. *Environ. Sci. Technol.* **2006**, *40*, 7903–7907.
22. Wang, J.; Uma, S.; Klabunde, K. J. Visible light photocatalytic activities of transition metal oxide/silica aerogels. *Microporous Mesoporous Mater.* **2004**, *75*, 143–147.
23. Rodrigues, S.; Uma, S.; Martyanov, I. N.; Klabunde, K. J. Visible light induced photocatalytic activity for degradation of acetaldehyde using transition metal incorporated Al-MCM-41 (aluminum doped silica zeolitic material). *J. Photochem. Photobiol., A* **2004**, *165*, 51–58.
24. Rodrigues, S.; Ranjit, K. T.; Uma, S.; Martyanov, I. N.; Klabunde, K. J. Single-step synthesis of a highly active visible-light photocatalyst for oxidation of a common indoor air pollutant: Acetaldehyde. *Adv. Mater.* **2005**, *17*, 2467–2471.
25. Rodrigues, S.; Uma, S.; Martyanov, I. N.; Klabunde, K. J. AgBr/Al-MCM-41 visible-light photocatalyst for gas-phase decomposition of CH_3CHO. *J. Catal.* **2005**, *233*, 405–410.
26. Yang, J.; Zhu, L.; Zhang, J.; Zhang, Y.; Tang, Y. Synthesis of nanosized TiO_2/SiO_2 catalysts by the ultrasonic microemulsion method and their photocatalytic activity. *React. Kinet. Catal. Lett.* **2007**, *91*, 21–28.
27. Boote, B.; Subramanian, H.; Ranjit, K. T. Rapid and facile synthesis of siliceous MCM-48 mesoporous materials. *Chem. Commun.* **2007**, *43*, 4543–4545.
28. Wang, L.; Zhang, J.; Chen, F.; Anpo, M. Fluoride-Induced Reduction of CTAB Template Amount for the Formation of MCM-48 Mesoporous Molecular Sieve. *J. Phys. Chem. C* **2007**, *111*, 13648–13651.
29. Kang, M.; Ko, Y.-R.; Jeon, M.-K.; Lee, S.-C.; Choung, S.-J.; Park, J.-Y.; Kim, S.; Choi, S.-H. Characterization of Bi/TiO_2 nanometer sized particle synthesized by solvothermal method and CH_3CHO decomposition in a plasma-photocatalytic system. *J. Photochem. Photobiol., A* **2005**, *173*, 128–136.
30. Yang, X.; Cao, C.; Hohn, K.; Erickson, L.; Maghirang, R.; Hamal, D.; Klabunde, K. Highly visible-light active C-doped and V-doped TiO_2 for degradation of acetaldehyde. *J. Catal.* **2007**, *252*, 296–302.

31. Hamal, D. B.; Klabunde, K. J. Synthesis, characterization, and visible light activity of new nanoparticle photocatalysts based on silver, carbon, and sulfur-doped TiO_2. *J. Colloid Interface Sci.* **2007**, *311*, 514–522.
32. Marugan, J.; Lopez-Munoz, M.-J.; van Grieken, R.; Aguado, J. Photocatalytic Decolorization and Mineralization of Dyes with Nanocrystalline TiO_2/SiO_2 Materials. *Ind. Eng. Chem. Res.* **2007**, *46*, 7605–7610.
33. Wilhelm, P.; Stephan, D. Photodegradation of rhodamine B in aqueous solution via SiO_2@TiO_2 nanospheres. *J. Photochem. Photobiol., A* **2007**, *185*, 19–25.

Chapter 12

An Inorganic Oxide TiO_2-SiO_2-Mn Aerogel for Visible-Light Induced Air Purification

Kennedy K. Kalebaila† and Kenneth J. Klabunde*

Chemistry Department, Kansas State University, Manhattan, KS 66506
***E-mail: kenjk@ksu.edu; Phone: 785-5326849; Fax: 785-532-6666**
†E-mail: kkalebai@ksu.edu

An eco-friendly high surface area TiO_2-SiO_2-Mn aerogel degrades acetaldehyde (CH_3CHO) under visible-light illumination (>420 nm) as evidenced by production of CO_2 detected by GC-MS. When CH_3CHO vapors come into contact with TiO_2-SiO_2-Mn, about 50 % of the initial CH_3CHO is destroyed with formation of nearly 5 % CO_2 (initial rate of 0.40 mmols/min). The rate of reaction increases 6-fold (2.57mmols/min) when the pollutant and catalyst are physically mixed and kept under constant agitation. Furthermore, 80 % of CH_3CHO is consumed under agitation with a 2-fold increase in detectable CO_2. ESR indicates that Mn exists as Mn^{+4} prior to reaction with CH_3CHO and is reduced to Mn^{2+} during the reaction. The catalytic circle is regenerated by reoxidation of Mn^{2+} in air under visible-light. Calcination of used TiO_2-SiO_2-Mn in air regenerates the photocatalytic activity. Aerogel samples of TiO_2-SiO_2-Fe did not photooxidize CH_3CHO despite having visible-light absorption.

1. Introduction

There is a growing trend globally to develop new materials and technologies in environmental remediation in particular the purification of air. Air purification is essential in removing contaminants such as mold spores, smoke particulates and volatile organic compounds (VOCs) from the air. Researchers have looked to the development of novel materials for decomposition of both indoor and outdoor pollutants. These materials are designed not only to adsorb pollutants but also to bring about a near or complete mineralization of the adhered molecules by utilizing solar energy in the presence of air via a process known as photocatalysis. Titanium dioxide (TiO_2), a UV semiconductor with a bandgap of 3.2 eV, has been extensively utilized as a photocatalyst in environmental remediation (*1–4*). However, the need for higher UV energy hampers widespread use of this material, and since UV light accounts for only about 4 % of sunlight the systems are less efficient in the utilization of the solar energy.

Extension of the photocatalytic activity of TiO_2 in the visible region has attracted remarkable research interest as visible-light constitutes about 45 % of the solar energy. In 1986 Sato reported that nitrogen-doped TiO_2 displayed visible-light photocatalysis, and since then several research groups have shown that N 2p states play a role by mixing with O 2p states leading to a red shift in the wavelength (*5–7*). Other nonmetals such as S and C have also been incorporated into a TiO_2 matrix for visible-light activated materials with the nonmetals replacing some of the O in the TiO_2 lattice. In a report by this group, the incorporation of S and C into TiO_2 was achieved by sol-gel reaction using thiocyanate as sources of S and C (*8*). Another group has shown that S incorporation into TiO_2 can be obtained by annealing TiS_2 to form TiO_2 at 600 °C. The residual S was shown to occupy O sites in TiO_2 and the band gap lowering was attributed to the mixing of S 3p and O 2p states (*9*, *10*).

Transition metal doping into the TiO_2 framework to achieve visible-light absorption has been another avenue that has received attention. Anpo and Takeuchi have recently prepared Cr-doped TiO_2 which was photocatalytic in the decomposition of NO into O_2 and N_2O above 450 nm (*11*). Furthermore, a visible-light responsive Fe-TiO_2 material prepared by chemical vapor deposition has been reported and was found to destroy azo dyes in wastewater (*12*). Our research group has been very active in this area with reports of Cr-doped TiO_2 supported on MCM-48 matrix as well as Ag-TiO_2 with incorporated C and S (*8*, *13*). Although, we have studied a number of transition metals including Mn doped on either TiO_2 or SiO_2 as the support for both UV and visible-light photocatalysis, only Cr, V and Co showed the best results. Moreover, the photoactivity of V and Cr-doped TiO_2 under visible-light was enhanced by preparing these materials via the sol-gel method to produce high surface area aerogels (supercritically dried gels). Additionally, the high oxidation states of the cations (V^{5+} and Cr^{6+}) played a role in the destruction of acetaldehyde, a model pollutant (*14–16*).

Presently, we are interested in the design of novel nanostructured mixed metal oxides that will not only have high capacity for adsorption of toxins but also utilize visible-light to photooxidize the adsorbents in air. More importantly, this material should be nontoxic to the environment. The reported V- and Cr- doped TiO_2 are effective in the adsorption and destruction of acetaldehyde using visible-light, but these metal dopants are toxic. Chromium (VI) is absorbed into the body as a soluble chromate (CrO_4^{-2}) which is reduced to the harmful Cr^{+5} which is believed to damage DNA via phosphate links. Vanadium (V) in the form of V_2O_5 enters the body by inhalation and is also known to cause damage to tissue cells (*17*, *18*). Thus, in order to avoid toxicity, and to increase surface area while still preserving the anatase phase, we sought to develop a visible active photocatalyst based on environmentally friendly inorganic oxide gels. We chose to incorporate either Fe or Mn ions into a TiO_2-SiO_2 framework prepared by a modified sol-gel process and supercritical drying to produce aerogel materials. In this study, we describe the synthesis, characterization and visible photocatalytic behavior of TiO_2-SiO_2-Mn aerogels as well as the role of the doped Mn ions. In addition, the results from Fe doped and the reference V-doped TiO_2-SiO_2 aerogels are presented.

2. Experimental

2.1. Materials

With the exception of iron(III) nitrate ($Fe(NO_3)_3.9H_2O$, Alfa Aesar), Titanium(IV) isopropoxide, ($Ti(O^iPr)_4$, 98 %), tetraorthosilicate (TEOS, 99 %), manganese(III) acetylacetonate (Mn(III) acac) and vanadium(V) oxytriisopropoxide ($VO(O^iPr)_3$, were all purchased from Aldrich and the chemicals were used as received. Acetylacetoacetate, methanol (MeOH), and ethanol (EtOH) all from Fisher Chemicals were used without further handling. Distilled water and ammonium hydroxide (36 %) were targeted to induce gelation of titania-silica as well as the metal doped titania-silica matrices.

2.2. Synthesis of TiO_2-SiO_2 Gels

Titania-silica (TiO_2-SiO_2) gel composites were prepared by dispersing 32 mL (0.11 mols) of $Ti(O^iPr)_4$ in 90 mL of MeOH and 14 mL of acetylacetoacetate under vigorous stirring producing a yellowish solution. To this solution was added 24 mL (0.11 mols) of TEOS previously dispersed in 90 mL MeOH. It was necessary to add acetylacetoacetate to MeOH before the addition of $Ti(O^iPr)_4$ during the synthesis, otherwise precipitation of TiO_2 occurred. The resultant mixture was allowed to stir at room temperature for 1 hour prior to addition of 4 mL of 3.72 mL NH_4OH (36 %) in 23 mL of H_2O to induce gelation. After 24 hours a thick gel had formed in the entire volume of the solution mixture and was left to age for 3 days.

2.3. Synthesis of Metal Incorporated TiO_2-SiO_2

Iron-doped TiO_2-SiO_2 gels were achieved by dissolving 4.36g (0.011 mols) of $Fe(NO_3)_3.9H_2O$ in MeOH followed by addition of this solution to a freshly prepared solution of $Ti(O^iPr)_4$ and TEOS as described in Section 2.2. After stirring for 30 minutes, an aqueous solution of NH_4OH was slowly added to initiate the formation of gels. Similarly, gel composites of TiO_2-SiO_2 doped with Mn and V were obtained by using 3.80g (0.011 mols) of Mn(III) acac or 2.55 mL of $VO(O^iPr)_3$, previously dissolved in 10 mL of MeOH, respectively. These salt solutions were then added to separate solutions of $Ti(O^iPr)_4$ and TEOS in methanol. When a basic solution was added to the vanadium precursor solutions, precipitation occurred immediately and hence only water was used to form TiO_2-SiO_2-V gels. The wet gels were aged for 3 days in the mother liquor before further processing. The calculated amounts of transition metal precursors used represent a doping level of 10 mol % of the metal ions in the TiO_2-SiO_2 matrix.

2.4. Supercritical Drying and Calcination of the Wet Gels

The wet gels (undoped TiO_2-SiO_2 and TiO_2-SiO_2 doped with Fe, V and Mn) were separately loaded into a cylindrical glass tube and loaded into a PARR 4843 High Pressure Autoclave and sealed under air for supercritical drying to make aerogels. The autoclave was set to reach 265 °C in 4 hours, after which the crude products were removed and ground to a powder and then calcined at 500 °C for 2 hours using a CARBOLITE CWF 1100 Furnace. A white product was obtained in the case of undoped TiO_2-SiO_2 aerogels, while the calcined aerogels for TiO_2-SiO_2 doped with Fe, V and Mn oxides were orange, purple and brown, respectively. The products were further ground to a fine powder and stored for further use. In future discussions, the doped aerogel samples will be denoted to as TiO_2-SiO_2-Fe, TiO_2-SiO_2-V, TiO_2-SiO_2-Mn and TiO_2-SiO_2 for undoped TiO_2-SiO_2.

2.5. Characterization Techniques

Powder X-ray diffraction (PXRD) was used to identify the crystallite phase present in TiO_2-SiO_2 metal doped aerogels and to determine the approximate particle size using the Scherer equation. PXRD data was obtained on a Scintag XDS 200D8 diffractometer equipped with a copper anode (Cu Kα radiation of wavelength 0.15406 nm) from 2-75° (2θ). Samples for PXRD study were prepared by spreading the powdered aerogels on a quartz background. The BET (Brunauer-Emmett-Teller) surface area and the BJH (Barret-Joyner-Halenda) pore size distribution of the aerogels were obtained from nitrogen sorption isotherms acquired at 77 K using a 30 sec equilibrium interval on a NOVA 1000 series Quantachrome Instrument. The samples were degassed at 100 °C for about 10 hrs prior to analysis (using Nova software version 1.11) to remove any adsorbed molecules (*19*).

Solid state UV-Vis absorption features of the TiO_2-SiO_2 and metal doped aerogels were studied on a CARY 500 UV/VIS/NIR spectrophotometer equipped with an integrating sphere over a wavelength range of 200-800 nm. The instrument was calibrated with 1 micron polytetrafluoroethylene (PTFE, Aldrich), a light reflecting powder as a reference material prior to sample analysis. Electron Spin Resonance (ESR) was conducted at 77 K to determine the active species in the doped TiO_2-SiO_2 aerogels. ESR spectra were collected using a Bruker EMX spectrometer equipped with an ER041XG microwave bridge at a frequency of 9.34 GHz and power of 2.02mW.

2.6. Photocatalysis of Acetaldehyde

The photocatalytic degradation experiments were carried out in a cylindrical glass reactor of a total volume of 305 mL. The powdered aerogels (100 mg) were placed on a circular glass dish supported on a circular stand and placed in the reactor. Typically, 100 μL of liquid acetaldehyde (CH_3CHO) was injected into the sides of the reactor through a rubber septum onto the bottom of the reactor. The reactor and its contents were kept at 25 °C using a ThermoFisher Temperature bath to allow the gas-solid equilibration of CH_3CHO with the solid catalysts under

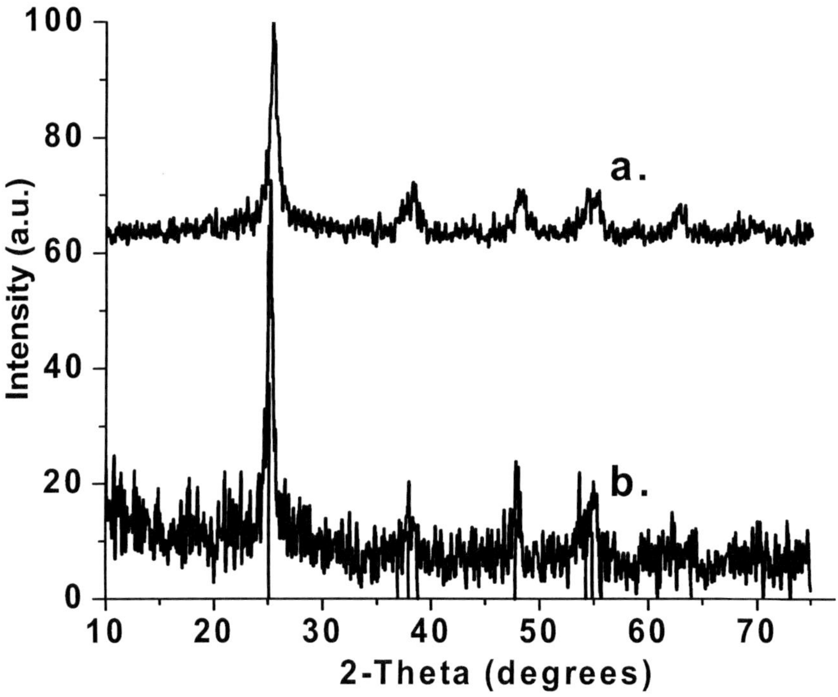

Figure 1. PXRD diffraction patterns of (a) metal doped TiO_2-SiO_2-Mn (b) TiO_2-SiO_2 aerogels. The diffraction lines match those of crystalline anatase TiO_2 (PDF # 86-1157).

Table I. Surface area and pore size distribution data for the TiO_2-SiO_2, TiO_2-SiO_2-Mn, Fe and V aerogels. Note that TSO denotes TiO_2-SiO_2

10 mol % metal-doped	Surface area (m^2/g)	Pore diameter (nm)	Pore volume (cm^3/g)
TSO	490	2.0	1.2
TSO-Fe	551	6.8	2.9
TSO-Mn	390	8.3	1.3

constant stirring. Initially the catalyst and CH_3CHO are kept in the dark for about 2 hours during which time an aliquot (35μL) was extracted from the headspace of the reactor every 10 minutes and injected into a gas chromatograph-mass spectrometer (GC-MS QP5000, Shimadzu) for product analysis. Then visible-light was turned on using an Oriel 1000W high pressure Hg lamp in combination with colored glass filters (>420 nm) to cutoff UV light and collection of the aliquots continued to assess the effect of visible-light on CH_3CHO in the presence of air and the aerogels.

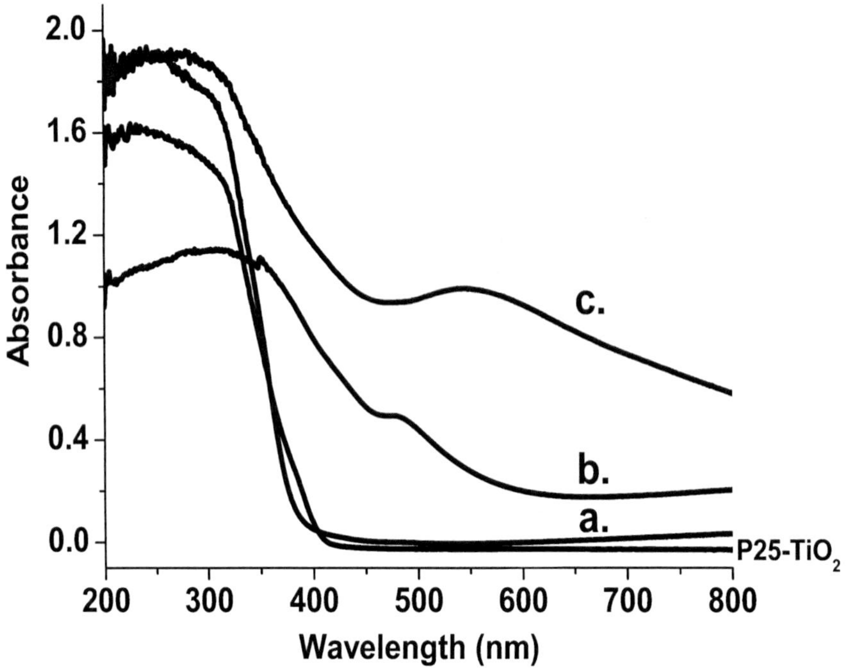

*Figure 2. UV-Vis absorption spectra of **(a)** TiO_2-SiO_2 **(b)** TiO_2-SiO_2-Fe **(c)** TiO_2-SiO_2-Mn aerogel. Absorption from commercial P25-TiO_2 is shown for reference.*

In order to estimate the efficiency of the TiO_2-SiO_2-Mn aerogels in the photodegradation process, the turnover frequency (TOF) and quantum yields (QY) were determined. TOF is the amount of product formed or reactant consumed per amount of active catalyst in a defined time interval. Quantum yield measurements are indicative of how efficient the catalyst utilizes the absorbed photons in relation to the number of molecules of product or reactant processed. The apparent quantum yield was calculated from the number of molecules of CH_3CHO consumed divided by the number of photons during a specified time interval. The number of photons was calculated from the intensity of visible-light reaching the catalyst and CH_3CHO mixture. The intensity of light was 147mW/cm^2 measured by a LabMax-TO laser power meter (Coherent Inc.).

3. Results

3.1. Structure and Composition

PXRD patterns (Figure 1) of TiO_2-SiO_2 and TiO_2-SiO_2-Mn show that the anatase phase of TiO_2 is preserved with a crystallite size of 11 nm, while amorphous SiO_2 can only be seen at low angles (~2 Theta, not shown). The observed diffraction lines were indexed to the anatase phase of TiO_2 (PDF 3# 87-1157.) No peaks from doped metal oxides are observed in the diffraction pattern. The PXRD pattern of Fe and V-doped TiO_2-SiO_2 was similar to that of Mn-doped aerogels. The surface area and pore size distribution of TiO_2-SiO_2 and metal-doped aerogels were calculated from the BET and BJH models, respectively. A summary of the BET and BJH data for the aerogel samples is shown in Table I. The BET surface areas of the materials are quite high exceeding 320 m^2/g compared to 50 m^2/g of commercial TiO_2-P25 Degusa.

The UV-Vis absorption data (Figure 2) shows a red shift (absorption in the visible region) in the metal doped TiO_2-SiO_2 aerogels compared to undoped TiO_2-SiO_2 aerogels and commercial TiO_2-P25 Degusa. SiO_2 does not affect absorption features since pure TiO_2 and TiO_2-SiO_2 show similar features (UV absorption). Infrared analysis of the aerogel samples is shown in Figure 3, where the presence of Ti-O-Ti and Si-O-Si is observed at 455, 1090 cm^{-1}, while the Si-O-Ti band is located at 950 cm^{-1} (*20, 21*). The presence of the Si-O-Ti group has been attributed to the substitution of Ti into a SiO_2 network. The vibrational features at 1630 and 3420 cm^{-1} are due to the presence of surface OH groups. The vibrations due to the doped metals (Fe, V and Mn) are not observed in the spectra. However, the metal oxide vibrations would be expected to fall in the same region as the vibrations from Ti and Si. Moreover, the intensity of the OH absorptions in TiO_2-SiO_2-Mn aerogels is greatly diminished compared to plain TiO_2-SiO_2 as well as TiO_2-SiO_2-Fe aerogels, while noting the Si-O-Si peak as an internal standard.

3.2. Products of Photocatalysis of Acetaldehyde

The photocatalytic degradation of acetaldehyde using previously calcined metal doped TiO_2-SiO_2 samples was followed by detection of CO_2 using GC-MS. For TiO_2-SiO_2-Fe aerogels, no discernible amount of CO_2 was produced (i.e. no degradation of CH_3CHO) over the course of the reaction (Figure 4a), despite the Fe sample having visible-light absorption. The rise in the amount of CH_3CHO in Figure 4a is attributed to the vaporization of adsorbed CH_3CHO when light is turned on (*22*). On the other hand, TiO_2-SiO_2-Mn samples showed steady consumption of CH_3CHO with an increase in CO_2 production over time as illustrated in Figure 4b. Under dark conditions, there is no formation of CO_2 or degradation of CH_3CHO but when visible-light is turned on there is a definite reaction taking place. About 40-50 % of initial CH_3CHO is decomposed by TiO_2-SiO_2-Mn while the amount of CO_2 produced is less than 5 % (in practice other intermediates such as acetic acid and methane are formed but these were not measured). Moreover, only CO_2 that desorbed is measured, it is likely that some CO_2 is still adsorbed on the surface of the catalyst.

Figure 5 shows a comparison of the activity of the V, Mn and Fe doped TiO_2-SiO_2 aerogels in terms of the production of CO_2 under visible-light. Firstly, the V-doped aerogel showed a higher apparent initial rate (0.72 mmols/min) followed by Mn doped aerogel (0.42 mmols/min) while no CO_2 was produced from the Fe doped sample as described earlier. Secondly, the reactivity of the doped materials may be dependent on the oxidation state as can be seen V^{5+} $>Mn^{4+}>Fe^{3+}$ i.e. a higher oxidation state Mn^{4+} (active) vs. Fe^{3+} (inactive) may be needed for degrading the organic pollutant in the visible region. This idea will be discussed further in a later section. Since the intended Fe doped TiO_2-SiO_2 did

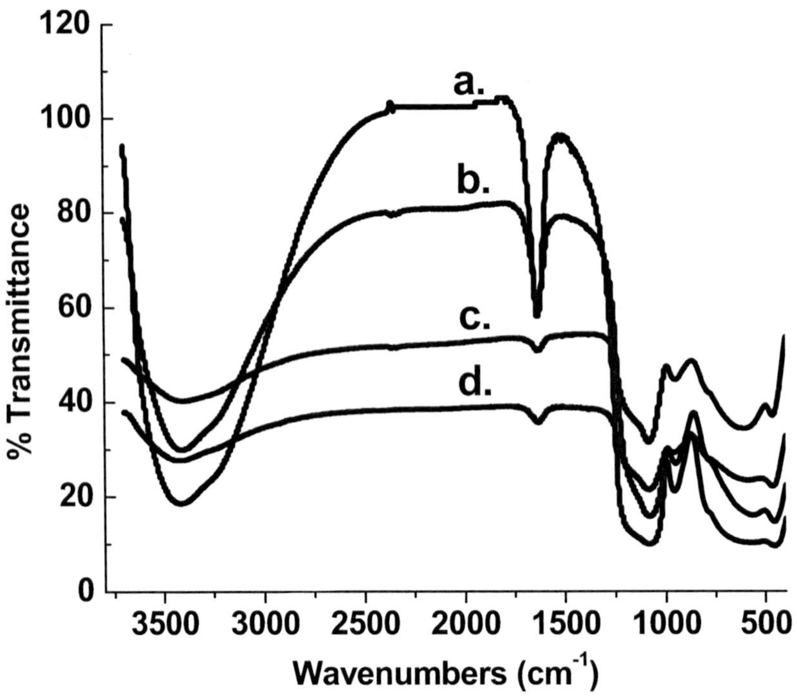

*Figure 3. Infrared spectra of **(a)** TiO_2-SiO_2 **(b)** TiO_2-SiO_2-Fe **(c)** TiO_2-SiO_2-V **(d)** TiO_2-SiO_2-Mn aerogels.*

not show any visible-light photocatalysis, attention was shifted to TiO_2-SiO_2-Mn samples and will be explored in detail.

3.3. ESR Analysis of Unused and Used TiO_2-SiO_2-Mn Aerogel

ESR studies performed on a diluted (2 mol % Mn) TiO_2-SiO_2-Mn- sample before and after exposure to CH_3CHO under visible-light are shown in Figure 6. Before photocatalysis of CH_3CHO, a sextet hyperfine structure (Mn, I=5/2) with g value of 1.97 is observed and is assigned to Mn^{4+}. After visible-light degradation of acetaldehyde, additional peaks appear between the major sextet lines and these were assigned to be Mn^{2+} (since Mn^{3+} is ESR silent) (*23*, *24*). The data obtained for the 2 mol % Mn doped sample was correlated to the 10 mol % TiO_2-SiO_2-Mn as the samples were prepared in the same way except for mol % loading of Mn. ESR peaks of TiO_2-SiO_2-Mn (10 %) were strong and overlapping, so little fine structure was observed.

Table II shows the apparent TOF and QY at two different times and concentrations of CH_3CHO. In our calculation the amount of active catalyst was determined to be 5.85 mg of MnO_2 (MnO_2 based on ESR data) in 100 mg of TiO_2-SiO_2-MnO_2. Moreover, it was assumed that MnO_2 covers the surface of TiO_2-SiO_2 since undoped TiO_2-SiO_2, with a similar surface area, does not degrade CH_3CHO under visible-light. We assume, for our TOF calculations, that incoming visible light was accessible to the inner surface area of the catalyst

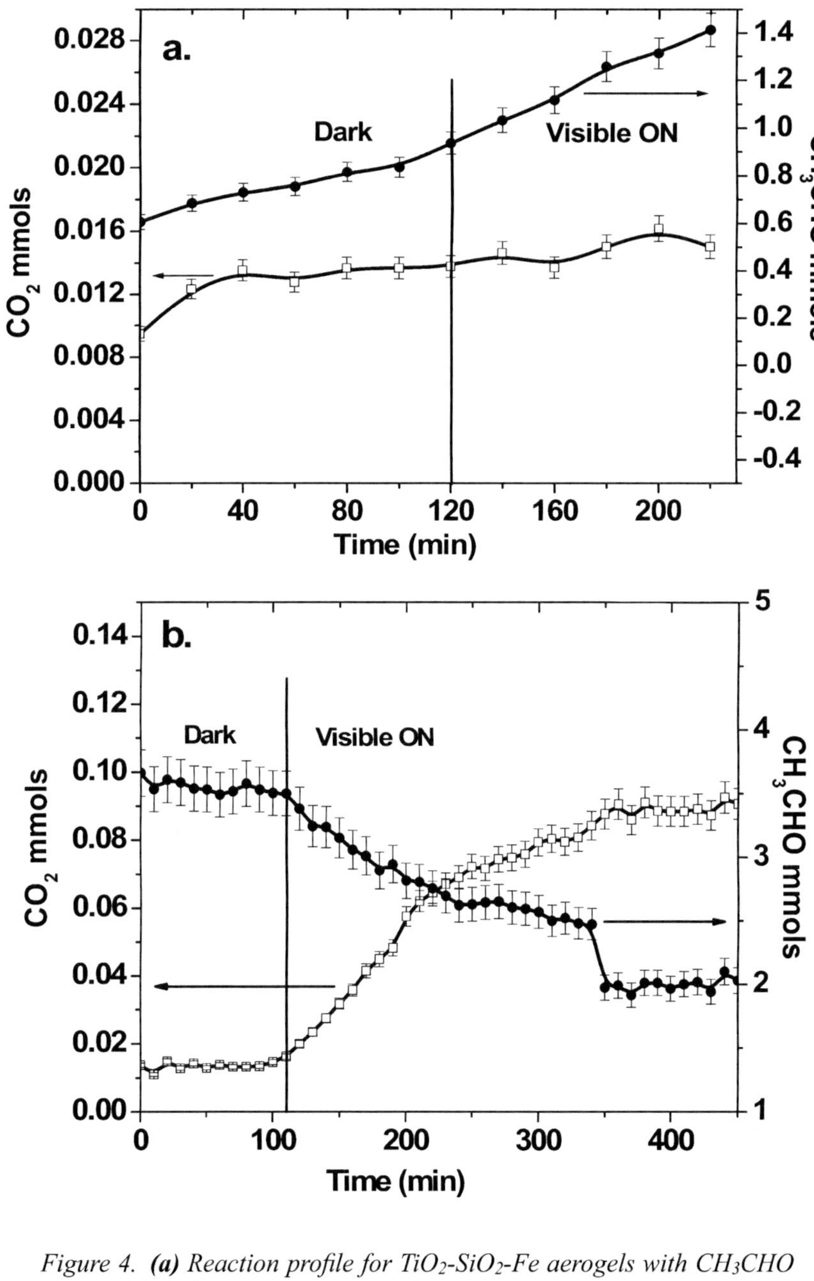

Figure 4. ***(a)*** *Reaction profile for TiO_2-SiO_2-Fe aerogels with CH_3CHO under visible-light.* ***(b).*** *Consumption and production of CH_3CHO and CO_2, respectively, after photocatalysis of TiO_2-SiO_2-Mn frameworks with acetaldehyde.*

due to the open network structure and is available to all reactive sites that are generated. The determination of TOF and QY were based on the disappearance of CH_3CHO as this represented about 50-80 % loss of initial CH_3CHO, and since we assumed that all of the MnO_2 was on the surface and active.

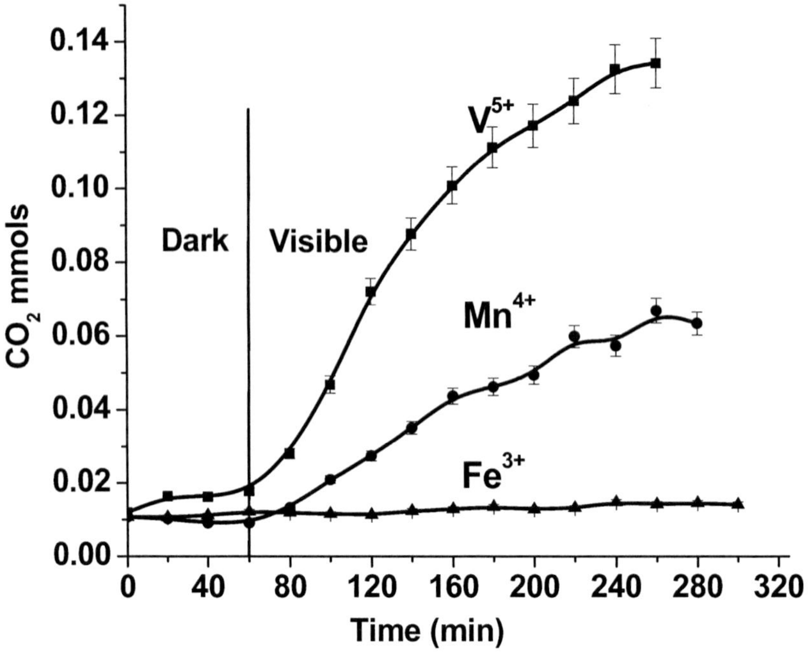

Figure 5. A comparison of the relative production of CO_2 from the photochemical reactions of V, Mn and Fe doped TiO_2-SiO_2 aerogels with CH_3CHO as a function of the oxidation state of the metal ions.

4. Discussion

4.1. Synthesis and Characterization

The metal (Fe, V and Mn) doped TiO_2-SiO_2 aerogels obtained via the sol-gel process and supercritical drying, exhibited high surface areas and mesopores compared to commercial TiO_2. The aerogels had the anatase phase which is the most photocatalytic active form of TiO_2 under UV light. The absence of diffraction lines from the doped metal oxides could be due to the fact that some metal ions replace Si^{4+} or Ti^{4+} in the TiO_2-SiO_2 lattice. Moreover, the low concentration (10 mol %) of the doped metals may have led to a homogeneous dispersion of the transition metal ions in the TiO_2-SiO_2 matrix, leading to diminished intensity of the metal-oxide (M-O) diffraction peaks. The main reason for incorporating SiO_2 into the composites was to impart high surface areas (> 300 m^2/g) to the aerogels in order to increase the adsorption capacity of these materials and hence bring into contact the adsorbed acetaldehyde with the generated radicals. The incorporation of metal dopants into TiO_2-SiO_2 framework does not significantly alter the surface area 490 m^2/g for TiO_2-SiO_2 compared to 390 m^2/g for TiO_2-SiO_2-Mn aerogels.

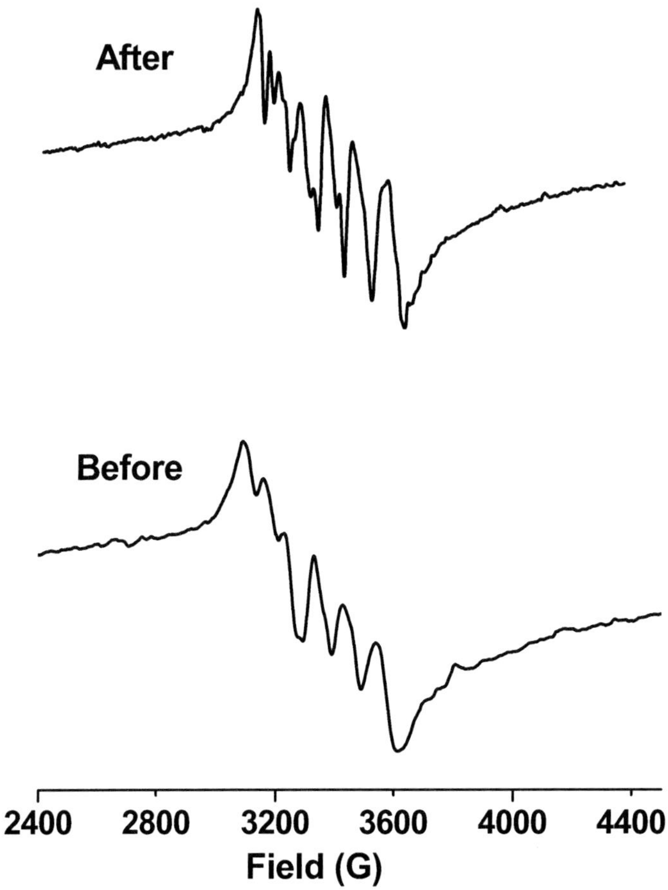

Figure 6. ESR spectra of diluted (2 mol %) Mn doped TiO_2-SiO_2 aerogels before and after photocatalysis with CH_3CHO. New peaks seen after photocatalysis are assigned to Mn.

ESR was used to probe the nature of the oxidation state of the active species before and after photocatalysis and after a careful analysis the sextet structure seen before reaction with CH_3CHO is assigned to Mn^{4+}. When the reaction with CH_3CHO was stopped the Mn existed in the +2 state in addition to the original +4 state. From this result the original Mn(III) ion in the Mn(III) acac precursor used is oxidized to this +4 state upon calcination in air at 500 °C. Iron exists as Fe^{3+} since the starting precursor ($Fe(NO_3)_3.9H_2O$) was in the highest oxidation state of the metal and is expected to stay the same upon calcination. The Mn component facilitates absorption of visible-light enabling excitation of electrons from the valence band of TiO_2 into unoccupied d-orbitals of Mn^{4+} ions and then transfers the electron into the conduction band of TiO_2. The overlap of the conduction band of Ti^{4+} (d-orbital) and the doped metal d-orbital leads to a decrease in the band gap of TiO_2 leading to visible-light absorption (*25*). Furthermore, it has been reported that the presence of a positive charge on the TiO_2 network facilitates

Table II. Apparent turnover frequency and quantum yield for the destruction of CH_3CHO at 25 °C by TiO_2-SiO_2-Mn calculated after a period of time

Initial CH_3CHO	Active catalyst[a] (MnO_2) mmol	Time (hr)	CH_3CHO used mmol	TOF (hr^{-1})	% QY
200μL non-agitated	0.067	4	1.54	5.7	1.6
200μL agitated	0.067	4	2.57	9.5	3.0
100μL non-agitated	0.067	3	0.46	2.3	0.5

a. TiO_2-SiO_2-MnO_2 (100 mg): 5.85 mg is MnO_2 which represents 0.067 mmols.

chemical adsorption of anionic molecules and contributes in breaking down of the adsorbed species (*26*, *27*). Thus, the Mn^{4+} will play a similar role in helping to trap and weaken the C=O bond of adsorbed CH_3CHO.

4.2. Photocatalytic Reactions and Role of Mn Ions

The initial goal was to create TiO_2-SiO_2-Fe aerogels but upon testing for its ability to degrade acetaldehyde there was no production of CO_2 even upon prolonged exposure to visible-light (Figures 4a and 5). To test whether the synthetic process had anything to do with the lack of photocatalytic activity of TiO_2-SiO_2-Fe, TiO_2-SiO_2-V was prepared as a reference since it is known that TiO_2-V is visible-light photoactive (*16*). Indeed, there was degradation of CH_3CHO accompanied by formation of CO_2 (Figure 5) upon shining visible-light. So TiO_2-SiO_2-Mn was prepared and as observed in Figures 4b and 5, this material does destroy acetaldehyde (200 μL) with production of CO_2 similar to the reference material (V-doped TiO_2-SiO_2). The color of the TiO_2-SiO_2-Mn aerogel of the initial sample is purple and after the reaction with CH_3CHO appeared to have stopped, the surface of the material had changed from purple to a pale brown. This discoloration may be a deactivation effect of the catalyst attributed to formation of reduced Mn species. Moreover, when the used TiO_2-SiO_2-Mn is removed from the reactor, the inner powder is still purple suggesting that the diffusion of CH_3CHO to access unreacted catalyst is slow and thus there is need for mechanical agitation to allow for exposure of unreacted surface to pollutant molecules.

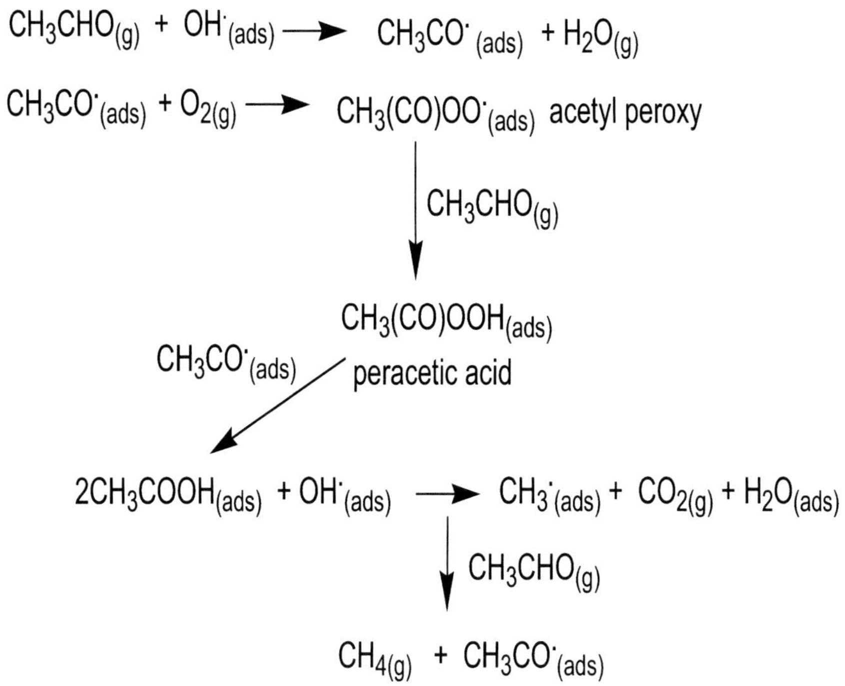

Figure 7. Degradation of acetaldehyde via hydroxyl radical attack generated from visible-light absorption by Mn doped TiO_2-SiO_2 aerogels. Species exist in either gaseous or adsorbed (ads) form.

Manganese plays two roles in the destruction of acetaldehyde; firstly it allows for visible-light absorption leading to promotion of electrons into the conduction band of TiO_2 and leaves holes in the valence band. The generated charge carriers are well known to react with air forming superoxides (O_2^-) and hydroxyl radicals ($OH^·$) and these radicals will initiate attack on the adsorbed acetaldehyde molecules (*10*, *28*). Although both O_2^- and $OH^·$ radicals can bring about destruction of acetaldehyde, the mechanism of $OH^·$ (a very strong oxidizing agent) attack is adopted here to discuss the visible-light photocatalytic action on CH_3CHO. As depicted in Figure 7, the reaction proceeds through a series of radical formation producing intermediate species such as peracetic acid, acetic acid, methane, water and eventually CO_2 (a marker for destruction of CH_3CHO).

Apart from facilitating the absorption of visible-light, the Mn^{4+} can accept the generated electron undergoing reduction to Mn^{2+} (from ESR) via Mn^{3+} formation. However, Mn^{3+} is unstable and will oxidize CH_3CHO forming Mn^{2+}. In the presence of peracetic acid, $CH_3(CO)OOH$, Mn^{2+} will be oxidized to either Mn^{3+} or Mn^{4+} thereby regenerating the active Mn component in TiO_2-SiO_2-Mn aerogel. In fact it has been reported that peracetic acid oxidizes Mn^{2+} to Mn^{3+} in solution (*25*), and in this case the presence of moisture in air can facilitate this oxidation. Because of the incomplete mineralization of CH_3CHO, only about 5 %

CO_2 was detected. However, nearly half of the starting CH_3CHO is converted to into less harmful intermediates.

Figure 5 reveals that the activity of the doped TiO_2-SiO_2-metal aerogels maybe dependent on the oxidation state of the transition metal ions: Mn^{4+} >> Fe^{3+}. A comparison of the redox potentials (*17*), shows that Mn^{4+} has a higher propensity to be reduced to Mn^{3+} which is highly unstable and is reduced to Mn^{2+} by attacking CH_3CHO. Despite both Fe and Mn ions attracting CH_3CHO, the Mn has the added advantage of having more oxidizing power and thus destroys acetaldehyde, compared to Fe^{3+} that is reluctant to be reduced to Fe^{2+}. In a related study, UV and visible photooxidation of yperite, bis(2-chloroethylsulfide) a chemical warfare agent, using xerogels (low surface area) similar to the aerogels (high surface area) has been reported. The authors studied TiO_2-SiO_2-doped with Fe, V and Mn and reported the destruction of this warfare agent into various products with the detection of SO_2 and CO_2 as the products of complete mineralization (*29*).

When no more CO_2 formation was observed, undecomposed CH_3CHO was allowed to escape over 12 hrs without disturbing the brown surface of the exposed TiO_2-SiO_2-Mn (initially purple in color). The used catalyst was reactivated by exposing visible light to the discolored surface of the catalyst for over 10 hrs, before sealing the reactor and injecting fresh CH_3CHO as before. When degradation experiments were repeated, formation of CO_2 was slower than before but some activity was obtained by the reactivation procedure. The results showed that the active species in the catalyst (Mn^{4+}) can be regenerated in the presence of air under visible-light illumination.

4.3. Effect of Agitation on Photoactivity of TiO_2-SiO_2-Mn Aerogel

To determine whether or not agitation of the catalyst would help in complete mineralization of acetaldehyde, the experiment was carried out in the manner described in Section 2.6. Here TiO_2-SiO_2-Mn was physically mixed with 200 μl of CH_3CHO and kept under constant stirring and intermittent shaking. The results are shown in Figure 8 and it was found that the initial rate of formation of CO_2 was 6 times faster (2.57 mmols/min) compared to 0.40 mmols/min (Figure 4b) without agitation. When the formation of CO_2 seemed to be over, a fresh amount of CH_3CHO was added to the sealed reactor and CH_3CHO is either immediately adsorbed or continues to degrade. However, the rate of degradation is slow and air was injected to facilitate further decomposition, Figure 8, until no significant change in CO_2 formation was observed. Furthermore, the efficiency of the catalyst is improved as evidenced by destruction of 83 % of initial CH_3CHO, with nearly twice (9 %) CO_2 detected. The TOF and QY are also improved on from 5.7 hr^{-1} and 1.6 % for the non-agitated system to 9.5 hr^{-1} and 3.0 % for the agitated system (Table II). The reactivity of used TiO_2-SiO_2-Mn could be restored as described in Section 4.3 and when experiments were repeated with fresh CH_3CHO under agitation, the amount of CO_2 and CH_3CHO converted were 10 % and 83 %, respectively. Regardless, of whether the catalyst was reactivated in the dark or light, the photoactivity was the same. Moreover, whether the reaction

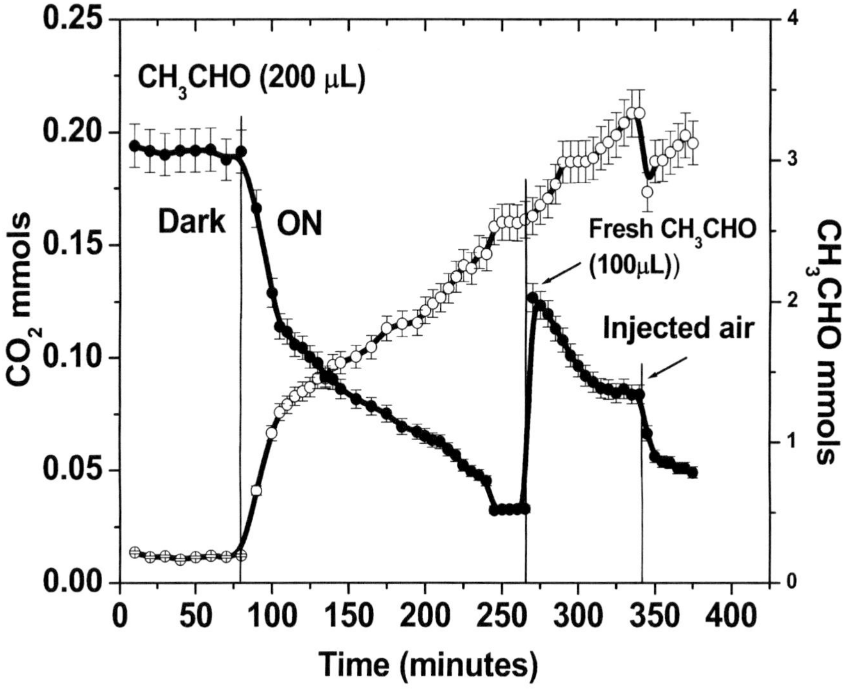

Figure 8. Reaction profile between TiO_2-SiO_2-Mn and CH_3CHO under constant agitation. Fresh CH_3CHO and air where injected at appropriate times when the reaction seemed to have stopped.

system is agitated or not, TiO_2-SiO_2-Mn catalyst is photoactive for up to 6 hours, Figures 4b and 8, in air under visible-light.

These results indicate the need to design a system that will allow for an intimate mixing of the catalyst with the VOCs to ensure near or complete destruction of these air pollutants. In addition, excess air is needed to facilitate continued supply of radicals needed for the reaction even when light is turned off as well as to oxidize Mn^{2+} to Mn^{4+}. The need for a continuous supply of oxygen has been demonstrated in the literature during the decomposition of yperite by a system analogous to that reported here as is the oxidation of Mn^{2+} to Mn^{3+} by adsorbed oxygen (*29*, *30*).

5. Conclusion

Visible-light active TiO_2-SiO_2-Mn has been prepared utilizing eco-friendly manganese ions. TiO_2-SiO_2-Mn demonstrates good photoactivity as evidenced by destruction of CH_3CHO to about 50-83 %. Fe-doped samples did not show visible activity. It appears that a high oxidation state (Mn^{4+}) is required for visible photocatalysis compared to the inactive Fe^{3+} (highest oxidation state possible) species. PXRD and FTIR show the presence of the anatase phase of TiO_2 with no Mn oxide vibrations. It is likely that Mn^{4+} substitutes some Ti^{4+} in the TiO_2 lattice. Furthermore, the photocatalytic efficiency is increased when the materials reacting are under constant motion as opposed to a static system. This work shows the significance of bandgap engineering in TiO_2 as well as a choice of high oxidation state, eco-friendly transition metal ions for novel materials that demonstrate unique photo- and photocatalytic properties in the destruction of biological pollutants under visible-light.

Acknowledgments

We acknowledge financial support from the Army Research Office (GOCH00493) and Targeted Excellence (NOCH-283002) at Kansas State University. ESR measurements were performed by Trenton Parsell in Prof. Andrew Borovik's laboratory, Department of Chemistry, University of California, at Irvine.

References

1. Hoffmann, M. R.; Martin, S. T.; Choi, W.; Bahnemann, D. W. *Chem. Rev.* **1995**, *95*, 69–96.
2. Muggli, D. S.; Ding, L. *Appl. Catal., B* **2001**, *32*, 181–194.
3. Deng, X.; Yue, Y.; Gao, Z. *Appl. Catal., B* **2002**, *39*, 135–147.
4. Kim, W.; Suh, D. J.; Park, T.; Hong, I. *Top. Catal.* **2007**, *44*, 499–505.
5. Sato, S. *Chem. Phys. Lett.* **1986**, *123*, 126–128.
6. Ozaki, H.; Iwamoto, S.; Inoue, M. *J. Mater. Sci.* **2007**, *42*, 4009–4017.
7. Asahi, R.; Morikawa, T.; Ohwaki, T.; Aoki, K.; Taga, Y. *Science* **2001**, *293*, 269–271.
8. Hamal, D. B.; Klabunde, K. J. *J. Colloid Interface Sci.* **2007**, *311*, 514–522.
9. Umebayashi, T.; Yamake, T.; Itoh, H.; Asai, K. *Appl. Phys. Lett.* **2002**, *81*, 454–456.
10. Ohno, T.; Mitsui, T.; Matsumura, M. *Chem. Lett.* **2003**, *32*, 364–365.
11. Anpo, M.; Takeuchi, M. *J. Catal.* **2003**, *216*, 505–516.
12. Zhang, X.; Lei, L. *Science* **2008**, *62*, 895–897.
13. Rodrigues, S.; Ranjit, K. T.; Uma, S.; Martyanov, I. N.; Klabunde, K. J. *Adv. Mater.* **2005**, *17*, 2467–2471.
14. Wang, J.; Uma, S.; Klabunde, K. J. *Microporous Mesoporous Mater.* **2004**, *75*, 143–147.
15. Kapoor, P.N.; Uma, S.; Rodriguez, S.; Klabunde, K. J. *J. Mol. Catal. A* **2005**, *229*, 145–150.
16. Yang, X.; Cao, C.; Hohn, K.; Erickson, L.; Maghirang, R.; Hamal, D.; Klabunde, K. J. *J. Catal.* **2007**, *252*, 296–302.
17. Cotton, F. A.; Wilkinson, G.; Murillo, C.; Bochmann, M. *Advanced Inorganic Chemistry*, 6th ed.; John Wiley and Sons, Inc.: New York, 1999.
18. Altamirano-Lozano, M.; Valverde, M.; Alvarez-Barrera, L.; Molina, B.; Rojas, E. *Teratog., Carcinog., Mutagen.* **1999**, *19*, 243–255.
19. Briggs, D.; SeahM. P. *Practical Surface Analysis*; John Wiley: New York, 1990.
20. Samantaray, S. K.; Parida, K. *Appl. Catal., B* **2005**, *57*, 83–91.

21. Zhang, X.; Zhang, F.; Chan, K. *Appl. Catal., A* **2005**, *284*, 193–198.
22. Tanizaki, T.; Murakami, Y.; Hanada, Y.; Ishikawa, S.; Suzuki, M.; Shinohara, R. *J. Health Sci.* **2007**, *53*, 514–519.
23. Keeble, D. J.; Li, Z.; Poindexter, E. H. *J. Phys.: Condens. Matter* **1995**, *7*, 6327–6333.
24. Velu, S.; Shah, N.; Jyothi, T. M.; Sivasanker, S. *Microporous Mesoporous Mater.* **1999**, *33*, 61–75.
25. Yamashita, H.; Harada, M.; Misaka, J.; Takeuchi, M.; Ichihashi, Y.; Goto, F.; Ishida, M.; Sasaki, T.; Anpo, M. *J. Synchrotron Radiat.* **2001**, *8*, 569–571.
26. Choi, W.; Termin, A.; Hoffmann, M. R. *J. Phys. Chem.* **1994**, *98*, 13669–13679.
27. Ranjit, K. T.; Willner, I.; Bossmann, S. H.; Braun, A. M. *J. Catal.* **2001**, *204*, 305–313.
28. Schwitzgebel, J.; Ekerdt, J. G.; Gerischer, H.; Heller, A. *J. Phys. Chem.* **1995**, *99*, 5633–5638.
29. Neaţu, Ş.; Pârvulescu, V. I.; Epure, G.; Preda, E.; Şomoghi, V.; Damin, A.; Silvia Bordiga, S.; Zecchina, A. *Phys. Chem. Chem. Phys.* **2008**, *10*, 6562–6570.
30. Xu, Y.; Lei, B.; Guo, L.; Zhou, W.; Liu, Y. *J. Hazard. Mater.* **2008**, *160*, 78–82.

Chapter 13

Comparative Pulmonary Toxicity of Metal Oxide Nanoparticles

J. A. Pickrell,[1,*] D. van der Merwe,[1] L. E. Erickson,[2] K. Dhakal,[1,3] M. Dhakal,[1,4] K. J. Klabunde,[5] and C. Sorensen[6]

[1]Comparative Toxicology Laboratories, Diagnostic Department of Medicine/Pathobiology, Kansas State University, Manhattan, KS
[2]Department of Chemical Engineering, Kansas State University, Manhattan, KS
[3]Doctoral graduate student, Department of Occupational and Environmental Health, University of Iowa, Iowa City, IA
[4]Doctoral graduate student, Department of Infectious Disease, University of Missouri, Columbia, MO
[5]Department of Chemistry, Kansas State University, Manhattan, KS
[6]Department of Physics, Kansas State University, Manhattan, KS
*pickrell@ksu.edu

Comparative pulmonary toxicity of metal oxide nanoparticles is addressed for several different materials. The size of the particles is an important variable which depends on both the fundamental particle size and the aggregation of nanoscale particles into larger clusters. The solubility of each material in lung fluids is also significant because it impacts the fate of particles that enter the lung. The effects of metal oxide nanoparticles on cultured lung epithelial cells are described.

Introduction

Different lung responses, progressing from normal to abnormal, result from inhaling progressively increasing doses of particles or noxious gases. High doses of particles can cause acute pulmonary edema. Intermediate doses can cause healing by secondary intent, fibrosis, emphysema – loss of lung tissue and airway enlargement, metaplasia – barely controlled lung growth, and neoplasia – tumors or uncontrolled lung tissue growth. Finally, progressively lower doses will cause mild responses healing by primary intent, activation of lung tissue, or escape to the interstitial lung space or into the blood stream and to other organs such as the liver, heart. Alternatively, particles can enter the brain by deposition in the olfactory bulb (*1–14*), where they can remain stationary or move by antero- or retro-grade movement to deep brain (*1–14*). This review will target solubility, crossing cell barriers of non-infectious lung diseases after inhalation of particles in the fine particle range (mass median aerodynamic diameters – MMAD; < 2,500 nanometers - nm) or untrafine particle-ranges (MMAD < 100-200 nm).

Particle dimensions can be described differently by different scientists, because different aspects are most important to them. Aerosol scientists are most interested in describing physical dimensions (MMAD), particle number, particle volume, particle surface area and motion (particle movement). Alternatively, engineers are more interested in how particles are delivered – rate of deposition. Health scientists of different types study the response of animals or persons. Specifically, health effects of animals or persons exposed to fine or ultrafine particles. To examine and interpret broad health implications of different types of particles, such as nanoparticles it is often necessary to form teams among health scientists, aerosol scientists and engineers. Each scientist on this team develops measurements they consider most important, because such measurements (aerosol size, surface area, mass and the distributions of these parameters) are often technically difficult (*5*).

As each team specialist interprets their data, they and the group, collectively, can infer the tendency of particles to cause health effects. For example, individual animals or people are not advised to live or work in environments that contain more than the toxic mass of particles, or small (fine or ultrafine) particles because such inhabitants are expected to have an increased possibility of developing either pulmonary or systemic disease. If the person is working in that environment, such levels are called occupational limits (REL – [recommended exposure limits, levels recommended we stay below]; and PEL – (permissible exposure limits, levels we are not permitted to exceed without penalty by the Environmental Protection Agency - EPA]). For purposes of regulation, PEL are similar to TLV (Threshold limit values). Standardized instruments are available to measure the chemical concentration of substances in the particles. Aerosol scientists describe particle size, shape and environmental reactivity, while health scientists use such data to estimate risk and the resulting health effects.

Aerosol Particle Size

Physical aerosol size is important because it strongly influences where in the respiratory tract such particles will be deposited, and enter the epithelial lining fluid. Location determines which cells will be responding to the deposition of such particles. For example, if very small singly dispersed particles are deposited in the deepest recesses of the gas-exchange units (alveoli; sacs) of the lung, they may injure lung epithelial cells, go to the lung interstitium or penetrate into the vessel and translocate to other organs such as the liver or heart. Particles may deposit directly on the olfactory bulb and possibly go to deep brain (*1*).

Alternatively, particles can enter the epithelial lining fluid (ELF) and either persist, or be dissolved (*2*). Physical aerosol size can vary depending on solvent content (monomers that were just formed), in epithelial lining fluid (aggregated) or in air (agglomerated). Aerosol size in air is most important to deposition after inhalation of particles (*2–5*).

Particle Size in the Atmosphere

Aggregated and agglomerated nanoparticles in air are often quite large, sometimes > 2,500 nm, although the individual monomers may have been < 50 nm in diameter. An interesting property of manufactured nanoparticles is that aggregation and subsequent agglomeration does not greatly reduce the total surface area of the agglomerate, although it may limit the size of the substance which can interact with the agglomerate.

Particle size in the ELF determines the interaction intensity with different types of lung parenchymal cells. Of perhaps greatest interest are interactions with epithelial cells (type I and type II epithelial cells) and phagocytes (pulmonary alveolar macrophages) in the lung parenchyma. Increasing particle surface area predicts increasing ELF or cellular interaction, if that surface area allows access to the ELF, the cells or both. Access can have very different outcomes. Access to ELF favors particle dissolution and limits interaction with epithelial cells or phagocytes. Alternatively, access predominantly to epithelial cells and phagocytes may limit dissolution, if both occur at the same rates. Certainly, access to both ELF and epithelial cells will allow both processes to occur. However, the process that is most rapid will tend to predominate over the slower process. Manufactured nanoparticles are approximately the same size as single, natural ultrafine particles (*2*, *5*).

Hydrodynamic Particle Size

Particle size in ELF is intermediate between agglomerated particles in air, and single metal oxide nanoparticles. Thus, particles in ELF are aggregated to some degree, but most were less than 500 nm (*4*). Variable, but lower percentages fell in the fine particle range, and thus only minimally aggregated (in distilled water ~ 20%; in tissue culture media ~45%) (*6*). By comparison, when measured by dynamic light scattering in our laboratory (*7*), in tissue culture, if particles were centrifuged, the remaining suspended particles were less than 300 nm in diameter, only minimally aggregated. If not centrifuged some particles were as large as 3,500 nm in diameter, and thus more agglomerated (*7*). This tendency was greater in the minimally soluble titanium dioxide particles (TiO_2) and experimental aerogel particles with a covalent compound containing oxides of Mn (aerogel-Mn). Particles after centrifugation most likely constituted < 1-2% of the mass (*2–5*), possibly reflecting mechanical agitation. We would expect that the smaller diameter metal oxide nanoparticles would dissolve more rapidly than would the larger conventional bulk particles. The more soluble MgO would be more likely to demonstrate this trend than would the virtually insoluble TiO_2 or aerogel-Mn (*2–5*).

Solubility and Dissolution Rate in Simulated Epithelial Lining Fluid (ELF)

To examine the effect of crystalline size on solubility in ELF (*2*), we compared the solubility of 2 nanocrystalline nanoparticles of MgO (nanoactive® MgO – crystallite size 5 nm, and > 230 m^2/g surface area; nanoactive® MgO Plus – crystallite size 4 nm, and > 600 m^2/g surface area, and 1 conventional macrocrystalline MgO – surface area 30 m^2/g in a simplified tissue culture media (Hanks Basic Salt Solution – HBSS) near that in ELF at the end of expiration and a more complex supportive media (Dulbecco's Modified Eagles Medium – DMEM) with bicarbonate near that in ELF during expiration. Dissolution was measured as total amount and as rate of dissolution of magnesium (Mg) using an Inductively Coupled Plasma Atomic Emission Spectrophotometry (ICP-AES) (*2*), and calculated as MgO. The results related most closely to the bicarbonate content of the media (*2, 5*). Thus, most of the MgO solubility in lungs would be expected to have occurred during expiration. There was a small effect of reduced particle size on total solubility and a modest effect on the rate of solutility at an early time, when nanoactive® MgO Plus, premium grade > nanoactive® MgO = conventional macrocrystalline MgO. Added 5% CO_2 to culture media increased total solubility an additional > 30% for nanoactive® MgO Plus, premium grade for HBSS, but not DMEM, suggesting that this was near maxiumum solubility and solubility rate. Under these conditions, > 75% of the MgO was dissolved in the first 1-3 hours, presumably as magnesium carbonate (*2, 5*). This amount exceeded the expected maxiumum inhaled dose of MgO several-fold, suggesting little residual MgO remained to injure lung. Moreover, since MgO is non-toxic *Milk of Magnesia* we would expect no toxicity from the minimum residual MgO. Mg is used in numerous metabolic reactions throughout the lung and body.

Similar solubility studies were made on TiO_2 and aerogel containing oxides of Mn and solubility measured as soluble or suspended Ti or Mn (ICP-AES) (*2, 8, 9*). For these metals, minimal to no solubility or suspension (< 0.5-1.5%) resulted. In many cases, suspended Ti or Mn could not be distinguished from no suspension or dissolution (*9*). Thus, all, or nearly all of the inhaled TiO_2 or the aerogel particles containing coordinate compound aerogel with oxides of Mn, Ti and Si would be expected to interact with the epithelium lining and the ELF where this material is deposited. As insoluble particles, they would have the possibility of either injuring or causing inflammation to epithelial cells so contacted (*9*).

Nanoparticle Penetration through Human Skin (in Vitro)

The dermal absorption potential of a mixture of nanocrystalline magnesium oxide (MgO) and titanium dioxide (TiO_2) through human skin with intact, functional stratum corneum was estimated using Bronaugh-type flow-through diffusion cells (*8*). The skin surface was exposed to the particles over a period of 8 hours, at a 50 mg/cm^2 dose rate in a water suspension, a water/surfactant suspension with sodium lauryl sulfate, and as a dry powder (*8*). The nanomaterials were not detectable in the receptor fluid using the presence of Mg and Ti in the receptor fluid as an indicator of absorption. Particle penetration into the skin was not detectable by transmission electron microscopy, except for particles penetrating loose outer stratum corneum layers (*8*). These data suggest that the exposure of healthy skin to TiO_2 , which is widely used in consumer products for the skin, and nanocrystalline MgO/TiO_2 mixtures, which are used for chemical spill containment, are unlikely to be the cause of adverse effects under conditions that do not break down the stratum corneum barrier, independent of the ability of these particles to cause harm if absorbed into the body (*8*).

Summarizing

Particle Size in air is an important determinant of deposition in the pulmonary tree – upper or lower airways. Larger, more heavily agglomerated particles would be expected to deposit in the upper airways, while the much smaller singly dispersed manufactured metal oxide nanoparticles would penetrate deeply into the lower airways and alveoli. These could migrate to the pulmonary interstitium, or enter the vessels and deposit in other organs (liver and heart etc.) and tissues throughout the body (*2*, *5*, *6*, *10*). Deposition site determines the epithelial cells that the nanopaticles would interact with, and the point that they would enter the ELF. Once in the ELF, the large strongly agglomerated large particles, modeled by the uncentrifuged particle size *in vitro*, would be expected to be phagocytized, and the smaller minimally aggregated, metal oxide nanoparticles would be expected to interact with epithelial cells by diffusion (Brownian motion) (*3–6*). We expect that the less soluble TiO_2 and aerogel particles containing a covalent compound with oxides of Mn would have greater chances to be phagocytized and to persist than would the more soluble MgO particles. Those that persisted would either remain in the lung intersitiium to injure lung epithelium or penetrate into the blood and migrate to the liver, heart and other tissues throughout the body (*5*, *10*).

Effects of Metal Oxide Nanoparticles to Cultured Lung Epithelial Cells

Introduction

Inhaled coarse metal oxide nanoparticles interact with upper airway fluid (nasal and ELF) and disperse, dissolve or travel upward on the mucociliary escalator and are often coughed up and swallowed (*7*). Smaller fine and ultrafine particles evade the upper respiratory tract physical defenses and are deposited on the bronchiolar, alveolar epithelium or enter the interstitial lung spaces or enter the blood and distribute throughout the liver, heart or other body tissues (*9–11*). Production of reactive oxygen substances (ROS) formed the basis of toxicity and injury of lung epithelial cells from metal oxide nanoparticles. Although the exact mechanisms are not known, what is known indicates the importance of ROS to lung epithelial cell injury (*9*, *12*, *13*). In fact, Nel et al. (*12*, *13*), have proposed a heirarchial stress model where nuclear regulatory factor (Nrf)-2 drives the lowest level (Tier 1) responses of lung epithelial defense when glutathione and ability to heal are relatively plentiful. At this level lung epithelium can often be healed spontaneously (by primary intent). At an intermediate level (Tier 2), pro-inflammatory tumor necrosis factor (TNF) alpha, interleukin (IL) 1 and

IL6 are triggered, inflammatory change occurs and epithelial cells are healed by secondary intent (fibrosis). Strength of Tier 2 responses were measured as released IL-6 by radioimmunoassay.

Tier 3 Responses related to mitochondrial toxicity, mitochondrial pore extent and size. Tier 3 responses were found to be proportional to the fraction of lung cells that had low mitochondrial membrane potential (JC-1 cell fraction). Strong Tier 3 responses (many JC-1 cells) (*12*, *13*) suggested a good possibility that apoptosis or necrosis would occur. Tier 3 responses occur at the highest level of oxidative stress when glutathione and ability to relieve oxidative stress are relatively low (*12–16*).

Man-made ultrafine (nano) particles of TiO_2 were cleared more slowly than their natural counterparts (*12*). Slowed clearance was associated with impaired phagocytosis, sequestration of particles in the interstitium (*9*, *14*, *15*), or escape from the lung into the blood and distribution in the liver, heart or throughout body tissues (*5*, *9–13*).

Tier 2 Responses (Release of IL 6 from BEAS 2B Lung Epithelial Cells)

To examine the strength of tier 2 responses, we cultured BEAS 2B lung epithelial cells in DMEM with 1% serum and compared the release of IL 6 in response to 50-1,000 micrograms/ml of MgO, TiO2 and Aerogel-Mn particles. The highest exposure levels were projected to be > 20 times the nuisance dust level, to allow projection of safety of these nanoparticles. These were compared to manure feedlot dust positive controls that increased strength of Tier 2 responses as indicated by IL 6 by 2-10 fold above unexposed BEAS 2B lung epithelial cell controls. Manure dust controls did not affect cell viability, but were > 10-fold above the nuisance dust level. At these high exposure levels to metal oxide nanoparticles, no increases in Tier 2 responses as IL 6 were noted. In fact, at the highest levels, IL 6 appeared to be reduced, especially with the aerogel-Mn particles (*9*).

Veranth et al (*16*) reported that at such high exposure levels (TiO_2), IL 6 was reduced, probably due to the interference of nanodust with IL 6 activity measurements. Our experiments revealed no similar trend toward reduction with MgO, a similar trend with TiO_2, extended below 50% of control at the highest level of TiO_2 and a greater reduction (below 25%) with aerogel-Mn particles, confirming and extending their observations with our aerogel-Mn particles. We believe that this reduction, at least in the case of TiO_2, was due to nanoparticles getting in the way of cell and enzyme function (*5*, *14–17*).

Cell Killing

An additional reason was cell killing as indicated by vital dye staining and flow cytometry (propidium iodide 40-60% reduction in live cells TiO_2 at 250 and 1,000 micrograms/ml. our highest exposure levels and 5-20 times the nuisance dust level) constitutes a trend toward reductions in cell numbers; $P > 0.05$ ANOVA that fails to attain statistical significance when analyzed by ANOVA, or student T taking into account the number of comparisons (*9*). From a public health standpoint, cell killing is suggestive, but not significant only at > 5 times the nuisance dust level. Such dust levels are difficult to attain in aerosol exposures. Clearly, further work will need to be done to verify the extent to which each factor contributes to this reduction.

TiO_2 had no measureable increase in BEAS 2B lung epithelial cells with low mitochondrial membrane potential (Tier 3) responses, or increased release if IL 6 (Tier 2 response). The data of Nel et al.'s laboratory (*12, 13*), modeled at 1-2 times the nuisance dust level showed no Tier 2 or 3 responses for TiO_2, agreeing with our data that was collected at 5-20 times the nuisance dust level. The degree (fidelity) of this model remains to be determined by future work. We believe that single high level exposures of BEAS 2B lung cells are better to rank single acute exposures than to predict chronic exposures. Since TiO_2 does not clear significantly, chronic exposures may be of more significance. Future work will be needed to determine utility, if any, beyond ranking for relatively high level single acute exposures. TiO_2 is regarded as a nuisance dust which requires a combined inflammatory trigger (endotoxin) to cause significant lung injury.

Tier 3 Responses (Fraction of BEAS 2B Lung Epithelial Cells with a Low Mitochondria Membrane Potential; JC 1 Cells)

Tier 3 responses were measured in BEAS 2B lung epithelial cells exposed to the highest levels (250 and 1,000 micrograms/ml) of metal oxide (TiO_2 and Aerogel-Mn) nanoparticles. Measurements were made as cell population with low mitochondrial membrane potential (JC-1 red fluorescence), believed to be a more significant indicator cell function and health than cell viability (*9*).

The aerogel-Mn, but not TiO_2 (*9*), or manure dust, even at the highest concentrations were significantly above (65-85% of the positive CN control) relative to the negative (unexposed) BEAS 2B cell control (< 5% of the positive CN control) $p< 0.05$ when corrected for multiple comparisons (Tuckey) or analyzed by ANOVA (*9*). The increased fraction of BEAS 2B lung epithelial cells with lowered mitochondrial membrane potential suggested that they might be or become less functional or apoptic.

How Might Tier 2 and Tier 3 Responses Relate to a Lung Epithelial Cell?

Data reviewed (*3*) from laboratories of Wyatt and Von Essen (*18*) and Unfried and Sydlik (*19, 20*), suggested that healthy lung epithelial cells undergoing normal amounts of phosphorylation and not damaged from inflammation or oxidative stress relate to PI3 kinase/Akt/ERK 1/2 and protein kinase (PK) C ε. More recently, data suggested that rat cerebral cortex cells might be sensitized to iron-induced oxidant stress damage by a related pathway (PI3K/Akt/glycogen synthesis kinase (Gsk) 3B) (*21*). Could lung epithelial cells or cells from related tissues carry the same sensitivity? Future work must decide. Additionally this might be an important change if manmade nanoparticles are exposed to the brain cells by way of the ophthalmic bulb in sufficient quantities to cause oxidant stress. Finally, it is clear (*3, 5, 14–21*), that inflammation and oxidative stress (Tier 2 responses) relate to cells with low potential in their mitochondrial membrane by depletion of glutathione. However, we sense that other mechanisms will emerge and we are only looking at the tip of the iceberg.

Comparatively, TiO_2 and manure dust, even at our highest tested levels, produced no functional decrement in BEAS 2B lung epithelial cells (*14*). Thus, neither of these dusts would be expected to affect mitochondrial membrane potential. Thus, only aerogel-Mn has the potential to lower mitochondrial membrane potential, and cause apoptosis, and only at extremely high levels.

Public Health Analysis

We must consider the question, "*Do aerogel-Mn nanoparticles constitute a danger if used in residences"?* The concentrations we tested *in vitro* modeled 5-20 fold of the nuisance dust level. *From a public health standpoint*, the aerogel is designed to be painted on walls of rooms. Thus, it is extremely unlikely that dust levels would attain even a few percent of the nuisance dust level, let alone exceed it. However, future measurements of aerosol levels in the room with walls to which it was applied, will be needed to confirm this interpretation.

Acknowledgments

We acknowledge support by M2 Technologies through contracts with The Marine Corps Systems Command, U.S. Marine Corps; Department of Defense, Quantico, VA, and the support of NanoScale Corporation, Manhattan, KS.

References

1. Kim, J. S.; Yoon, T. J.; Yu, K. N.; Kim, B. G.; Park, J. P.; Kim, H. W.; Lee, K. H.; Park, S. B.; Lee, J.-K.; Cho, M. H. *Toxicol Sci* **2006**, *89*, 338–347.
2. Pickrell, J. A.; Dhakal, M.; Castro, S. D.; Klabunde, K. J.; Gakhar, G.; Erickson, L. E.; Hayden, E.; Hazarika, S.; Oehme, F. W. *Biomed. Environ. Sci.* **2009**, accepted for publication.
3. Pickrell, J. A. Biomedical responses and toxicity of nanoparticles. In *Veterinary Toxicology, Basic and Clinical Principles*; Gupta, R. C., Ed.; Academic Press: New York, NY, 2007; pp 305–312.
4. Pickrell, J. A. Respiratory Toxicity. In *Veterinary Toxicology, Basic and Clinical Principles*; Gupta, R. C., Ed.; Academic Press: New York, NY, 2007; pp 177–192.
5. Pickrell, J. A.; Erickson, L. E.; Dhakal, K.; Klabunde, K. J. Part IX Biological and Environmental Aspects of Nanomaterials; Chapter 22; Toxicity of inhaled nanoparticles. In *Nanoscale materials in Chemistry*, 2nd ed.; Klabunde, K. J., Richards, R. M., Eds.; Wiley-Interscience: New York, 2009; pp 729–769.
6. Sayes, C. M.; Reed, K. L.; Warheit, D. B. *Toxicol. Sci.* **2007**, *97*, 163–180.
7. Sorensen, C. M.; Hageman, W. B.; Rush, T. J.; Huang, H.; Oh, C. *Phys. Rev. Lett.* **1998**, *80*, 1782–1785.

8. Van Der Merwe, D.; Tawde, S.; Pickrell, J. A.; Erickson, L. E. *Cutaneous Ocul. Toxicol.* **2009**, *28*, 78–82.
9. Dhakal, K. Ph.D. thesis, Kansas State University, Manhattan, KS, 2009.
10. Nemmar, A.; Vanbilloen, H.; Hoyaerts, M. F.; Verbruggen, A.; Nemery, B. *Am. J. Respir. Crit. Care Med.* **2001**, *164*, 1665–1668.
11. Kemp, S. J.; Thorley, A. J.; Gorelik, J.; Seckl, M. J.; O'Hare, M. J.; Acaro, A.; Korchev, Y.; Goldstraw, P.; Tetley, T. D. *Am. J. Respir. Cell Mol. Biol.* **2008**, *39*, 591–597.
12. Nel, A. Toxic potential of materials at the nanolevel. *Science* **2006**, *311*, 622.
13. Xia, T.; Kovochich, M.; Liong, M.; Meng, H.; Kabehie, S.; George, S.; Zink, J. L.; Nel, A. E. *ACS Nano* **2008**, *2*, 2121–2134.
14. Ferin, J.; Oberdorster, G.; Penney, D.P. *Am. J. Respir. Cell Mol. Biol.* **1992**, *6*, 535–542.
15. Renwick, L. C.; Donaldson, K.; Clouter, A. *Toxicol. Appl. Pharmacol.* **2001**, *172*, 119–127.
16. Veranth, J. M.; Kaser, E. G.; Veranth, M. N.; Koch, M.; Yost, G. S. *Part. Fiber Toxicol.* **2007**, *27* (4), 2.
17. Moss, O. R.; Wong, V. A. *Inhalation Toxicol.* **2006**, *18*, 711–716.
18. Wyatt, T. A.; Slager, R. E.; DeVasure, J.; Auvermann, B. W.; Mulhern, M. L.; Von Essen, S.; Mathiesen, T.; Floreant, A. A.; Romberger, D. J. *Am. J. Physiol Lung Cell Mol Physiol* **2007**, *293*, L1163–L1170.
19. Sydlik, U.; Bierhals, K.; Soufi, M.; Abel, J.; Schins, R.P.; Unfried, K. *Am. J. Physiol. Lung Cell Mol. Physiol.* **2006**, *291*, L725–L733.
20. Unfried, K.; Sydlik, U.; Weissenberg, A.; Abel, J. *Am. J. Physiol. Lung Cell Mol. Physiol.* **2008**, *294*, L358–L367.
21. Uranga, R. M.; Giusto, N. M.; Salvador, G. A. *Toxicol. Sci.* **2009**, *111*, 331–344.

Chapter 14

The Development of FAST-ACT® by NanoScale Corporation

Olga B. Koper*

NanoScale Corporation, Inc., Manhattan, Kansas
***okoper@nanoscalecorporation.com**

Nanocrystalline metal oxides possess very large surface areas, defect rich morphology, large porosities, and small crystallite sizes. This combination of properties results in extremely high chemical reactivity including both enhanced reaction kinetics and large capacities. Based on this core technology, originally developed at Kansas State University, NanoScale has developed FAST-ACT® (First Applied Sorbent Treatment Against Chemical Threats), a family of products for containment and neutralization of a wide range toxic chemicals including chemical warfare agents. Significant advantages over other chemical release mitigation measures include (1) non-reversible destruction of reactive toxic compounds leading to less toxic byproducts, (2) safe utilization on any release without the need for prior identification, and (3) effectiveness against vapor hazards as well as liquids at a broad range of environmental conditions; (4) proven safety and long shelf-life. This product is being utilized by first responders, HAZMAT teams, as well as industry and academia around the world.

Nanocrystalline metal oxides exhibit very high reactivity due to increased surface area, unique morphology, additional functional groups on the surface, large porosities, small crystallite sizes and altered electronic state. In the 1980s Prof. Kenneth J. Klabunde, from Kansas State University (KSU), developed manufacturing processes and started exploring applications of high surface metal oxides. In 1995 he founded a company, NanoScale Corporation (originally known as Nantek), that licensed the patented technologies from

KSU in order to commercialize them in a variety of applications, primarily related to environmental remediation and decontamination. Throughout the years NanoScale has compiled considerable data on the performance of these metal oxides, trademarked as NanoActive® materials, towards many hazards including acidic and caustic gases, chlorocarbons, organophosphorus compounds, simulants as well as actual chemical warfare agents (*1–8*). The nanocrystalline metal oxides not only adsorb toxic compounds but also chemically decompose them in a nonreversible manner, which inhibits the possibility of desorption of captured agents. In addition, decomposition products are much less toxic, or even nontoxic, and typically are in an easily disposable solid form.

During the last few years the company entered large scale production of specialized chemicals including the branded chemical hazard containment and neutralization system, FAST-ACT (First Applied Sorbent Treatment Against Chemical Threats) family of products (Figure 1), which is the focus of this chapter. The technology was developed with support from the United States military to target the destruction of chemical warfare agents. Building on this knowledge, the product's utility was extended to include commonly encountered spills and releases.

FAST-ACT is offered in pressurized cylinders (analogous to fire extinguishers) capable of addressing both liquid and vapor hazards as well as manually dispersed containers for liquid hazard treatment. Due to the fine size of the powder, the material remains suspended in the air for minutes, enabling capture of the volatile species. The formulation is non-toxic, non-flammable and non-corrosive, can operate in a broad range of environmental conditions, acts rapidly upon contact reducing the threat to life and safety, as well as on-site management time and costs, Figure 2. The product, as well as the underlying technology, has very strong U.S. and international patent protection (*9–12*).

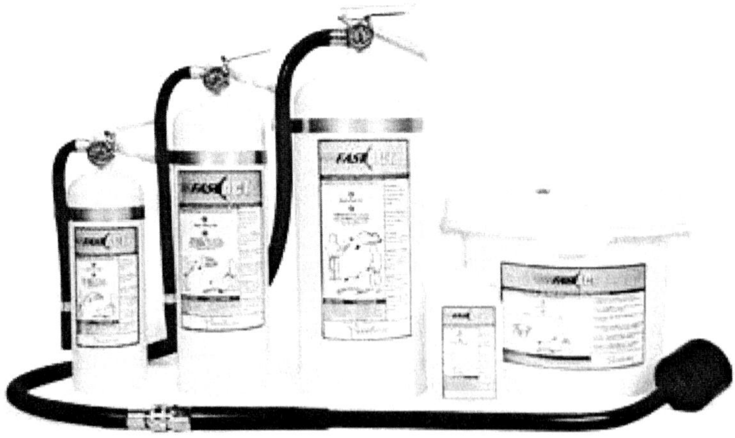

Figure 1. FAST-ACT family of products.

Figure 2. Utilization of FAST-ACT.

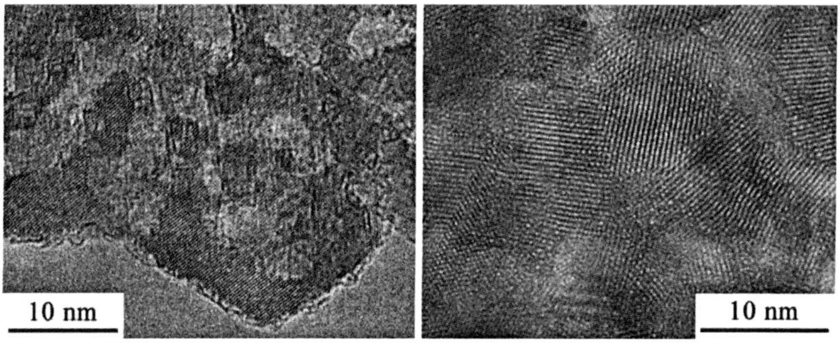

Figure 3. High resolution TEM images of NanoActive MgO (a) and NanoActive TiO_2 (right)

FAST-ACT has been widely recognized, receiving numerous awards such as: nomination by the United States Army as the Invention of the Year, recognition by Popular Mechanics as one of "20 inventions that will change the world" (*13*), and receipt of the prestigious Tibbetts Award, given to small businesses that successfully commercialized technologies originating from government sponsored basic research. Sections below describe in more detail the technology, efficacy, safety and operational/manufacturing aspects of the product.

Nanocrystalline MgO and TiO_2

FAST-ACT is a proprietary blend of NanoActive Magnesium Oxide and NanoActive Titanium Dioxide. Both of these materials are nanocrystalline, i.e. they are composed of crystals that are less than 100 nanometers in at least one dimension. However, the large surface energy of these nanocrystals causes them to agglomerate into particles that are significantly larger. While the crystallite size of the FAST-ACT formulation averages less than 10 nanometers, as measured using the powder X-ray diffraction and calculated by Scherrer Equation, the median particle size is approximately 3-5 microns. The agglomerated particles are very strongly bound and do not deaggregate to individual crystallites, even if high sheer forces are applied. Figure 3 shows high resolution Transmission Electron Microscopy (TEM) images of NanoActive MgO, and NanoActive TiO_2, where the small crystallites can be seen. Figure 4 presents a Scanning Electron

Figure 4. SEM image of FAST-ACT formulation.

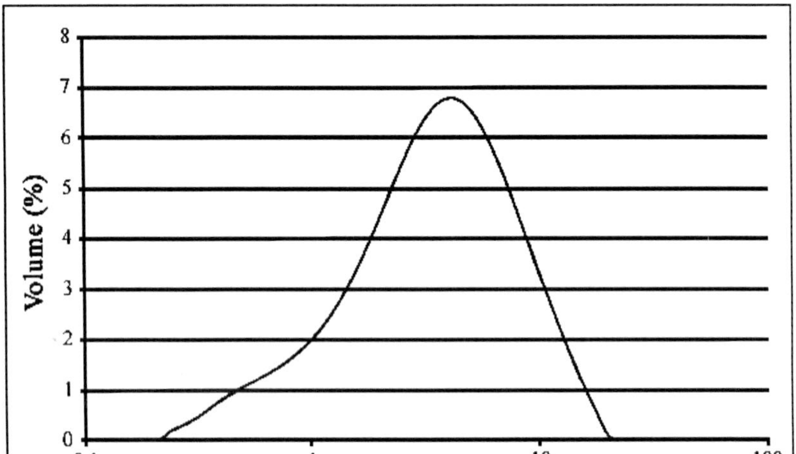

Figure 5. Particle size distribution of FAST-ACT formulation.

Microscopy (SEM) image of the FAST-ACT formulation, where the micron size aggregates are visible. A particle size distribution graph is shown in Figure 5 indicating there are no particles smaller than 200 nm in the formulation.

The average surface area of the FAST-ACT formulation, as determined by the conventional Brunauer, Emmett, Teller (BET) method that assumes multilayer gas adsorption behavior, is around 280 m^2/g, with the porosity of 0.3 cc/g and the average pore radius of 22 Å.

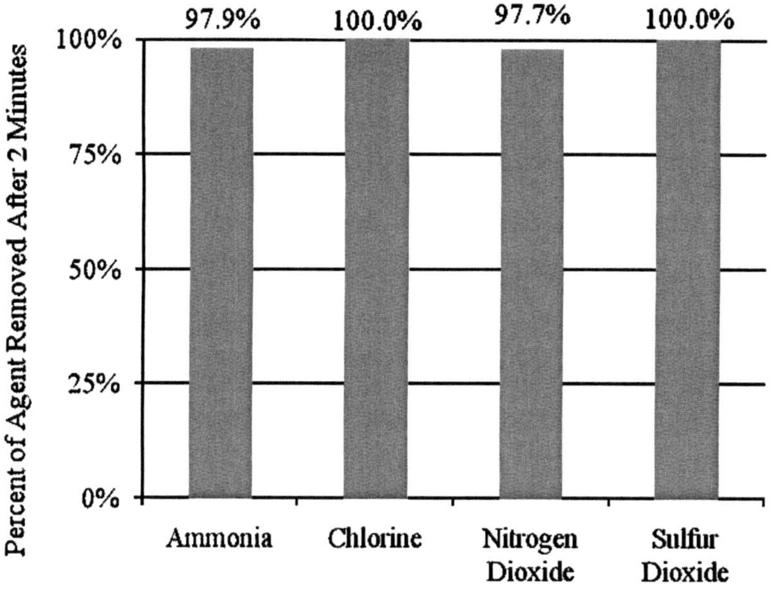

Figure 6. Removal of hazardous vapors by FAST-ACT from the air.

Efficacy of FAST-ACT

Hazardous chemicals are being manufactured, transported, stored and utilized by industry, academia, government, and individuals in their daily activities. Releases of these materials are generally accidental; however, a limited number of them, usually involving the most toxic chemicals (i.e., chemical warfare agents), can be intentional. The capabilities of FAST-ACT have been thoroughly tested with various toxic industrial chemicals including acids, bases, halogenated solvents, organic compounds, etc. In most cases, the decontamination efficacy was at least 95% after 2-minute contact. For uniformity, all tests were conducted at 1 to 50 agent to sorbent ratio by weight. However, for many hazards complete neutralization could be realized at much lower (up to stoichiometric) sorbent to agent ratios. Adsorption of gases by FAST-ACT was determined using Fourier Transform Infrared spectroscopy (FT-IR) and gravimetric measurements. In the IR experiments, the disappearance of the agent, upon exposure to FAST-ACT, was monitored based on the decrease in the intensity of the most prominent agent peaks. Since chlorine is a symmetrical molecule, IR spectroscopy could not be employed. Therefore, the amount of adsorbed chlorine was calculated based on the weight gain of the FAST-ACT formulation upon exposure to chlorine gas. FAST-ACT effectively removes acidic gases (i.e., chlorine, hydrogen chloride, nitrogen dioxide, sulfur dioxide) as well as basic gases, such as ammonia, Figure 6.

Table 1. Examples of FAST-ACT efficacy towards various classes of compounds

Neutralization		*Adsorption*	*Containment*
Corrosive Materials		*Vapor Hazards*	*Liquid Solvent Spills*
Acids Inorganic and Organic Hydrochloric Acid Hydrofluoric Acid Hydrobromic Acid Nitric Acid Phosphoric Acid Sulfuric Acid Acetic Acid Methanesulfonic Acid Ethanesulfonic Acid Benzenesulfonic Acid Toluenesulfonic Acid **Phosphorus** Pesticides Dimethylmethyl Phosphonate Paraoxon Parathion **Sulfur** 2-Chloroethyl Ethyl Sulfide Methyl Mercaptan **Phenols** Nitrophenols Chlorophenols	**Carbonyl Compounds** Aldehydes Ketones Carboxylic Acids **Nitrogen Compounds** Acetonitrile Sodium Cyanide (aq) 4-vinylpyridine **Halogens/Halides** Acetyl Chloride Chloroacetyl Chloride Chlorine Chloroform Hydrogen Bromide Cyanogen Chloride Methylene Chloride Carbon Tetrachloride TCE, PCE Bis-(2-Chloroethyl) Sulfide Pinacolyl methyl-phosphonofluoridate O-ethyl S-(2-diisopropylaminoethyl) methylphosphonothioate	**Acidic and Caustic Gases** Hydrogen Chloride Hydrogen Fluoride Hydrogen Bromide NOx/N_2O_4 Sulfur Dioxide Hydrogen Sulfide Diborane Hydrogen Selenide Phosphine Ammonia Carbonyl Sufide Hydrogen Cyanide **Chlorinated Organics** Acetyl Chloride Chloroacetyl Chloride Chloroform Methylene Chloride **Halogens** Chlorine Bromine Iodine **Volatile Organics** Methyl Mercaptan Ethylene Oxide Formaldehyde Phosgene Arsine	**Alcohols/Phenols** Ethanol Methanol Allyl Alcohol Nitrophenols Chlorophenols **Caustics** Anhydrous Ammonia Metal Hydroxides (aq) **Petrochemicals** Diesel Gasoline Oils **Others** Acrylonitrile Benzene Hydrazine Toluene Acrolein Methylhydrazine Methylisocynate
LIQUID & VAPOR CHEMICAL SPILLS AND RELEASES			

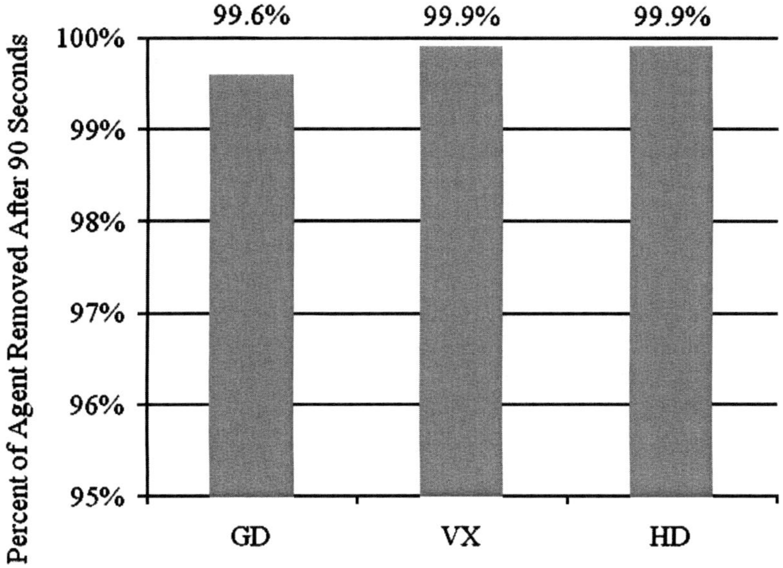

Figure 7. Effectiveness of FAST-ACT towards removal of chemical warfare agents from surfaces.

The formulation is also very effective at absorption, and/or neutralization of liquid hazards including acids, i.e., hydrochloric acid, nitric acid, phosphoric acid, sulfuric acid; alcohols, (methanol, ethanol); aldehydes, (acetaldehyde); fuels, (diesel, gasoline); aromatics, (p-cresol), as well as organic compounds containing heteroatoms such as: sulfur (mercaptans), phosphorous (paraoxon, dimethyl methyl phosphonate) and nitrogen (4-vinyl pyridine, acetonitrile), Table 1. Generally, if the compound is non-reactive, such as gasoline, FAST-ACT will sorb it; however, if the compound has reactive groups, such as parathion, FAST-ACT formulation will convert it to more benign byproducts. Since it is a dry powder system, the cleanup requires only sweeping or vacuuming of the decontaminated residue and disposing of as regulated. Upon treatment, the products are chemically inert and considerably less toxic, therefore, they may often be landfilled.

Chemical warfare agents can only be destroyed if their chemical structure is modified. Nerve agents, such as sarin, soman, tabun and VX all are phosphorus containing compounds, which can be chemically detoxified through hydrolysis with the formation of surface bound metal phosphonates. Blistering agents, such as distilled mustard contain sulfur and chlorine groups that are also neutralized by FAST-ACT through hydrolysis and dehydrochlorination reactions.

The FAST-ACT formulation has been tested by Battelle Memorial Institute in West Jefferson, OH, to confirm its detoxification versatility towards chemical warfare agents VX, GD and HD. An inert surface (glass) was spiked with CWA, FAST-ACT was applied to the surface and mixed/agitated with the agent. After 90 seconds the powder was removed from the surface, the surface extracted with an organic solvent, and the amount of agent in the extract determined by gas chromatography with flame ionization detector (GC/FID). FAST-ACT was capable of removing 99%+ of all agents tested under 90 seconds, as shown in Figure 7. After 10 minutes for VX and GD, and 60 minutes for HD, the powders were extracted and the amount of extracted agent quantified by gas chromatography-mass spectrometry (GC-MS). The nerve agents were completely destroyed under 10 minutes and the majority of HD was destroyed in under 60 minutes. To elucidate the destruction mechanisms solid state nuclear magnetic resonance (NMR) was employed (*14–17*).

In the case of GD (Soman, 1,2,2-trimethylpropyl methylphosphonofluoridate) destruction, the product is pinacolyl methylphosphonic acid (GD-acid) that converts to surface bound methylphosphonic acid (MPA); eliminated HF is also neutralized by FAST-ACT.

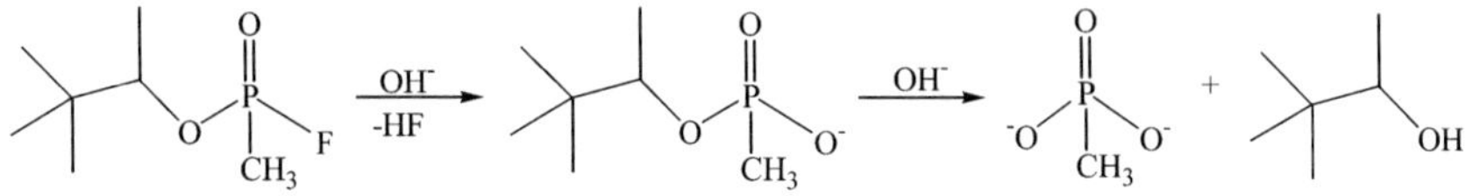

GD **GD-acid** **MPA**

For VX (O-ethyl S-diisopropylaminomethyl methylphosphono-thiolate), the major destruction product is ethyl methylphosphonic acid (EMPA) that converts into surface bound methylphosphonic acid (MPA).

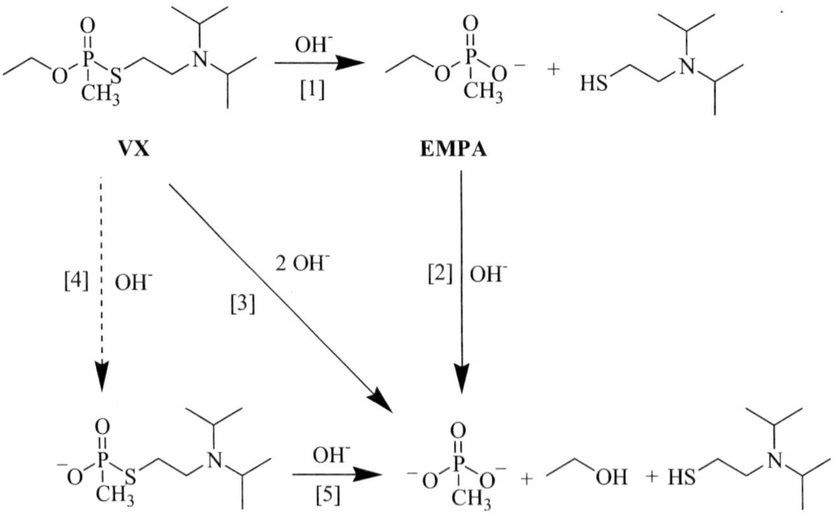

VX **EMPA** **EA-2192** **MPA (major)**

During HD (bis(2-chloroethyl) sulfide) destruction, the hydrolysis products are thioxane and chlorohydrin (CH) that forms surface bound thiodiglycol (TG).

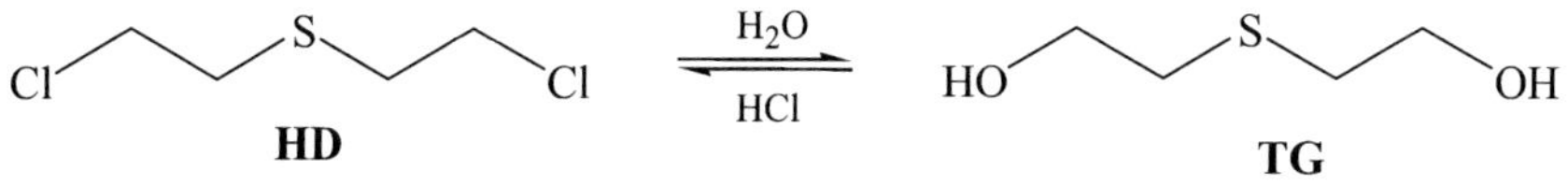

Decomposition products through elimination are 2-chloroethyl vinyl sulfide (VHD) and divinyl sulfide (DVHD).

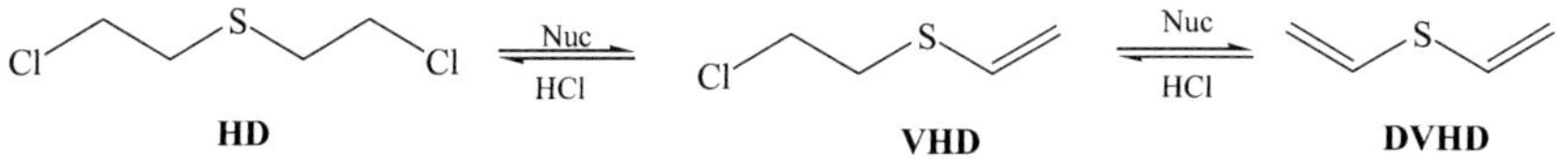

Product Manufacture, Shelf Life and Quality Control

To ensure FAST-ACT meets specifications all suppliers and subcontractors go through a review and approval process and incoming materials and finished goods are inspected and tested in NanoScale's Analytical Services laboratories before use or shipment. To simulate long storage times at ambient storage conditions shelf life studies were conducted using a thermodynamic temperature coefficient rule that is known as the Von't Hoff rule which states that a 10 °C rise in temperature doubles the rate of chemical reaction. Samples were placed in a calibrated environmental chamber at 50 +/- 2 °C for nine months. Real time studies at ambient temperature and humidity were also performed to validate accelerated shelf life results. Surface area of the material, as well as its chemical reactivity towards 2-chloroethyl ethyl sulfide and dimethyl methyphosphonate, were employed for determining the shelf life of the material. This data is shown in Figure 8 and indicates that packaged FAST-ACT (in cylinders, shakers bottles, or pails) can be stored for years, since the packaging prevents interaction with the environment (humidity).

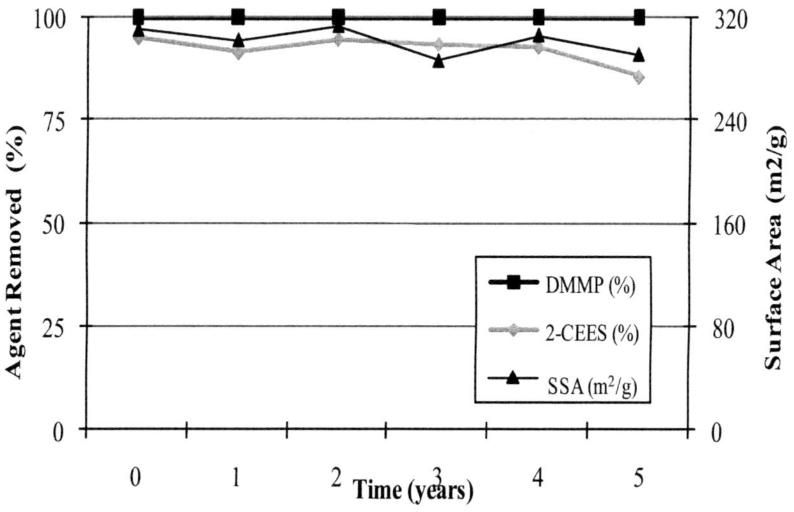

Figure 8. Shelf-life of FAST-ACT formulation.

Table 2. Summary of toxicity data for FAST-ACT and its components

	NanoActive TiO_2	*NanoActive MgO*	*FAST-ACT*
Acute Oral Toxicity	$LD_{50} > 2$ g/kg 928-006*	$LD_{50} > 5$ g/kg 85-MA-5302-01	-
Acute Dermal Toxicity	$LD_{50} > 5$ g/kg 85-XC-03GG-06	-	-
Acute Dermal Irritation	EPA Category IV Non-irritating 85-XC-03GG-05	EPA Category IV Non-irritating 85-XC-5302-01	-
Skin Sensitization	Non-sensitizer 85-XC-03GG-05	Non-sensitizer 85-MA-5302-02	-
Acute Eye Irritation	EPA Category IV Non-irritating (minimal affects clearing in less than 24 hours) 85-XC-03GG-05	EPA Category III Slightly irritating 85-XC-5302-01	-
Acute Inhalation	EPA Category IV Practically Nontoxic 2334 mg/m^3 for 4 hours 85-XC-03GG-05	Nontoxic 259 mg/m^3 for 4 hours 85-XC-5302-03	Nontoxic 825 mg/m^3 for 4 hours 85-XC-01BN-03
Mutation Assay: Ames Test	-	-	Not a Potential Mutagen 200532203-01**
Cytotoxicity	-	-	Non toxic to human immune cells***

* Testing conducted at MPI Research, Inc. ** Testing conducted at Nelson Labs. *** Testing conducted at RMIT University and Nanosafe Australia.

NanoScale maintains an independent Quality *Assurance* Department to ensure the integrity of the products. The company and processes are in compliance with current Good Manufacturing Practices, ISO 9001:2000 and Company standards. Production operations are guided by standard operating procedures (SOP) and work instructions to ensure controls are maintained and products are manufactured consistently. At various strategic points safe guards/tests have been established to ensure that the production of NanoActive materials consistently meet or exceed established quality standards throughout the entire manufacturing process. Product traceability is maintained by documenting lot numbers of raw materials, packaging material, and finished goods throughout the production and shipping processes.

Toxicity

Extensive third party safety and toxicity testing has been conducted on FAST-ACT and its components, that documents that FAST-ACT is safe and will not cause undue harm to the end-user. Table 2, summarizes these results. The testing was conducted at CHPPM (Center for Health Promotion and Preventive Medicine), unless indicated otherwise.

One of the possible routes of exposure to FAST-ACT is through inhalation; therefore, acute inhalation toxicity in rats was studied, where the animals were exposed to 825mg/m^3 FAST-ACT particle aerosol for 4 hours. The toxic signs, body weight measurements, and gross necropsy findings indicated that inhalation exposure to FAST-ACT under these conditions is nontoxic to rats. Preliminary data on subchronic inhalation testing indicate the same findings. FAST-ACT has also undergone Ames Mutagenicity testing by Nelson Laboratories, and the results indicate that FAST-ACT does not impart any mutagenicity. In a cytotoxicity assessment by *in vitro* exposure to human immune cells FAST-ACT was incubated *in vitro* with cultured human immune cells. The viability of these cells was assessed after 24 and 72 hours of exposure, using a standard method employing a compound that is converted to a colored product by living cells. FAST-ACT was not directly toxic to the cells under these *in vitro* exposure conditions. At this printing, ecotoxicity and environmental kinetics testing has not been conducted, however, the composition of the materials (magnesium oxide and titanium dioxide), large agglomerate size (micron size), and the conversion of magnesium oxide to magnesium carbonate due to prolonged exposure to humidity and carbon dioxide, lead to the conclusion that these materials will not pose environmental risks.

Comparison to Other Spill Countermeasure Technologies

Current solutions for counteracting toxic chemical spills or vapor releases can be divided into four categories: (1) Spill kits; (2) Procedures (evacuation); (3) Commodity absorbents/neutralizers; and (4) Military products. There are hundreds of products on the market that address specific types of hazards. There are separate spill controls for acids, caustics, and organic compounds with many countermeasures geared only towards a single chemical. There are very logistically burdensome solutions that address vaporous threats, or specially designed products that counteract the most toxic chemicals. Before any action can be undertaken a lengthy process, usually carried out by certified responders, takes place to identify and classify the hazard and then to utilize the proper countermeasure. However, if there is a life threatening situation, that involves a mixture of several hazards, or there is no identification equipment or proper countermeasure available, the need for one solution that addresses all of these scenarios is evident.

Table 3 shows a comparison between FAST-ACT and spill kits. Each spill kit contains three components, one to counteract acids, one for caustics (bases) and one for organic compounds. As can be seen, FAST-ACT is capable of addressing all of these chemicals as one product. In addition, spill kits do not address gaseous threats, nor do they neutralize the organic hazards, rather, they simply absorb them. Also, they do not address chemical warfare agents, and may cause an undesirable effect when a wrong countermeasure is applied. For example, when a standard silica based organic absorbent is utilized on a hydrofluoric acid spill, an adverse reaction will occur with the formation of a highly poisonous silicon fluoride gas. In contrast, application of FAST-ACT on a hydrofluoric acid spill will generate magnesium fluoride, a benign solid byproduct.

Table 3. Comparison of FAST-ACT and Spill Kits

	FAST-ACT		*Spill Kits*					
			Acid		*Caustic*		*Solvent*	
Effectiveness	*Contain*	*Neutr*	*Contain*	*Neutr*	*Contain*	*Neutr*	*Contain*	*Neutr*
Acids	Yes	Yes	Yes	Yes	No	No	No	No
Halogenated Compounds	Yes	Partial	Yes	No	Yes	No	Yes	No
Phosphorus Compounds	Yes	Yes	Yes	No	Yes	No	Yes	No
Acidic and Caustic Gases	Yes	Yes	No	Partial	No	Partial	No	No
Organic Compounds	Yes	Partial	Partial	No	Partial	No	Partial	No
Chemical Warfare Agents	Yes	Yes	Yes	No	Yes	No	Yes	No
Liquid & Vapor Hazards	Yes		No		No		No	
Pretreatment Identification	No		Yes		Yes		Yes	
Safe for All Spills	Yes		No		No		No	
Packaging	Pressurized canisters		Carton box, shaker		Carton box, shaker		Carton box, shaker	
Material Testing Program	Yes		No		No		No	

Overall, the FAST-ACT solution is competing against all current solutions based on performance across a much broader spectrum of utilization: acids, bases, toxic chemicals and chemical warfare agents. It is simple to apply, does not require any ancillary equipment of deployment, and is compatible with detection equipment. Most importantly, it has a significant differentiation regarding safety in that it can be applied to *any* spill and vapor release, including unknowns and mixtures, with actual destruction of the threat, as opposed to simple containment.

Marketing and Sales

The FAST-ACT market can be broadly defined as any place where accidental or intentional (terrorist attacks) hazardous chemical spills and vapor releases may be encountered. The market environment for FAST-ACT is stimulated by the fact that one reported hazardous material spill occurs every 3.5 minutes in the U.S., there is a heightened awareness of terrorism utilizing chemical warfare agents or toxic chemicals, and there is increased Homeland Security funding for weapons of mass destruction countermeasures. The primary market segments include First Responders, Government, Industry, Department of Transportation and Academia. The product is available through NanoScale's national and international distributors, and directly from the company website www.fast-act.com.

Conclusion

The active component of FAST-ACT is based on patented nanotechnology, originating from Prof. Kenneth Klabunde's laboratory at Kansas State Univerisity. It is enabled dry powder that has unsurpassed chemical reactivity due to its large surface area, high porosity, unique surface morphology and small crystallite sizes. These characteristics allow FAST-ACT to destroy toxic chemical hazards, as opposed to mere absorption, the mode of operation for the majority of other hazard control products. The powder is non-flammable and non-corrosive and can be utilized in any environmental condition. The logistical benefits of the system, combined with effectiveness against a wide range of toxic chemical threats, make it an attractive solution for markets where accidental or intentional hazardous chemical spills and vapor releases are encountered.

References

1. Klabunde, K. J.; Park, D. G.; Stark, J. V.; Koper, O. B.; Decker, S.; Jiang, Y.; Lagadic, I. In *Nanoscale Metal Oxides as Destructive Adsorbents. New Surface Chemistry and Environmental Applications, NATO Advanced Study Institute*; Pelizzetti, E., Ed.; Kluwer Academic Publishers: The Netherlands, Fine Particle Science and Technology, 1996; pp 691–706.
2. Klabunde, K. J.; Stark, J. V.; Koper, O.; Mohs, C.; Park, D. G.; Decker, S.; Jiang, Y.; Lagadic, I.; Zhang, D. *J. Phys. Chem.* **1996**, *100*, 12142–12153 (invited Feature Article).
3. Koper, O. B.; Lagadic, I.; Volodin, A.; Klabunde, K. J. *Chem. Mater.* **1997**, *9*, 2468–2480.

4. Khaleel, A.; Kapoor, P.; Klabunde, K. J. *Nanostruct. Mater.* **1999**, *11*, 459–468.
5. Klabunde, K. J.; Decker, S.; Lucas, E.; Koper, O. In *Cluster and Nanostructure Interfaces*; Jena, P.; Khanna, S. N.; Rao, B. K., Eds.; Proceedings of the International Symposium, Richmond, VA, USA, October 25–28, 1999; pp 577−582.
6. Lucas, E.; Decker, S.; Khaleel, A.; Seitz, A.; Fultz, S.; Ponce, A.; Wi, L.; Carnes, C.; Klabunde, K. J. *Chem.−Eur. J.* **2001**, *7*, 2505–2510.
7. Rajagopalan, S.; Koper, O.; Decker, S.; Klabunde, K. J. *Chem.−Eur. J.* **2002**, *8*, 2602–2607.
8. Kim, S.; Wang, X.; Buda, C.; Neurock, M.; Koper, O. B.; Yates, J. T., Jr. *J. Phys. Chem. C* **2009**, *113*, 2219–2227.
9. Klabunde, K. J. U.S. Patent 5,990,373, 1999.
10. Koper, O.; Klabunde, K. J. U.S. Patent 6,057,488, 2000.
11. Lanz, B. E.; Allen, T. U.S. Patent 7,279,129, 2007.
12. Mulukutla, R. S.; Malchesky, P. S.; Maghirang, R.; Klabunde, J. S.; Klabunde, K. J.; Koper, O. U.S. Patent 7,276,640, 2007.
13. Ward, L. *Pop Mech.* **2005** (November), 75.
14. Wagner, G. W.; Koper, O. B.; Lucas, E.; Decker, S.; Klabunde, K. J. *J. Phys. Chem. B* **2000**, *104*, 5118–5123.
15. Wagner, G. W.; Bartram, P. W.; Koper, O.; Klabunde, K. J. *J. Phys. Chem. B* **1999**, *103*, 3225–3228.
16. Wagner, G. W.; Procell, L. R.; Munavalli, S. *J. Phys. Chem. C* **2007**, *111*, 17564–17569.
17. Wagner, G. W.; Procell, L. R.; O'Connor, R. J.; Munavalli, S.; Carnes, C. L.; Kapoor, P. N.; Klabunde, K. J. *J. Am. Chem. Soc.* **2001**, *123*, 1636–1644.

Chapter 15

Nanoscale Catalysts and In Room Devices To Improve Indoor Air Quality and Sustainability

Steve Eckels,[1,*] Olga B. Koper,[3] Larry E. Erickson,[2] and Lynette Vera Bayless[2]

[1]Institute for Environmental Research, [2]Department of Chemical Engineering, Kansas State University, Manhattan, Kansas 66506
[3]NanoScale Corporation, 1310 Research Park Drive, Manhattan, Kansas 66502
[*]eckels@ksu.edu

Reduced energy consumption and increased indoor environmental quality are the two main/major goals when constructing green buildings. Nanoscale catalysts and reactive sorbents have the potential to improve indoor air quality and operational efficiency. This work illustrates applications of these advanced materials for in room air cleaning devices. Catalysts attached to walls and those incorporated into packed beds have demonstrated potential performance in simulation studies. Operational results are also presented using the OdorKlenz-Air cartridge.

Introduction

There are many in room devices to improve air quality for health and comfort. There is a growing emphasis on energy conservation, efficiency and sustainable indoor environments. Nanoscale catalysts have been developed for air quality applications. Many of these developments can be used as in room devices or as part of heating, ventilation, and air conditioning systems. In this work, the emphasis will be on nanoscale catalysts that have application to in room devices for the improvement of indoor air quality (IAQ). The work is based on available literature (*1–22*), the recent work of Bayless (*23*), Klabunde (*24*), and work at NanoScale Corporation (*25*). Photocatalytic oxidation processes to oxidize volatile organic compounds have been developed through extensive research (*11*, *14*, *15*, *18–20*).

Progress has also been made in developing nanoscale catalysts for the destruction of biological airborne particulate matter including viruses, bacteria, and fungi (*11*, *24*).

Photocatalysis

The photocatalytic oxidation (PCO) process is effective in oxidizing most volatile organic compounds that are associated with indoor air at the concentrations normally found in homes, hospitals, and aircraft cabins (*11*, *23*). Some self-contained in room systems are available that make use of titanium dioxide catalysts and UV light (*10*). Photocatalytic oxidation can inactivate and destroy airborne pathogenic microorganisms such as viruses, bacteria, fungi, and their spores (*11*, *16*, *17*, *23*). Huang et al. (*8*) reported that viruses are the easiest to inactivate and that destroying spores is the most difficult. Chen et al. (*7*) have demonstrated that UV light emitting diodes can be used to save energy.

Yang et al. (*19*) have developed nanoscale photocatalysts that function with visible light by using carbon and vanadium dopants to modify the structure of titanium dioxide. The catalyst has comparable activity in the dark, with visible light, and with UV light. This development makes it possible to use room light and catalytic surfaces to improve air quality.

In Room Devices

There are a number of design options for in room devices that can be used to improve air quality. The fan, humidifier, and dehumidfier are common examples of plug in devices that are presently used. The concept of walls, windows and other surfaces that are covered with catalysts that are designed to improve air quality is a newer idea (*12*, *14*, *15*, *21*, *23*). A packed bed that contains nanoscale catalytic particles is another option (*11*, *23*). The OdorKlenz-Air® product of NanoScale Corporation contains nanoscale catalysts that are fixed in place; air is passed through the bed and volatile organic compounds are oxidized (*25*).

Conditions That Favor Use of In Room Devices

There are many indoor environments where in room cleaning devices are desirable to improve air quality. Applications include environments where odors are being generated (nursing homes, locker rooms, chemical laboratories, kitchens, etc) and where biological airborne particulate matter may be of concern (food processing, hospitals, nursing homes, medical waiting rooms and patient rooms). The common theme is that these spaces have either higher air quality needs or stronger contaminant sources than most indoor environments. Increasing ventilation rates are helpful but this often increases energy usage for the building as outdoor air must be conditioned before it enters the indoor environment.

Energy usage has become a bigger and bigger issue in recent years as many national organizations have led the charge to develop green building or net-zero building environments. It should be noted that the goal of most green buildings or net-zero buildings is not just reduced energy usage but also increased Indoor Environmental Quality (IEQ). This combination of reduced energy and increased IEQ has opened the door to new applications of in room devices that work synergistically with current building systems to deliver a better quality environment with lower energy consumption (*26–28*).

Understanding sustainability goals is an important part of understanding how new air cleaning technology will make a significant market contribution. ASHRAE (*26*) cites a number of facts that will play a significant role in future building designs.

- Heating, ventilation and air conditioning (HVAC) systems consume 1/3 of the global nonrenewable energy every year and thus will come under increased scrutiny.
- Significant reduction in CO_2 emissions must include addressing energy usage in buildings.
- ASHRAE soon to be adopted Standard 189 – Sustainable Buildings Standard will call for 30% reduction over current minimum energy standards.
- Current standard is ANSI/ASHRAE/IESNA 90.1 Energy Standard for Buildings (*27*).

It is easy to see that current movement in the Building Industry is putting significant pressure on energy usage in buildings. In room air cleaning devices can play a significant role in two ways. First, these devices can be used as source control at points where contaminant generation is significant. By eliminating the source it is possible to lower the overall ventilation rate in the room and thus save energy. Second, the use of such devices increases the overall air quality in the room and helps achieve not only energy reduction goals but also better overall IEQ.

The future building designs are being pushed in the direction of lower energy usage and better IEQ. Although current building designs span a wide spectrum in terms of energy usage it is interesting to look at the current green/sustainable build design push. It does represent one extreme but one that will see greater importance in the years to come. Sustainable or green building designs are currently judged based on a number of rating systems. Typically, this is done with a rating system developed by one of many "Green Councils". The most popular in the US is LEED developed by the U.S. Green Building Council. Buildings earn points based on sustainable features designed into the building. For the LEED rating systems, the following points can be earned in the minimum energy performance design category (ASHRAE 2008).

- Minimum Energy Performance Design
  - Sustainable Sites: 8 credits/14 points
  - Water Efficiency: 3 credits/5 points
  - Energy and atmosphere: 6 credits/17 points
  - Materials and resources: 8 credits/13 points
  - Indoor Environmental Qual.: 8 credits/15 points

Points and credits can be earned for sustainable building sites, efficient use of water, energy, and materials. It is important to see that indoor environmental quality is a significant feature of this rating system. Building designers typically see increased IEQ and reduced energy usage as conflicting goals but through tight integration and use of room specific cleaning devices such goals do not have to be mutually exclusive. In each subcategory points are won based on additional criteria. In the IEQ area the following are the stated goals.

- Prerequisites
  - Minimum IAQ performance (ASHRAE Standard 62.1)
  - Tobacco smoke control
- EQ 1: Outdoor Air Delivery Monitoring
- EQ 2: Increased Ventilation
- EQ 3: Construction IAQ Management Plan
- EQ 4: Low-Emitting Materials
- EQ 5: Indoor Chemical and Pollutant Source Control
- EQ 6: Controllability of Systems
- EQ 7: Thermal Comfort

Even within leading edge rating systems such as the LEEDS system one sees bias toward increased ventilation as the only solution for better air quality. It can be argued that through integration of in room specific air cleaning technology a reduced ventilation rate can be achieved. It should be noted that ventilation can never be eliminated because of the obvious need to supply oxygen and remove CO_2 due to human respiration. Minimum energy consumption can be achieved when we reduce the need for excess ventilation above this level. In room cleaning devices also are important in EQ 5 and EQ 6. In room devices that are either active or passive can give the building operator more controllability over the system and increase the ability to respond to IAQ problems in specific areas of the building. The LEEDS system highlights one system for rating new building designs and helps create the next generation of green/sustainable buildings.

In the effort to provide better indoor air quality it is important to identify the potential sources of problems in buildings. Although the list of potential chemicals and contaminations is almost endless ASHRAE has identified those issues that most commonly cause problems in buildings (*26*). The primary sources are

- Combustion sources
- Building materials
- Asbestos-containing insulation
- Wet or damp building materials (mold and mildew)
- Certain pressed wood
- Products for cleaning
- Wide range of industrial processes
- Occupants activities

Improving the air quality in new designs or solving current problems in buildings involves identifying the sources of contamination and controlling the spread or eliminating the source. In room cleaning devices can play an important role in solving such problems with reduced capital costs and increased effectiveness.

Photocatalytic Particles on Walls

Recently, Yang (*20*) reported the development of photocatalysts that are active both under visible light and dark conditions. The photocatalyst was made from TiO_2 doped with carbon and vanadium in varying amounts. The activity of the photocatalysts was tested for gaseous acetaldehyde. Results showed that the C- and V-doped TiO_2 containing 2% vanadium had comparable activity under visible light and dark conditions (*19*, *20*). This characteristic makes the 2% C-and V-doped TiO_2 potentially a great component of an energy-efficient PCO device that can clean the air both during the day and at night. Moreover, the light spectra of the fluorescent lights in homes and offices as well as the visible light component of the solar irradiation can be utilized to initiate the photocatalytic process. The potential use of the catalyst developed by Yang et al. (*19*) is considered further using simulation in this work.

PCO systems that incorporate thin films of photocatalytic materials have gained attention in recent years as an approach to treat gaseous pollutants (*18*, *21*). Thin film PCO designs have good commercialization potential (*14*). Materials coated with photocatalysts in the form of thin films can generate self-cleaning surfaces (*12*). Walls and ceilings can be made catalytically-active to treat and maintain low levels of air contaminants indoors. This configuration offers some advantages over other PCO configurations. Immobilizing the catalytic particles in building construction materials will enable them to be contained well, thus preventing the unintentional release of the particles in the air. Most importantly, this configuration offers large surface area for photocatalytic reaction.

Mass transfer is important in processes which involve diffusion. In thin films on walls, the overall rate of reaction may be limited by the rate of mass transfer of reactants (air contaminants) between the bulk gas and the catalytic surface. For instance, Noguchi et al. (*12*) studied the decomposition of formaldehyde and acetaldehyde using thin films and found that the rate was mass transfer-limited for acetaldehyde.

The modeling studies conducted here on systems involving catalytic surfaces on thin films involve two parts. The first is the estimation of the mass transfer coefficients ($k_m^{'}$) to obtain the rate of mass transfer. The second is the mass balance for IAQ modeling to determine the effect of mass transfer limitations on the overall contaminant removal process in a typical indoor environment.

Mass transfer coefficients , $k_m^{'}$, can be obtained through empirical expressions relating the Sherwood number ($N_{Sh,L}$) to the Reynolds and Schmidt numbers. For instance, Perry and Green (*13*) noted the following expression for a flat plate:

$$N_{Sh,L} = \frac{k_m^{'} L}{D} = 0.646 \left(N_{\mathrm{Re},L} \right)^{1/2} \left(N_{Sc} \right)^{1/3} \tag{1}$$

Reynolds number ($N_{Re,L}$) and Schmidt number (N_{Sc}) in equation (1) are defined by

$$N_{\mathrm{Re},L} = \frac{\rho_g V_o L}{\mu} \tag{2}$$

$$N_{Sc} = \frac{\mu}{\rho_g D} \tag{3}$$

where ρ_g and μ are the density and viscosity of the contaminated gas, V_o is the superficial velocity parallel to the catalytic surface and L is the characteristic length of the wall.

With the catalytic films immobilized on the walls and ceiling of the room with assumed dimensions of 3 m x 3m x 3 m, the characteristic length, L is set to be 300 cm. The mass transfer coefficients for the range of superficial velocities from 5 to 25 cm/s were obtained using equation (1). Results are shown in Table 1.

A mass balance inside the room for this system is presented below. The rate of contaminant removal (R_m) is due to the rate of mass transfer of the pollutant to the catalytically active wall.

$$\frac{VdC_i}{dt} = QC_0 + S_p p - QC_i - k_m' A_m \left(C_i - C_w\right) \tag{4}$$

$$R_m = k_m' A_m \left(C_i - C_w\right) \tag{5}$$

where QC_o is the flow of contaminant into the room, QC_i is the flow out where C_i is the concentration in the room. The term $S_p p$ is the contaminant generation rate in a room with p people. The contaminant concentration at the wall is assumed to be maintained at zero by the rapid catalytic reaction and the large surface area of the catalyst. The area, A_m, in equation (5) is the wall and ceiling surface area.

At steady state conditions, $V\frac{dC_i}{dt} = 0$ and with $C_w = 0$, equation (4) becomes

$$C_i = \frac{QC_o + S_p p}{Q + k_m' A_m} \tag{6}$$

Equation (6) can be used to estimate the indoor air concentration, C_i (in μg/m^3). The area of mass transfer, A_m, includes the four walls and ceiling of the room, thus, with a value of 4.5 x 10^5 cm^2. The emission rate, S_p, inside the room is steady at 35 μg/h-person (*4*). The influence of the following factors on the indoor air concentration (C_i) for the range of superficial velocities of 5 to 25 cm/s was studied: the mass transfer-limited rate of pollutant removal (R_m), ventilation rate (Q), magnitude of emission source (p), and outdoor air concentration (C_o). The variation of the levels of acetaldehyde inside the zone was studied for cases when there was no in-room device and with the device (thin film) installed inside the room. The contrasting effect on C_i of having a clean ventilation air versus outdoor air with a typical acetaldehyde concentration (C_o) of 20.12 μg/m³ (*3*, *4*) was also explored. Levels of acetaldehyde indoors were determined for conditions where there is no ventilation and with the ASHRAE standard ventilation rate (V_R) of 0.48 m³/min per person (*5*). The ventilation rate guidelines provided by ASHRAE (*5*) and the emission rate (S_p) are both dependent on the number of people (p) inside the room. The results for all cases are presented in Bayless (*23*).

The corresponding values of the mass transfer coefficient (k'_m) for air velocities of 5, 10, 15, 20 and 25 cm/s were calculated using equations (1), (2) and (3). The k'_m values (in cm/s) and Reynolds number for each flow velocity are tabulated in Table 1.

The mass transfer coefficients above were used to estimate the rate of removal term in equation (5). The concentration of the contaminant (C_i) in the room for various operating conditions was evaluated by utilizing the steady state mass balance in equation (6), wherein the rate of contaminant removal (described by equation 5) is mass transfer-controlled. The results are shown in Table 2.

Photocatalytic Particles in Packed Beds

In photocatalysis, a bed of photocatalytic material can be configured to clean the air indoors. Kowalski (*11*) noted that this is one of the more recent designs in PCO. Arabatzis et al. (*1*) and Ibhadon et al. (*9*) used packed bed photoreactors incorporating porous foaming titania photocatalysts in their study of the photocatalytic degradation of VOCs.

Figure 1 depicts a possible configuration of a packed bed in the upper part of the room that can utilize an effective surface area for photocatalytic oxidation of organic air contaminants.

The increased surface area resulting from the use of nanosized photocatalysts increases the reaction rate (*22*). However, the use of nanoparticles in a packed bed system can increase the pressure drop. Pelletization of the photocatalysts can reduce the pressure drop as well as prevent fine particles from becoming entrained in the air. An optimum pellet size is necessary in order to achieve good performance with minimal pressure drop. Along with this, the effect of reaction rate and mass-transport limitations in the overall PCO process must be considered in modeling and simulation studies for a packed bed photocatalytic reactor.

Equation (6) is a form of Ergun equation for spherical particles adapted from Bird et al. (*6*). The Ergun equation relates the pressure drop with depth of the bed, z, fluid flow, G_o, particle size, D_p, bed porosity, ε, viscosity, μ, and density of the fluid, ρ_g.

$$\frac{\Delta P}{z} = \frac{150\mu G_o}{g_c D_p^2 \rho_g}\frac{(1-\varepsilon)^2}{\varepsilon^3} + \frac{1.75 G_o^2}{g_c \rho_g D_p}\frac{(1-\varepsilon)}{\varepsilon^3} \tag{7}$$

Pressure drop (ΔP) was estimated using the Ergun equation (7). Values are shown in Table 3 for particles with diameter of 0.2 cm.

Table 1. Calculated mass transfer coefficients as a function of superficial velocity (*23*)

V_0 (cm/s)	$N_{Re,L}$	k'_m (cm/s)
5	9845	0.028
10	19691	0.039
15	29536	0.048
20	39381	0.056
25	49227	0.062

Table 2. Indoor air concentration with catalytic walls and ceiling, ventilation rate 0.48 m³/min-person, and 1 person* (*23*)

V_o (cm/s)	Indoor Air Concentration, C_i ($\mu g/m^3$)	
	$C_o = 0$	$C_o = 20.12\ \mu g/m^3$
5	0.47	8.34
10	0.38	6.66
15	0.33	5.76
20	0.29	5.18
25	0.27	4.76

* $C_i = 21.33\ \mu g/m^3$ without catalyst.

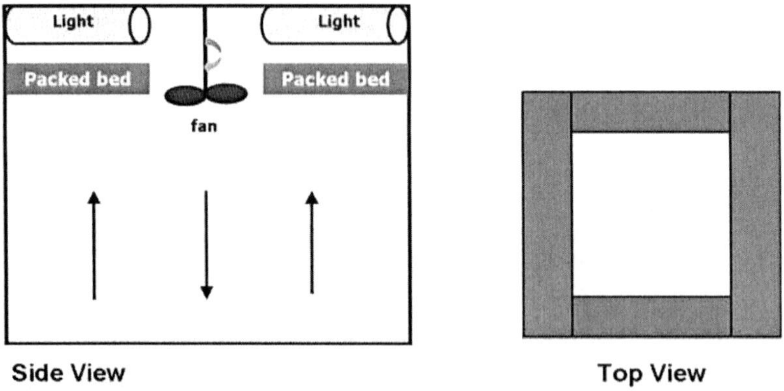

Figure 1. Packed bed in the upper part of the room (23).

Pellets of nanoscale catalyst with diameter of 0.2 cm were selected based on pressure drop and particle surface area.

Bayless (*23*) considered diffusion and reaction within the particles using the reaction rate observed by Yang et al. (*19*) and a zero order kinetic model. Mass

Table 3. Calculated pressure drop across a packed bed with spherical pellets (Dp = 0.2 cm) (*23*)

V_o (cm/s)	G_o (g/cm²-s)	$N_{Re,p}$	ΔP (dynes/cm²)
5	0.0060	11	22
10	0.0121	22	49
15	0.0181	33	80
20	0.0241	44	117
25	0.0301	55	159

transfer external to the particles was also considered. The values of the parameters used in the study are presented in Table 4. External mass transfer was found to be rate controlling.

For the case of external mass transfer limiting the rate of reaction, a plug flow model was used to describe the change of concentration through the packed bed over time due to mass transfer of the pollutant to the surface of the pellets.

$$\frac{dC}{dt} = -k_m^{'} C A_m \tag{8}$$

where C is the pollutant concentration at any residence time, t, $k_m^{'}$ is the mass transfer coefficient in cm/s and Am is the area of mass transfer. The acetaldehyde concentration at the pellet surface is assumed to be zero. Integration of equation (8) and applying the following boundary conditions in the packed bed yields equation (9): At t = 0, C = C_i and at t = t_R , C = C_{out}, where C_i is the concentration of the gas at the inlet of the packed bed, C_{out} is the concentration of the pollutant coming out of the packed bed and t_R is the residence time in the bed.

$$C_{out} = C_i e^{-k'_m A_m t_R} \tag{9}$$

The area of mass transfer (A_m) can be estimated by finding the total external surface area of the pellets over a differential volume of the bed. It is approximately 18 cm²/cm³.

The mass transfer coefficient values for the range of velocities of 5 to 25 cm/s are estimated using the correlation of the Sherwood number ($N_{Sh,p}$) for flow around a spherical particle shown in equation (10). This equation is appropriate for Reynolds numbers greater than 20 for flow in packed beds (*6*).

$$N_{Sh,p} = \frac{k_m^{'} D_p}{D} = 2.0 + 0.60\left(N_{\mathrm{Re},p}\right)^{1/2}\left(N_{Sc}\right)^{1/3} \tag{10}$$

$N_{Re,p}$ and N_{Sc} are the Reynolds number for pellets in packed beds and the Schmidt number, respectively. For a steady state balance around the packed bed,

Table 4. Values of parameters used in the modeling and simulation studies (*23*)

Parameter	*Value and Units*
Assumptions and general considerations	
Room Dimensions (*L* x *W* x *H*)	3m x 3m x 3m
Temperature, *T*	20 °C
Pressure, *P*	1 atm
Feed gas to the packed bed or catalytic walls and ceiling	
Air contaminant	Acetaldehyde (very dilute concentration)
Equivalent radius of the acetaldehyde molecule, r_{count}	0.28 nm
Density, ρ_g	1.206 x 10^{-3} g/cm^3 (air)
Diffusion coefficient of acetaldehyde in air, *D*	0.1202 cm^2/s
Viscosity, μ	1.81 x 10^{-4}g/cm-s (air) (*6*)
Range of superficial velocity, V_o	5 to 25 cm/s (*2*, *4*)
Emission rate, S_p in the room	35 μ g/h-person
Ventilation rate, V_R	0.48 m^3/min-person (*5*) and equivalent to *Q/p*
Packed bed	
Porosity, ε	0.4
Depth, *z*	1 cm
Pellets of 2% C-and V-doped TiO_2	
Rate, υ_o	1.4 x 10^{-8} gmoles/cm^3-s = 2.283 x 10^{-8} gmoles/gcat-s
Shape of pellet	
Spherical	
Diameter, D_p	0.2 cm
Range of pore radius (r_{pore})	2 – 10 nm
Pellet porosity parameter, ε_p	0.5
Bulk density of the catalyst, ρ_{bulk}	0.6 g/cm^3
Tortuosity factor, τ	2.0
Outdoor acetaldehyde concentration, C_o	20.12 μ g/m^3
Number of people in the room, *p*	One (1) or three (3)
Area of mass transfer, A_m	18 cm^2/cm^3 (packed bed)

$$qC_i - qC_{out} = qC_i\left(1 - e^{-k'_m A_m t_R}\right) \tag{11}$$

A mass balance for the room gives

$$V\frac{dCi}{dt} = QC_o + S_p p - QC_i - qC_i\left(1 - e^{-k'_m A_m t_R}\right) \tag{12}$$

where $q = V_o S$ and S is the area of the bed normal to the flow. At steady state conditions, equation (12) was rearranged to solve for C_i at various velocities in the bed.

$$C_i = \frac{QC_o + S_p p}{Q + V_o S\left(1 - e^{-k'_m A_m t_R}\right)} \tag{13}$$

The residence time (t_R) in a 1 cm length of the bed (z) for velocities of 5 to 25 cm/s is calculated as follows:

$$t_R = \frac{z\varepsilon}{V_o} \tag{14}$$

The indoor air concentrations for 1,000, 5,000 and 10,000 cm^2 area of the bed (S) were estimated using equation (13) and the parameter values in Table 4.

Table 5 shows the estimated mass transfer coefficients, residence time for a bed depth of 1 cm and indoor air acetaldehyde concentrations obtained by using equations (10) and (13) assuming that the rate is limited by the mass transfer exterior to the pellet.

To evaluate the effect of the addition of an in room device, the indoor air concentration results can be compared to the value of $C_o = 20.12$ μg/m^3 which is the estimated value in the entering air from outside the room. With the average generation rate of the people in the room, the estimated value is 21.33 μg/m^3. Thus, the in room device reduces the concentration significantly.

At low concentrations of volatile contaminants in rooms, mass transfer limited processes are expected for catalytic processes with reasonable reaction rates. Bayless (*23*) found that external mass transfer was rate limiting for both the packed bed and catalytic particles on the walls of the room.

OdorKlenz-Air® Product Field Experience

OdorKlenz-Air® technology (www.odorklenz.com) was developed by NanoScale Corporation in Manhattan, KS (*25*). This product utilizes a proprietary mixture of high surface area nanocrystalline metal oxides incorporated in a filtration cartridge (Figure 2) to remove and neutralize malodors and hazardous chemicals in enclosed airspaces. With the OdorKlenz-Air® system the odor causing compounds first are physisorbed on the surface of the metal oxides, and in a subsequent step are neutralized with the reaction byproducts bound to the

Table 5. Calculated indoor air concentrations (C_i) of acetaldehyde for external mass transfer control in a packed bed (*23*)

V_o (cm/s)	$k_m^{'}$ (cm/s)	t_R (s)	Indoor air concentration, C_i (µg/m³) 1,000 cm²	5,000 cm²	10,000 cm²
5	2.496	0.080	13.28	5.29	3.02
10	3.031	0.040	10.13	3.27	1.77
15	3.443	0.027	8.49	2.49	1.32
20	3.789	0.020	7.47	2.08	1.09
25	4.095	0.016	6.76	1.81	0.95

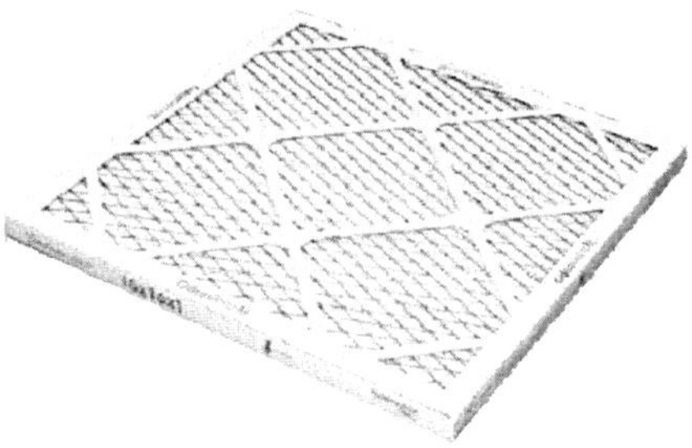

Figure 2. OdorKlenz-Air® cartridge (16"x16"x1").

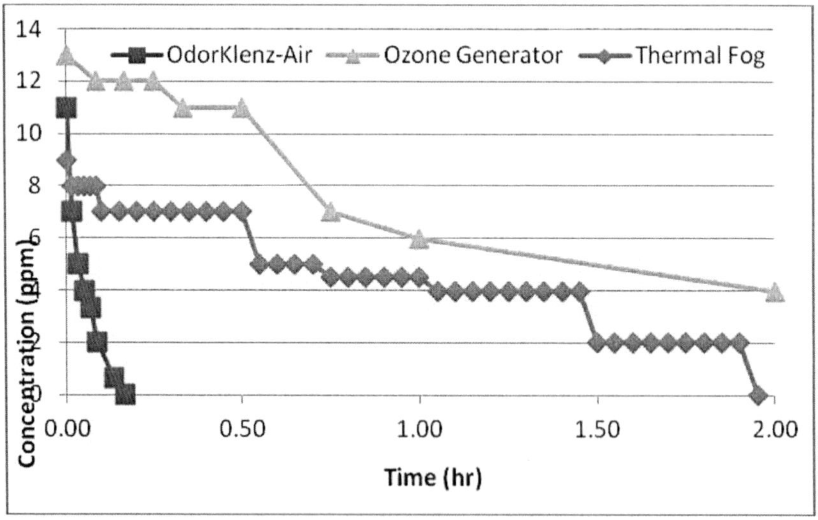

Figure 3. Removal of hydrogen sulfide using OdorKlenz-Air®, Ozone Generator and Thermal Fogging.

surface. The formulation was optimized to target a broad range of malodors including organic acids (i.e., propionic acid, isovaleric acid, acetic acid), aldehydes and ketones (i.e., acetaldehyde, formaldehyde, acetone), thiols (i.e., methanethiol, ethanethiol, hydrogen sulfide), and amines: including aliphatics (i.e., cadaverine and putrescine) as well as heterocycles (i.e., skatole and indole), and a multitude of others. The primary application is in the disaster recovery market, where rapid removal of hazardous and/or malodorous compounds without off-gassing is of paramount importance.

To illustrate the efficacy of the system, Figure 3 shows removal of hydrogen sulfide over time utilizing the OdorKlenz-Air® cartridge, incorporated into an airscrubber, ozone, and thermal fogging. OdorKlenz-Air® was highly superior in both kinetics and capacity for removal and neutralization by comparison.

Conclusions

In room products have been used extensively for thermal comfort and humidity control. Recently, there has been greater use of in room products to reduce contaminant concentrations.

Nanoscale materials are being used to improve indoor air quality in products such as OdorKlenz. Nanoscale sorbents and catalysts have the potential to reduce energy operational costs and improve indoor environmental quality through development of additional in room devices. These advances are needed to help accomplish the goals and objectives of ASHRAE and the U.S. Green Building Council.

Acknowledgments

Partial support for this research was provided by the Targeted Excellence Program at Kansas State University.

References

1. Arabatzis, I. M.; Spyrellis, N.; Loizos, Z. Design and theoretical study of a packed bed. *J. Mater. Proc. Technol.* **2005**, *161* (1–2), 224–228.
2. ASHRAE. *ASHRAE standard 55-2004 -thermal conditions for human occupancy*; 2004.
3. ASHRAE. *ASHRAE handbook -Fundamentals*; American Society of Heating, Refrigerating and Air-Conditioning Engineers, Inc.: Atlanta, GA, 2005.
4. ASHRAE. *ASHRAE handbook -HVAC applications*; American Society of Heating, Refrigerating and Air-Conditioning Engineers, Inc.: Atlanta, GA, 2007.
5. ASHRAE. *ASHRAE standard 62.1-2007 -ventilation for acceptable indoor air quality*; 2007.
6. Bird, R. B.; Stewart, W. E.; Lightfoot, E. N. *Transport phenomena*, 2nd ed.; John Wiley & Sons, Inc.: New York, USA, 2002.

7. Chen, D. H.; Ye, X.; Li, K. Oxidation of PCE with a UV LED photocatalytic reactor. *Chem. Eng. Technol.* **2005**, *28* (1), 95–97.
8. Huang, Z.; Mannes, P. C.; Blake, D. M.; Wolfrum, E. J.; Smolinski, S. L.; Jacoby, W. A. Bactericidal mode of titanium dioxide photocatalysis. *J. Photochem. Photobiol., A* **2000**, *130* (2), 163–170.
9. Ibhadon, A. O.; Arabatzis, I. M.; Falaras, P.; Tsoukleris, D. The design and photoreaction kinetic modeling of a gas-phase titania foam packed bed reactor. *Chem. Eng. J.* **2007**, *133* (1−3), 317–323.
10. Kowalski, W. J. *Immune building systems technology*; The McGraw-Hill Companies, Inc.: USA, 2003.
11. Kowalski, W. J.. Photocatalytic oxidation. *Aerobiological engineering handbook*; The McGraw-Hill Companies, Inc.: USA, 2006; p 295.
12. Noguchi, T.; Fujishima, A.; Sawunyama, P.; Hashimoto, K. Photocatalytic degradation of gaseous formaldehyde using TiO2 film. *Environ. Sci. Technol.* **1998**, *32* (23), 3831–3833.
13. Perry, J. L. *Trace chemical contaminant generation rates for spacecraft contamination control system design*; National Aeronautics and Space Administration: Marshall Space Flight Center (MSFC), AL, 1995.
14. Tompkins, D. T.; Lawnicki, B. J.; Zeltner, W. A.; Anderson, M. A. Evaluation of photocatalysis for gas-phase air cleaning part 1: Process, technical and sizing considerations. *ASHRAE Trans.* **2005**, *111* (2), 60–84.
15. Tompkins, D. T.; Lawnicki, B. J.; Zeltner, W. A.; Anderson, M. A. Evaluation of photocatalysis for gas-phase air cleaning, part 2: Economics and utilization. *ASHRAE Trans.* **2005**, *111* (2), 85–95.
16. Vera, L.; Erickson, L.; Maghirang, R.; Klabunde, K. Photocatalytic oxidation of bioaerosols and volatile organic compounds in air by using titanium dioxide. *Proceedings of the 35th Annual Biochemical Engineering Symposium*; Rapid City, SD, 2007; pp 96−104.
17. Vera, L. A.; Erickson, L. E.; Maghirang, R. G.; Klabunde, K. J. Application of nanotechnology to agricultural air quality. *Proceedings: International Conference on the Future of Agriculture: Science Stewardship and Sustainability*; Sacramento, CA, 2007; pp 609−625.
18. Xu, M.; Lin, S.; Chen, X.; Peng, Y. Studies on characteristics of nanostructure of N-TiO2 thin films and photo-bactericidal action. *J. Zhejiang Univ., Sci., B* **2006**, *7* (7), 586–590.
19. Yang, X.; Cao, C.; Hohn, K.; Erickson, L.; Maghirang, R.; Hamal, D.; Klabunde, K. Highly visible-light active C-and V-doped TiO2 for degradation of acatldehyde. *J. Catal.* **2007**, *252*, 296–302.
20. Yang, X. Sol-gel synthesized nanomaterials for environmental applications. Ph.D. dissertation, Kansas State University, 2008.
21. Yu, H.; Zhang, K.; Rossi, C. Theoretical study on photocatalytic oxidation of VOCs using nano-TiO2 photocatalyst. *J. Photochem. Photobiol., A* **2007**, *188*, 65–73.
22. Zhang, Y.; Yang, R.; Zhao, R. A model for analyzing the performance of photocatalytic air cleaner in removing volatile organic compounds. *Atmos. Environ.* **2003**, *37*, 3395–3399.

23. Bayless, L. V. Photocatalytic Oxidation of Volatile Organic Compounds for Indoor Air Applications. M.S. Thesis, Kansas State University, Manhattan, KS, 2009; http://krex.k-state.edu/dspace/handle/2097/1496.
24. Klabunde, K. J. (2009). Visible Light Active Photocatalysts and Biocides Based on Transition Metal Titanium Dioxide and Silicon Dioxide Aerogels. In *Nanoscale Materials in Chemistry*; ACS Symposium Series; in press.
25. NanoScale Corporation, Manhattan, KA, 2009; http://www.nanoscalecorp.com.
26. ASHRAE. ASHRAE Green Guide: for Design Construction, and Operation of Sustainable Buildings, 2nd ed.; American Society of Heating, Refrigeration and Air-Conditioning Engineers, Inc.: Atlanta, GA, 2008.
27. ANSI/ASHRAE/IESNA Standard 90.1-2004U. Energy Standard for Buildings Except Low-Rise Residential Buildings, SI Edition. American Society of Heating, Refrigeration and Air-Conditioning Engineers, Inc.: Atlanta, GA.
28. ASHRAE. Standard for the Design of High-Performance Green Buildings Except Low-Rise Residential Buildings.American Society of Heating, Refrigeration and Air-Conditioning Engineers, Inc.: Atlanta, GA, 2007.

Indexes

Author Index

Subject Index

N

O

P

R

S

T

U

V

W

X

Z